Autodesk Architectural Desktop R3.3 Fundamentals: Laying a Sound Foundation

Elise Moss and Chris Fox

authorized author

PUBLICATIONS

Schroff Development Corporation
www.SDCACAD.com

Schroff Development Corporation
P.O. Box 1334
Mission, KS 66222
(913) 262-2664
www.schroff.com

Trademarks
The following are registered trademarks of Autodesk, Inc.: AutoCAD, Mechanical Desktop, Inventor, Autodesk, AutoLISP, AutoCAD Design Center, Autodesk Device Interface, pointA and HEIDI.
Microsoft, Windows, NetMeeting, Word, and Excel are either registered trademarks or trademarks of Microsoft Corporation.
All other trademarks are trademarks of their respective holders.

Copyright 2001 by Elise Moss

All rights reserved. No part of this book may be reproduced, stored in a retrieval system, or transcribed in any form or by any means – electronic, mechanical, photocopying, recording, or otherwise – without the prior written permission of Schroff Development Corporation.

> Moss, Elise
>   Autodesk Architectural Desktop R3 Fundamentals:
>   Laying a Sound Foundation
>   Elise Moss and Chris Fox
> ISBN: 1-58503-030-9

The author and publisher of this book have used their best efforts in preparing this book. These efforts include the development, research, and testing of material presented. The author and publisher shall not be held liable in any event for incidental or consequential damages with, or arising out of, the furnishing, performance, or use of the material herein.

Printed and bound in the United States of America.

Preface

I am not an architect by training. I apologize in advance for any architectural errors that may be found in this book. I have taught and used CAD for almost twenty years. I know that many students find CAD applications to be extremely frustrating to learn and master. Architectural Desktop is particularly difficult because it is such a powerful software. I hope that using this text will ease some of that frustration. Perhaps the most frustrating element of writing this text was not being able to cover EVERY command and function within ADT. This software is even deeper than AutoCAD and any textbook can only scratch the surface of all the capabilities of this powerful design tool.

I have partnered with Chris Fox for this text. This textbook marks my first joint venture with Chris. Chris is also a teacher and a working draftsman in the building industry. This text marks his debut in the world of textbooks and we hope that readers will be kind.

The files used in this text are accessible from the Internet at www.schroff1.com/adtr3. They are free and available to students and teachers alike.

We value customer input. Please contact us with any comments, questions, or concerns about this text.

Elise Moss
elise_moss@mossdesigns.com

Chris Fox
lcfox@archimagecad.com.

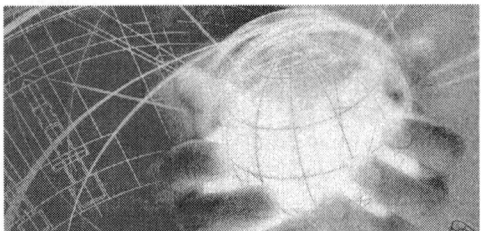

Acknowledgements from Elise Moss

This book would not have been possible without the support of some key Autodesk employees. A special thanks to Rebecca Bell, Lynn Allen, Melrose Ross, Bill Zhang, and Denis Cadu.

Additional thanks to the board and members of the Silicon Valley AutoCAD Power Users, a dedicated group of Autodesk users, for educating me about the needs and wants of CAD users.

My deep appreciation to Reid M. Addis of Addis Computer Consultants, Bob Callori, Lay Chris Fox, Shao-ying Cheng, Roy Salume, Jason Ellisor, and a host of other architects and facilities managers whose insight into the field of architecture and the use of Architectural Desktop has been invaluable.

The effort and support of the editorial and production staff of Schroff Development Corporation is gratefully acknowledged. I especially thank Stephen Schroff for his helpful suggestions regarding the format of this text.

Finally, truly infinite thanks to Ari for his encouragement and his faith.

- Elise Moss

Acknowledgements from Chris Fox

Heartfelt thanks to Elise Moss for inviting me to work on this book project, a sizable step up from my previous writing about AutoCAD. May this be the start of a rich relationship.

Thanks and appreciation also to David Harrington and John Clauson of AUGI (AutoCAD User Group International) and PaperSpace, the AUGI newsletter for local AutoCAD User Groups, for their encouragement of my efforts in writing articles.

Thanks to the members of the Rochester Area AutoCAD Users group for putting up with my enthusiastic but fumbling presentations and demonstrations.

Thanks to Vincenzo Buonomo of the Autodesk Training Center at Rochester Institute of Technology, for providing opportunities for all-important teaching experience.

Thanks also to Charles Traylor and James Turner of Archline CAD Services (www.archline.com) for their commitment to providing a work network for home-based architectural drafters, which gave me a start on working independently.

And, of course, all love and thanks to Sally, who bucks me up with smiles and tea.

--Lay Christopher (Chris) Fox

Table of Contents

Lesson 1
Desktop Features
    Exercise 1:   1-5
    Creating a New Drawing
    Exercise 2:   1-31
    Creating a New Geometric Profile
    Exercise 3:   1-38
    Creating a New Wall Style
    Exercise 4:   1-46
    Assigning Wall Properties
    Exercise 5:   1-51
    Downloading Styles from Autodesk's pointA
    Exercise 6:   1-53
    Creating a Fireplace/Chimney
    Layer Manager   1-67

Lesson 2
Site Plans
    Exercise 1:   2-2
    Creating Custom Line Types
    Exercise 2:   2-5
    Creating a Custom Text Style
    Exercise 3:   2-6
    Creating New Layers
    Exercise 4:   2-12
    Creating a Site Plan
    Exercise 5:   2-26
    Creating a Layer User Group

Quiz 1   Q1-1

Lesson 3
Foundations
    Exercise 1:   3-1
    Convert to Slab

Table of Contents

Lesson 4
Floor Plans

Exercise 1: 4-1
Layer Filters
Exercise 2: 4-3
Snapshots
Exercise 3: 4-7
Creating Exterior Walls
Exercise 4: 4-10
Convert to Walls
Exercise 5: 4-14
Wall Cleanup
Exercise 6: 4-17
Adding Closet Doors
Exercise 7: 4-22
Adding Interior Doors
Exercise 8: 4-24
Add Opening
Exercise 9: 4-34
Add Window Assemblies
Exercise 10: 4-43
Adding a Fireplace

Quiz 2 Q2-1

Lesson 5 5-1
Space Planning

Exercise 1: 5-1
Creating AEC Content
Exercise 2: 5-7
Furnishing the Bedrooms
Exercise 3: 5-13
Equipping the Bathrooms
Exercise 4: 5-17
Furnishing the Common Areas
Exercise 5: 5-20
Adding to the Service Areas
Exercise 6: 5-29
Adding Decorator Touches

Lesson 6
Roofs

    Exercise 1:   6-4
    Creating a Roof Using Existing Walls
    Exercise 2:   6-7
    Roof Slabs

Quiz 3   Q3-1

Lesson 7
Structural Members

    Exercise 1:   7-4
    Creating Member Sizes
    Exercise 2:   7-6
    Adding Members
    Exercise 3:   7-23
    Add Floorboards
    Exercise 4:   7-30
    Add Railing
    Exercise 5:   7-40
    Add Stairs

Lesson 8
Layouts

    Exercise 1:   8-2
    Creating Elevation Views
    Exercise 2:   8-7
    Creating Section Views
    Exercise 3:   8-21
    Creating 3D Section Views
    Exercise 4:   8-31
    Modifying Section Views

Quiz 4   Q4-1

Lesson 9
Documentation

    Exercise 1:   9-8
    Dimensioning a Floor Plan
    Exercise 2:   9-16
    Adding Wall Dimensions
    Exercise 3:   9-23
    Add Drawing Scale

Lesson 10
Schedules

Exercise 1: Adding a Schedule Table	10-11
Exercise 2: Adding Window Tags	10-14
Exercise 3: Editing a Schedule	10-16
Exercise 4: Creating a New Table Style	10-19
Exercise 5: Exporting a Schedule	10-29

Lesson 11
Creating a Video

Exercise 1: Creating a Path	11-1
Exercise 2: Adding A Camera	11-3
Exercise 3: Creating a Video	11-13

Final Exam   E-1

Appendix A: Toolbars   A-1
Appendix B: Exercise Time Chart   B-1
Index
About the Authors

Lesson 1:
Desktop Features

Architectural Desktop enlists object oriented process systems (OOPS). That means that ADT uses intelligent objects to create a building. This is similar to using blocks in AutoCAD. Objects in Architectural Desktop are blocks on steroids. They have intelligence already embedded into them. A wall in ADT is not just a collection of lines. It represents a real wall. It can be constrained, has thickness and material properties, and is automatically included in your building schedule.

The following table describes the key features of ADT:

Feature Type	Description
AEC Camera	Create perspective views from various camera angles. Create avi files.
AEC Profiles	Create AEC objects using polylines to build doors, windows, etc.
Anchors and Layouts	Define a spatial relationship between objects. Create a layout of anchors on a curve or a grid to set a pattern of anchored objects, such as doors or columns.
Annotation	Set up special arrows, leaders, bar scales
Ceiling Grids	Create reflected ceiling plans with grid layouts
Column Grids	Define rectangular and radial building grids with columns and bubbles
Design Center	Customize your AEC block library
Display System	Control views for each AEC object
Doors and Windows	Create custom door and window styles or use standard objects provided with ADT
Elevations and Sections	An elevation is basically a section view of a floor plan.
Layer Manager	Create layer standards based on AIA CAD Standards. Create groups of layers. Manage layers intelligently using Layer Filters.
Masking Blocks	Store a mask using a polyline object and attach to AEC objects to hide graphics
Model Explorer	View a model and manage the content easily. Attach names to mass elements to assist in design
Multi-view blocks	Blocks have embedded defined views to allow you to easily change view

Lesson 1
Desktop Features

Feature Type	Description
Railings	Create or apply different railing styles to a stair or along a defined path
Roofs	Create and apply various roof styles.
Floorplate slices	Generate the perimeter geometry of a building.
Spaces and Boundaries	Spaces and boundaries can include floor thickness, room height, and wall thickness.
Stairs	Create and apply various stair types
Tags and Schedules	Place tags on objects to generate schedules. Schedules will automatically update when tags are modified, added, or deleted.
Template Files	Use templates to save time. Create a template with standard layers, text styles, linetypes, dimension styles, etc.
Walls	Create wall styles to determine material composition. Define end caps to control opening and end conditions. Define wall interference conditions.

Architectural Desktop sits on top of AutoCAD. It is helpful for users to have a basic understanding of AutoCAD before moving to Architectural Desktop. Users should be familiar with the following:

- AutoCAD toolbars
- zoom and move around the screen
- manage blocks
- draw and modify commands
- model and paper space (layout),
- dimensioning and how to create/modify a dimension style

If you are not familiar with these topics, you can still move forward with learning ADT, but you may find it helpful to have a good AutoCAD textbook as reference in case you get stuck.

TIP: The best way to use Architectural Desktop is start your project with massing tools (outside-in design) or space planning tools (inside-out design) and continue through to construction documentation.

The AEC Project Process Model

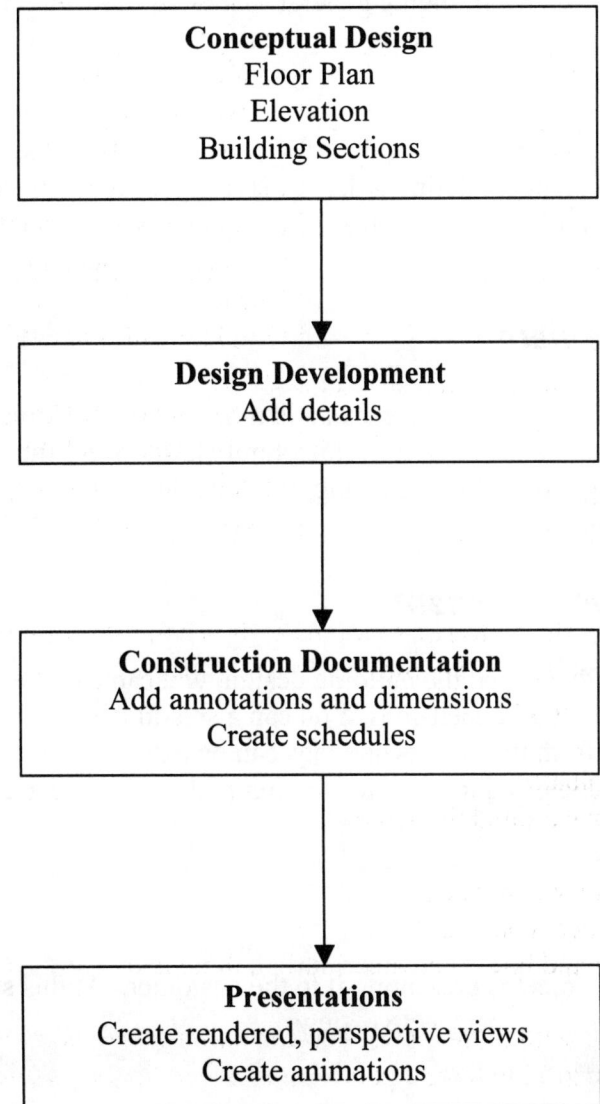

Conceptual Design

In the initial design phase, you can assemble Architectural Desktop mass elements as simple architectural shapes to form an exterior model of your building project. You can also lay out interior areas by arranging general spaces as you would in a bubble diagram. You can manipulate and consolidate three-dimensional mass elements into massing studies.

Later in this phase, you can create building footprints from the massing study by slicing floorplates, and you can begin defining the structure by converting space boundaries into walls. At the completion of the conceptual design phase, you have developed a workable schematic floor plan.

Design Development

As you refine the building project, you can add more detailed information to the schematic design. Use the features in Architectural Desktop to continue developing the design of the building project by organizing, defining, and assigning specific styles and attributes to building components.

Construction Development

After you have fully developed the building design, you can annotate your drawings with reference marks, notes, and dimensions. You can also add tags or labels associated with objects. Information from the objects and tags can be extracted, sorted, and compiled into schedules, reports, tables, and inventories for comprehensive and accurate construction documentation.

Presentations

A major part of any project is presenting it to the customer. At this stage, you develop renderings, animations, and perspective views.

Exercise 1:
Creating a New Drawing with an AEC Template

Drawing Name: New
Estimated Time: 15 minutes

This exercise reinforces the following skills:

- Use of templates
- Getting user familiar with tools and ADT environment

Menu	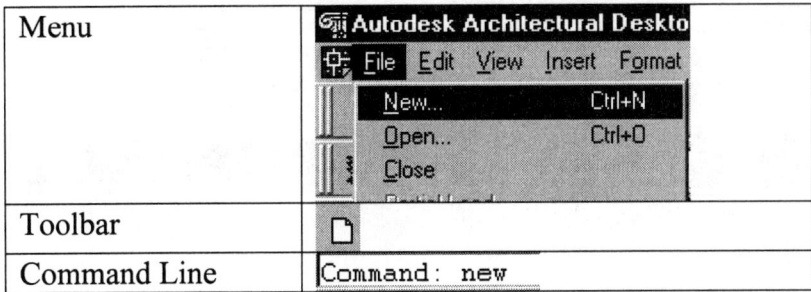
Toolbar	
Command Line	Command: new

Start a New Drawing. This can be done using the Menu, the Standard toolbar, or the command line as shown in the table above.

TIP: If you start a New Drawing using the toolbar, you will bypass the Start-Up Dialog box.

Lesson 1
Desktop Features

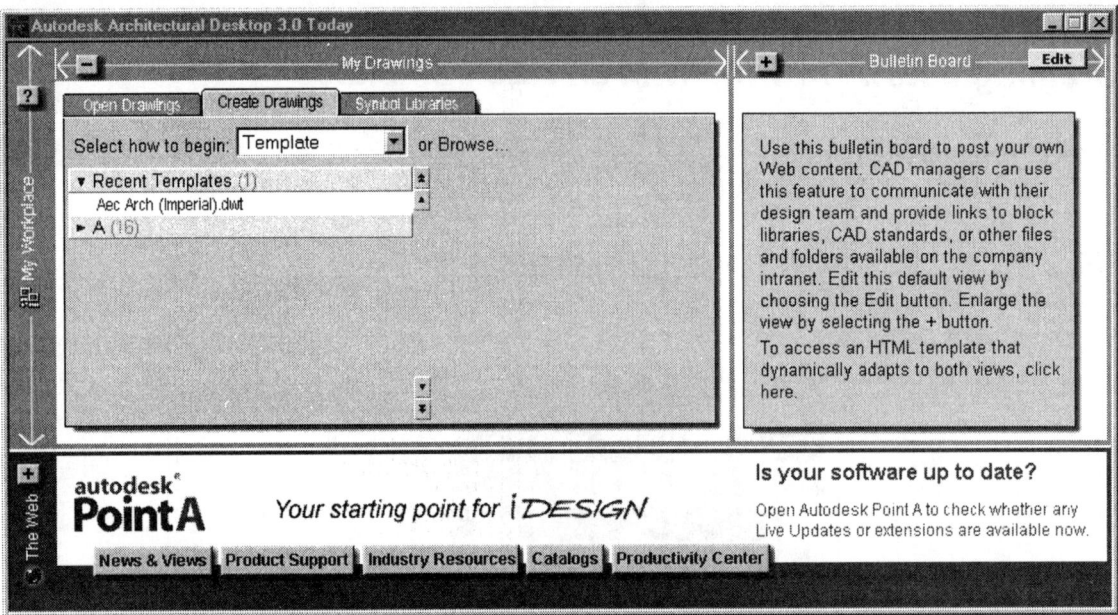

The Start-Up Dialog appears.

We will use the standard default template – Aec Arch (Imperial).dwt.

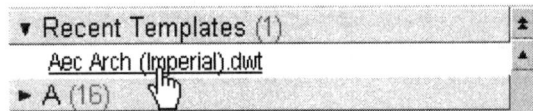

To select the template, we place our mouse over the template name as shown and left mouse button click.

At the bottom of the screen, we see eleven layout tabs. Each layout includes viewports set up to display the production type indicated.

Select each layout tab to see the differences.

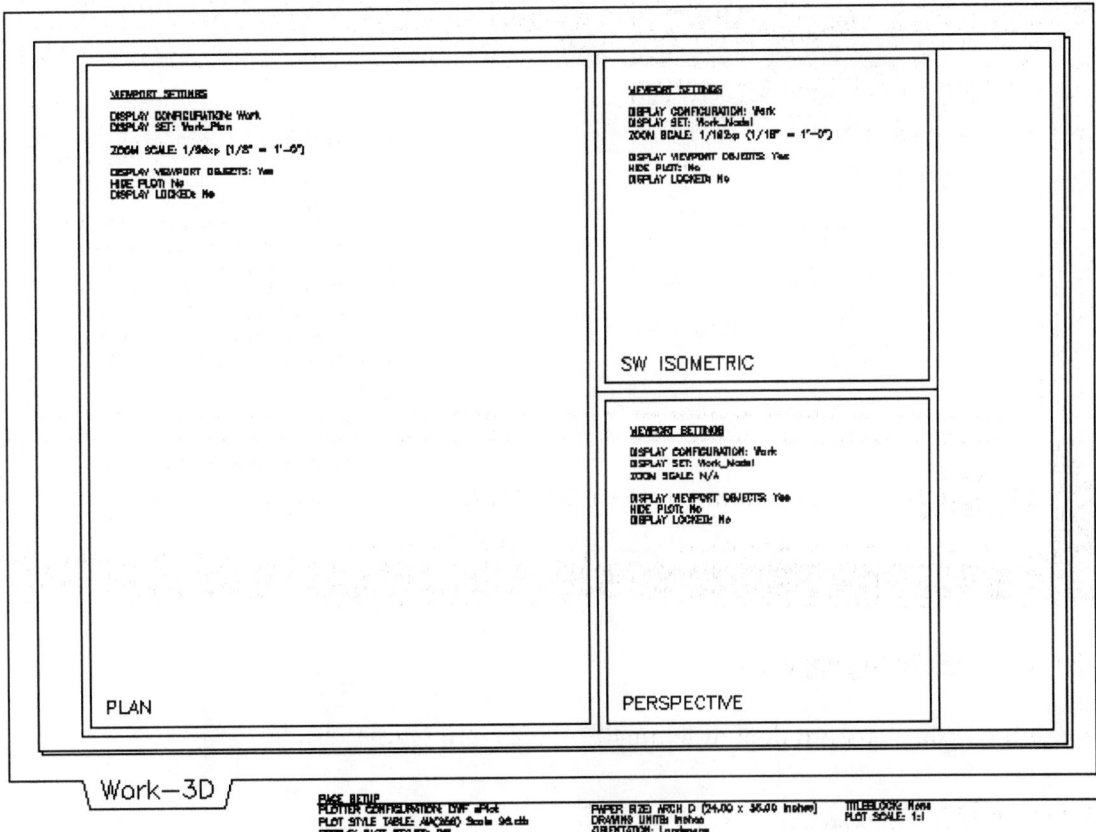

Use the Work-3D layout tab to work on everything in the building model except ceiling objects.

The template overview tab displays all the layout tabs and how they are set up. Use your Zoom tools to zoom in and read how each layout has been defined.

Setting AEC Drawing Options

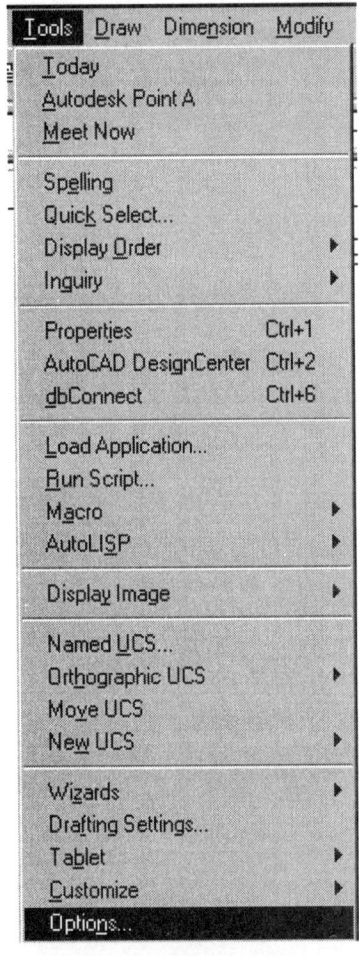
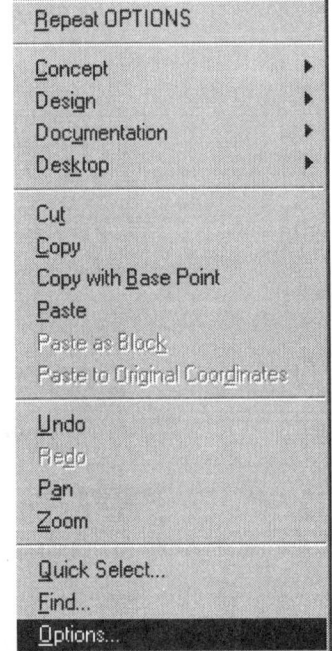
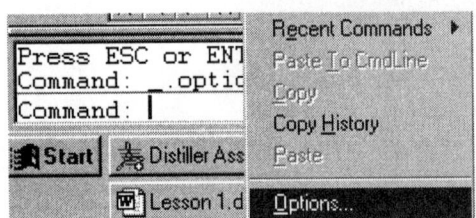

Menu	Tools->Options
Command line	Options
Context Menu->Options	Place Mouse in the graphics area and right click
Shortcut	Place mouse in the command line and right click

Access the Options dialog box.

Lesson 1
Desktop Features

| AEC Editor | AEC DwgDefaults | AEC Performance | AEC Stair Defaults | AEC Content | AEC Dimension |

ADT's Options includes six additional AEC specific tabs.
They are AEC Editor, AEC DwgDefaults, AEC Performance, AEC Stair Defaults, AEC Content, and AEC Dimension.

AEC Editor

Diagnostic Messages	
Diagnostic Messages	All diagnostic messages are turned off by default.
Object Relationship Graph	Displays data about the relationships between objects in the current drawing.
Display Management System	Turns on display control error messages
Geometry	Displays geometry-specific data.
Similar-Purpose Dialogs	Options for the position of dialog boxes and viewers
Use Common Position	Sets one common position on the screen for similar dialog boxes, such as door, wall, and window add or modify dialog boxes. Some dialog boxes, such as those for styles and properties, are always displayed in the center of the screen, regardless of this setting.
Use Common Viewer Position and Sizes	Sets one size and position on the screen for the similar-purpose viewers in Architectural Desktop. Viewer position is separately controlled for add, modify, style, and properties dialog boxes.
Block Properties of Exploded Objects	When you explode an AEC object, you create several objects grouped in a block definition. To maintain the layer, color, and linetype of AEC objects when you explode them, select Maintain Resolved Layer, Color, Linetype under Block Properties of Exploded Object. Any objects whose component layer, color, and linetype properties are set to ByBlock take the color of the parent object. If this option is cleared when you explode an object, then properties set to ByBlock remain ByBlock. Clear this option if you want to explode the block definition further.

1-9

Lesson 1
Desktop Features

Object Snap	Enable Optimize for Speed to limit certain display representations to respond only to the Node and Insert object snaps. This setting affects stair, railing, space boundary, multi-view block, masking block, slice, and clip volume result (building section) objects.
Layer Manager	Enable Optimize for Speed to load layers into the Layer Manager faster. After the layers are scanned and loaded into the Layer Manager, the status (in use/not in use) of each layer is indicated by the icon to the left side of the layer name in the Layer Manager. When you select Optimize for Speed, the Layer Manager does not scan the drawing to determine the status of the layers, speeding up the loading time. The icons that are displayed next to the layer name in the Layer Manager do not indicate whether or not the layer is in use.

TIP: Option Settings are applied to your current drawing and saved as the default settings for new drawings. Because ADT operates in a Multiple Document Interface, each drawing stores the Options Settings used when it was created and last saved. Some users get confused because they open an existing drawing and it will not behave according to the current Options Settings.

AEC DwgDefaults

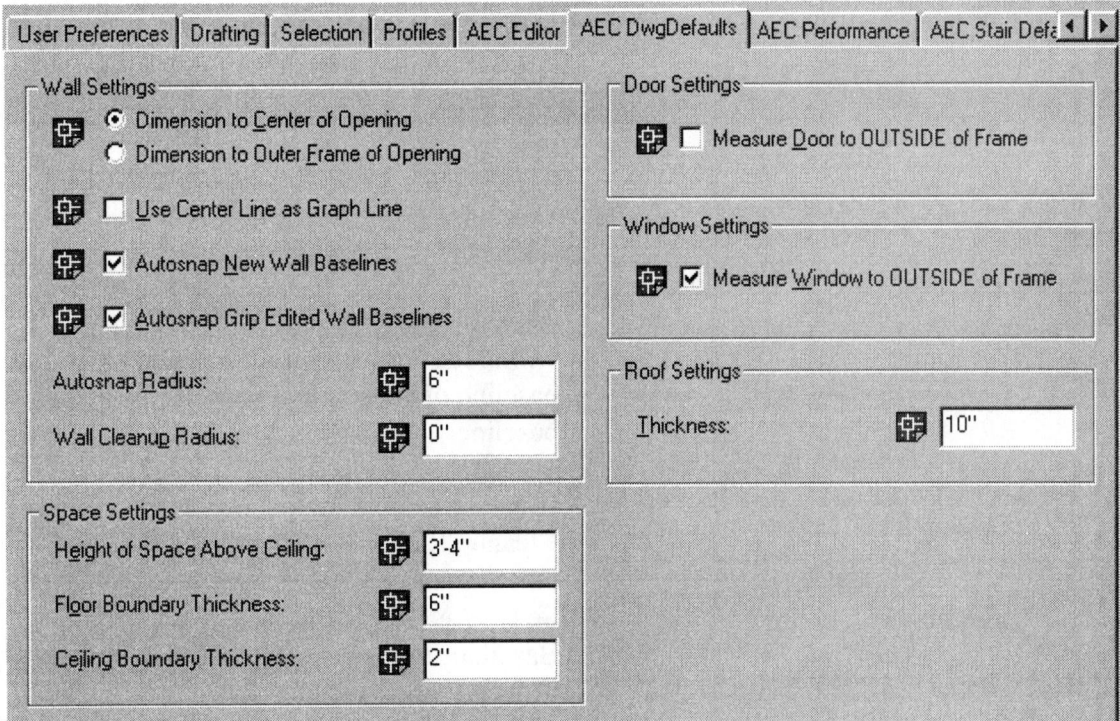

Wall Settings	
Dimension to Center of Opening	
Dimension to Outer Frame of Opening	
Use Center Line as Graph Line	Select Use Center Line as Graph Line to use wall center lines instead of the wall justification position to represent the walls in the Wall Graph display representation and calculate wall cleanups.

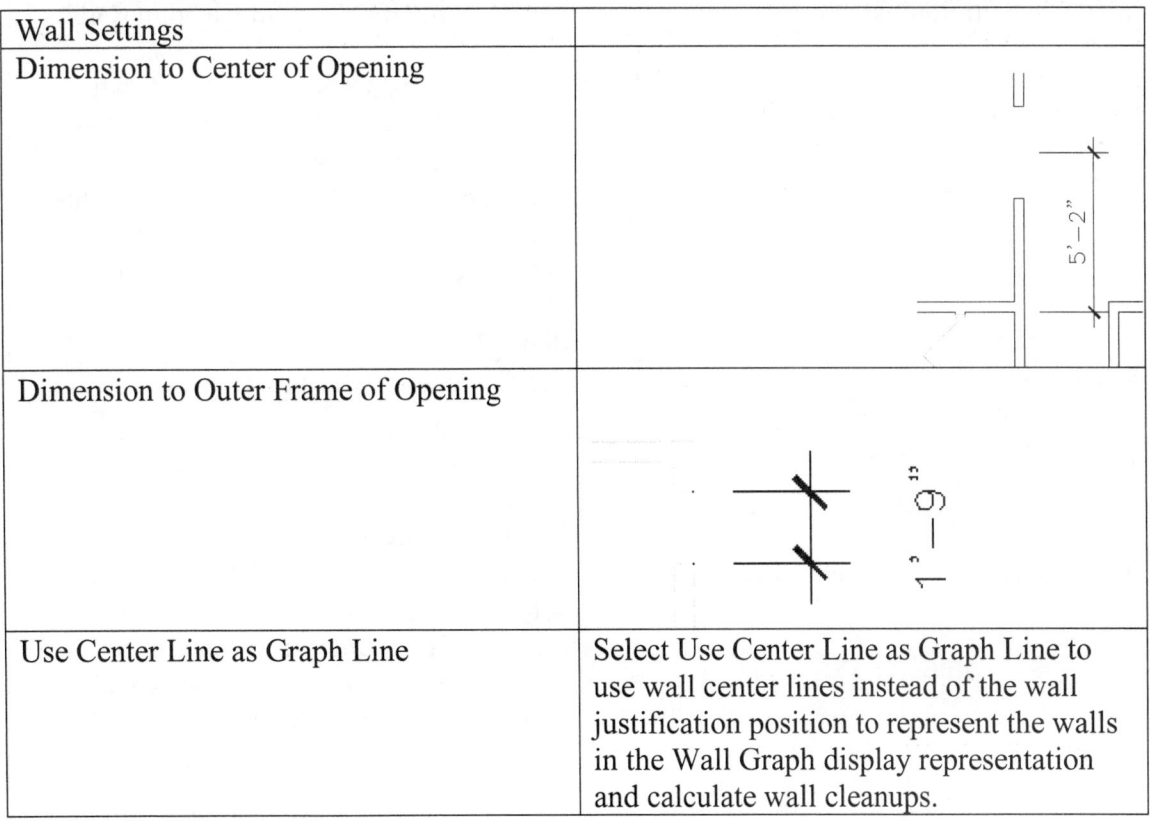

Lesson 1
Desktop Features

Autosnap New Wall Baselines	Select Autosnap New Wall Baselines to force the endpoint of a new wall drawn within the Autosnap Radius distance of the baseline of an existing wall to snap to that baseline. If you select this option and set your Wall Cleanup Radius to 0, then only walls that touch clean up with each other.
Autosnap Grip Edited Wall Baselines	Select Autosnap New Wall Baselines to force the endpoint of a new wall drawn within the Autosnap Radius distance of the baseline of an existing wall to snap to that baseline. If you select this option and set your Wall Cleanup Radius to 0, then only walls that touch clean up with each other.
Autosnap Radius	Select Autosnap Grip Edited Wall Baselines to force a wall that you grip edit to snap to the baseline of an existing wall within the Autosnap Radius distance.
Wall Cleanup Radius	Type a value for the Wall Cleanup Radius. The Wall Cleanup Radius controls whether or not walls clean up when they meet at a corner or intersection. You can set the default cleanup radius for all walls in this dialog box. You can also set different cleanup radii for individual walls on the Dimensions tab of the Wall Properties dialog box.
Space Settings	Set the default values to be used
Height of Space Above Ceiling	Type in a value
Floor Boundary Thickness	Type in a value
Ceiling Boundary Thickness	Type in a value
Door Settings	Set the defaults to be used
Measure Door to Outside of Frame	If Measure Door to OUTSIDE of Frame is selected then all measurements are from the outside of the frame. If Measure Door to OUTSIDE of Frame is not selected, all measurements are from the inside of the frame. All measurements to wall corners are to the nearest corner.

Window Settings	Set the defaults to be used
Measure Window to Outside of Frame	If Measure Window to OUTSIDE of Frame is selected then all measurements are from the outside of the frame. If Measure Window to OUTSIDE of Frame is not selected, all measurements are from the inside of the frame. All measurements to wall corners are to the nearest corner.
Roof Settings	Set the defaults to be used
Thickness	Type in a value

AEC Performance

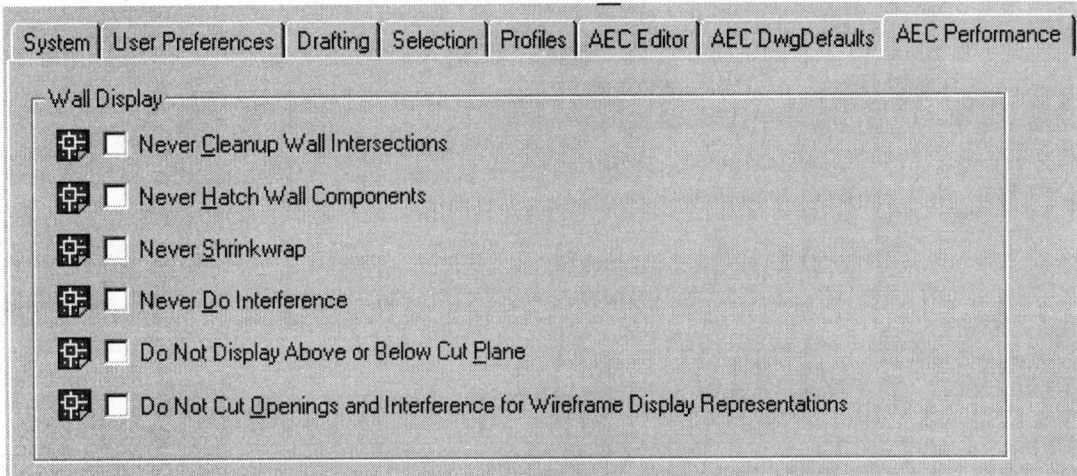

The settings in the AEC Performance tab enhance the display speed when drawing wall objects. The more Wall Display options enabled, the slower your regens.

AEC Stair Defaults

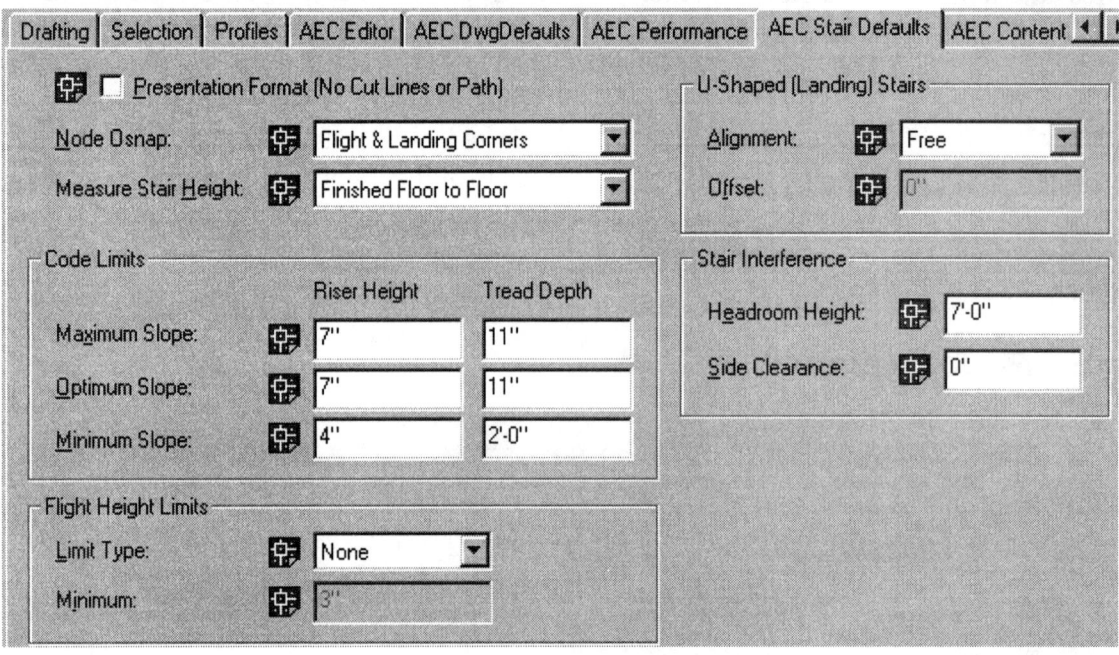

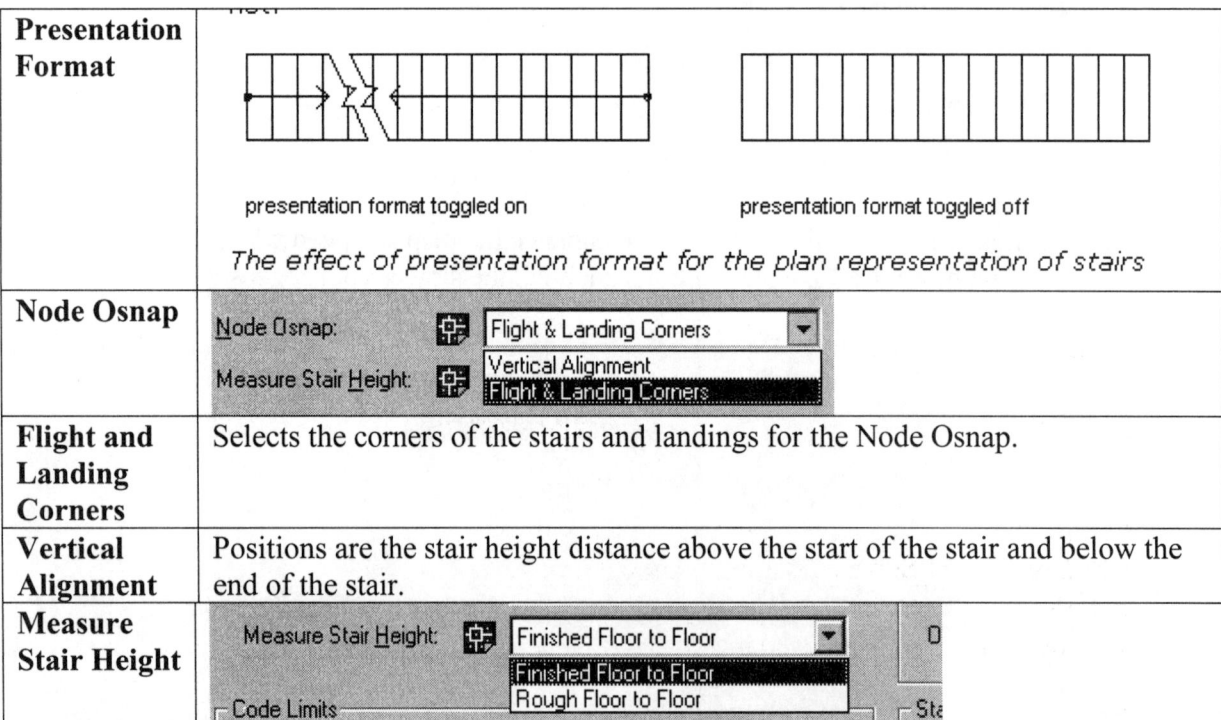

Presentation Format	
	The effect of presentation format for the plan representation of stairs
Node Osnap	
Flight and Landing Corners	Selects the corners of the stairs and landings for the Node Osnap.
Vertical Alignment	Positions are the stair height distance above the start of the stair and below the end of the stair.
Measure Stair Height	

Finished Floor to Floor	Includes top and bottom offsets
Rough Floor to Floor	Ignores top and bottom offsets.
Code Limits	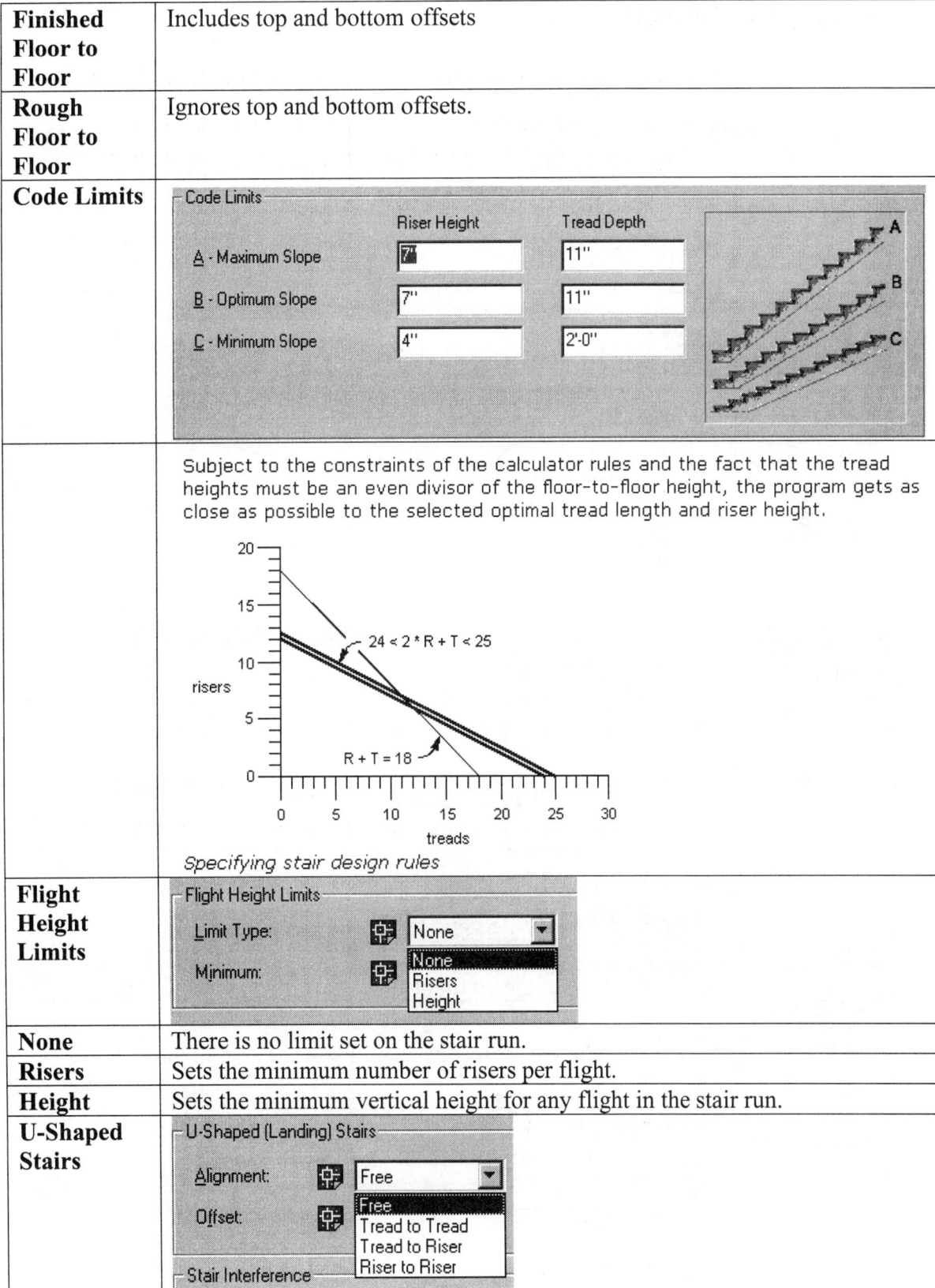

Subject to the constraints of the calculator rules and the fact that the tread heights must be an even divisor of the floor-to-floor height, the program gets as close as possible to the selected optimal tread length and riser height.

Specifying stair design rules |
Flight Height Limits	
None	There is no limit set on the stair run.
Risers	Sets the minimum number of risers per flight.
Height	Sets the minimum vertical height for any flight in the stair run.
U-Shaped Stairs	

Lesson 1
Desktop Features

Free	
Tread to Tread	Aligns the tread of the up flight to the tread of the down flight.
Tread to Riser	Aligns the tread of the up flight to the riser of the down flight.
Riser to Riser	

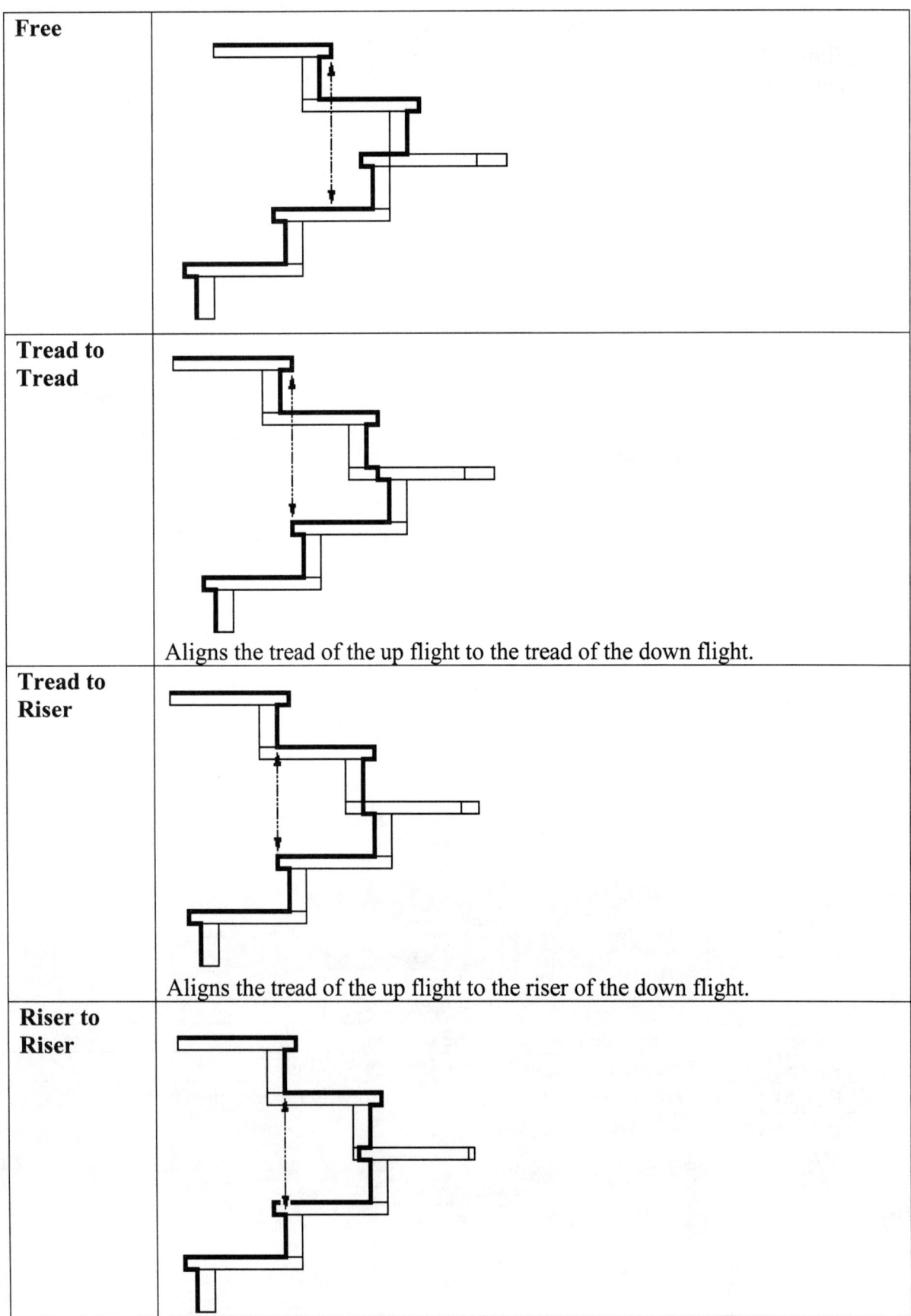

Offsets	Aligns the riser of the up flight to the riser of the down flight. *Specifying tread and riser alignment offsets* If you selected Tread to Tread, Tread to Riser, or Riser to Riser, you can set an offset value. A positive offset value means that the down flight tread location is closer to landing than the up flight tread location.
Stair Interference	The stair interference area, which is a box that is cut out from a space around the stair from the ceiling boundary based on the specified headroom height and side clearance.
Headroom Height	Sets the height above the treads for the interference condition.
Side Clearance	Sets the width of the interference condition on all sides of a stair run, other than where you step on and off at the first and last treads.

Lesson 1
Desktop Features

AEC Content

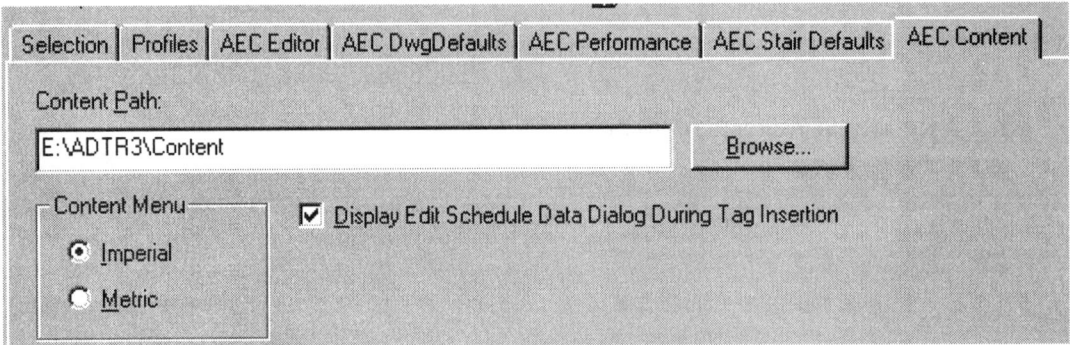

Content Path	Type the path and location of your content files, or click Browse to search for the content files.
Content Menu	The symbols in the Design and Documentation menus change to reflect the option that you select: Imperial (Feet and Inches) or Metric (meters)
Display Edit Schedule Data Dialog During Tag Insertion	To attach schedule data to objects when you insert a schedule tag in the drawing, this should be ENABLED.

TIP: Your Autodesk Architectural Desktop CD-ROM provides two imperial content menus: an Autodesk Architectural Desktop content menu and a CSI MasterFormat content menu. You can also locate FREE AEC Content on Autodesk's pointA.

Using AEC Content

Architectural Desktop uses several pre-defined and user-customizable content including:

- Architectural Display Configurations
- Architectural Profiles of Geometric Shapes
- Wall Styles and Endcap geometry
- Door styles
- Window styles
- Stair styles
- Space styles
- Schedule tables

Standard style content is stored in the AEC templates subdirectory. You can create additional content, import and export styles between drawings.

TIP: I highly recommend that you store any custom content in a separate directory as far away from AutoCAD Architectural Desktop as possible. This will allow you to back up your custom work easily and prevent accidental erasure if you upgrade your software application.

You should be aware that ADT currently allows the user to specify only ONE path for custom content, so drawings with custom commands will only work if they reside in the specified path.

AEC Dimension

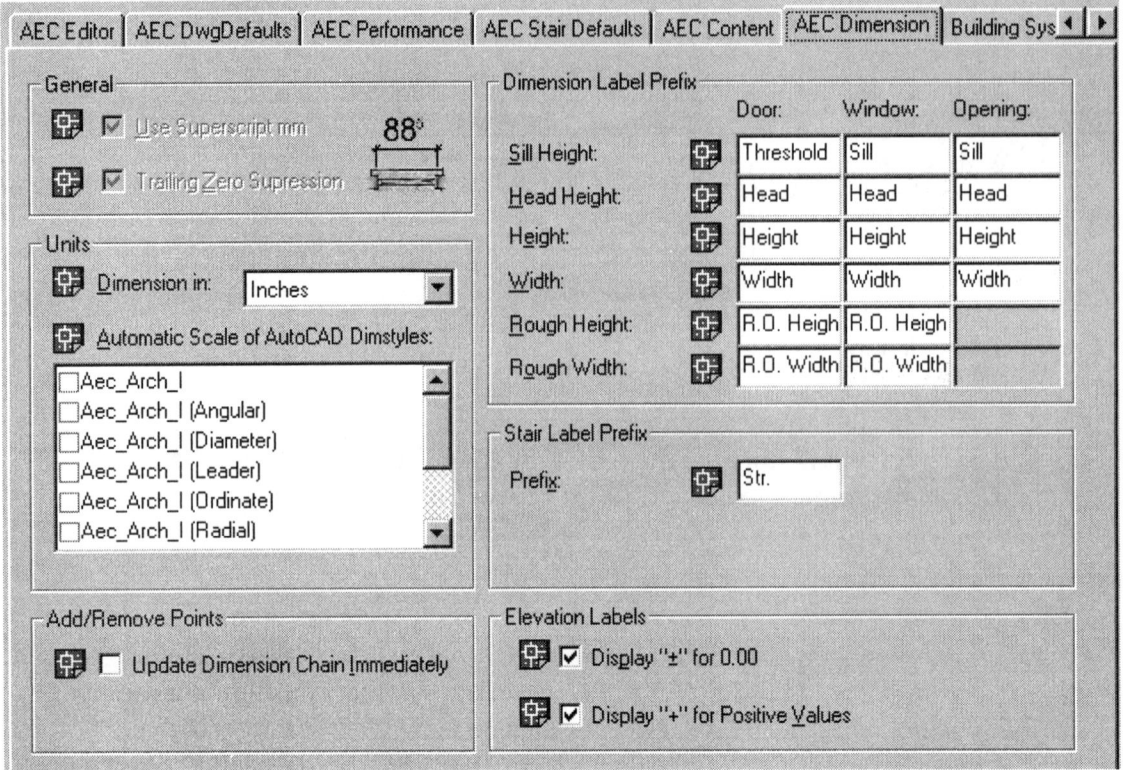

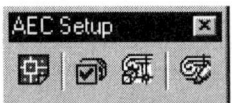

Desktop Display Manager

Menu	Desktop->Display Manager
AEC Setup Toolbar	
Command Line	DisplayManager

Access the Display Manager using the Menu.
Go to Desktop->Display Manager

The display system in Autodesk® Architectural Desktop controls how AEC objects are displayed in a designated viewport. By specifying the AEC objects you want to display in a viewport and the direction from which you want to view them, you can produce different architectural displays, such as floor plans, reflected plans, elevations, 3D models, or schematic displays.

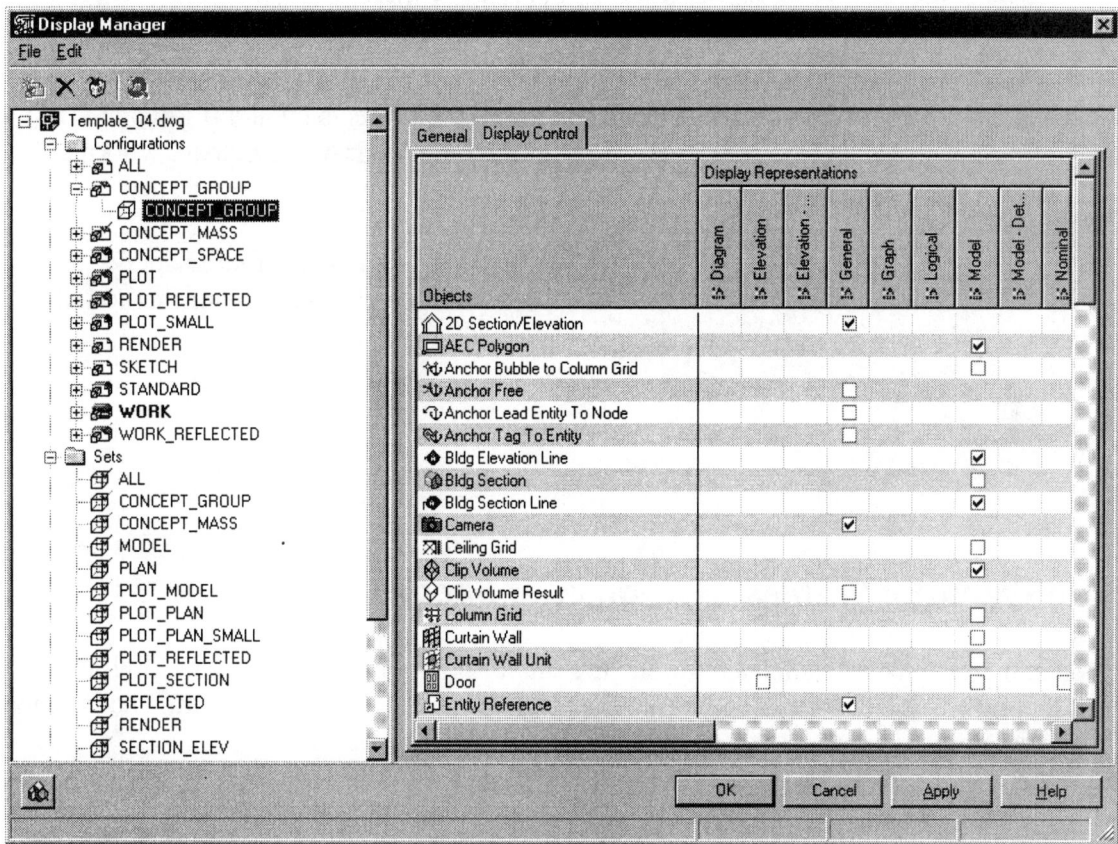

The Display Manager tool is located on the AEC Setup toolbar.

Lesson 1
Desktop Features

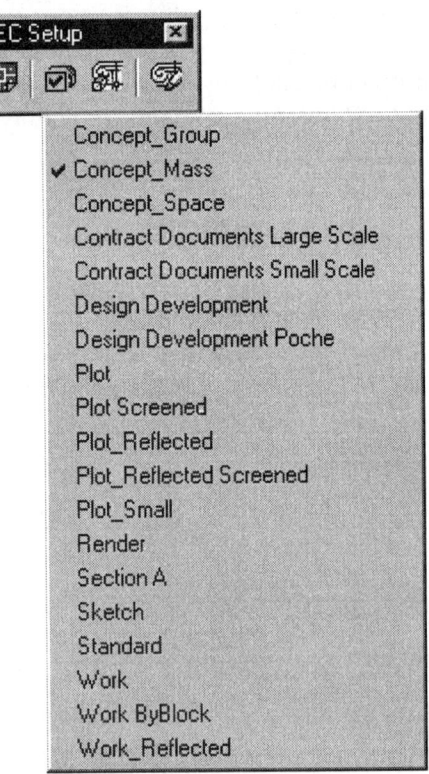

Left clicking the Check box on the AEC Setup toolbar allows the user to quickly switch between different Display Configurations.

ADT has several Display Configurations set up to make it easier for you to work. These configurations automatically turn visibility of objects on and off depending on the type of drawing you are working on.

You can also create custom Display Configurations based on your method of working.

Architectural Profiles of Geometric Shapes

You can create mass elements to define the shape and configuration of your preliminary study, or mass model. After you create the mass elements you need, you can change their size as necessary to reflect the building design.

- **Mass element**: A single object that has behaviors based on its shape. For example, you can set the width, depth, and height of a box mass element, and the radius and height of a cylinder mass element.

Mass elements are parametric, which allows each of the shapes to have very specific behavior when it comes to the manipulation of each mass element's shape. For example, if the corner grip point of a box is selected and dragged, then the width and depth are modified. It is easy to change the shape to another form by right-clicking on the element and selecting a new shape from the list.

Through Boolean operations (addition, subtraction, intersection), mass elements can be combined into a mass group. The mass group provides a representation of your building during the concept phase of your project.

- **Mass group**: Takes the shape of the mass elements and is placed on a separate layer from the mass elements.
- **Mass model**: A virtual mass object, shaped from mass elements, that defines the basic structure and proportion of your building. A marker appears as a small box in your drawing to which you attach mass elements.

As you continue developing your mass model, you can combine mass elements into mass groups and create complex building shapes through addition, subtraction, or intersection of mass elements. You can still edit individual mass elements attached to a mass group to further refine the building model.

To study alternative design schemes, you can create a number of mass element references. When you change the original of the referenced mass element, all the instances of the mass element references are updated.

The mass model that you create with mass elements and mass groups is a refinement of your original idea that you carry forward into the next phase of the project, in which you change the mass study into floorplates and then into walls. The walls are used to start the design phase.

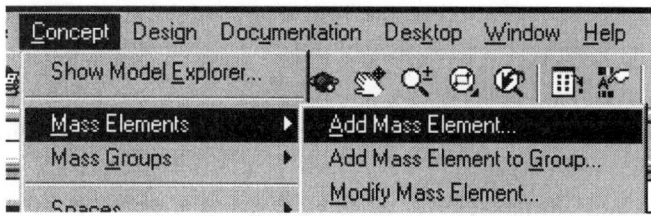

To add a Mass Element, use Concept->Mass Elements->Add Mass Element.

This will bring up the Add Mass Element dialog box.

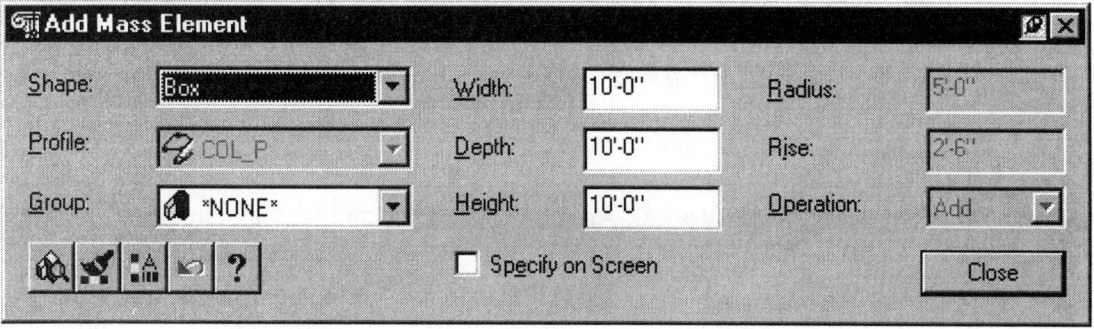

Mass Elements that can be defined include Arches, Gables, Doric Columns, etc.

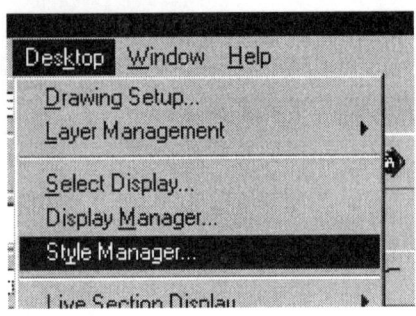 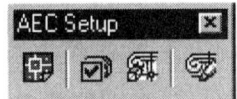

The Style Manager

Menu	Desktop->Style Manager
AEC Setup Toolbar	
Command Line	StyleManager

The Style Manager is a Microsoft® Windows Explorer-based utility that provides you with a central location in Autodesk Architectural Desktop where you can view and work with styles in drawings or from Internet and intranet sites.

Styles are sets of parameters that you can assign to objects in Autodesk Architectural Desktop to determine their appearance or function. For example, a door style in Autodesk Architectural Desktop determines what door type, such as single or double, bi-fold or hinged, a door in a drawing represents. You can assign one style to more than one object, and you can modify the style to change the all the objects that are assigned that style.

Depending on your design projects, either you or your CAD Manager might want to customize existing styles or create new styles. The Style Manager allows you to easily create, customize, and share styles with other users. With the Style Manager, you can:

- Provide a central point for accessing styles from open drawings and Internet and intranet sites

- Quickly set up new drawings and templates by copying styles from other drawings or templates
- Sort and view the styles in your drawings and templates by drawing or by style type
- Preview an object with a selected style
- Create new styles and edit existing styles
- Delete unused styles from drawings and templates

- Send styles to other Autodesk Architectural Desktop users by email

Objects in Autodesk Architectural Desktop that use styles include 2D sections and elevations, AEC Polygons, curtain walls, curtain wall units, doors, endcaps, railings, roof slab edges, roof slabs, schedule tables, slab edges, slabs, spaces, stairs, structural members, wall modifiers, walls, window assemblies, and windows.

Additionally, layer key styles, schedule data formats, and cleanup group, mask block, multi-view block, profile, and property set definitions are handled by the Style Manager.

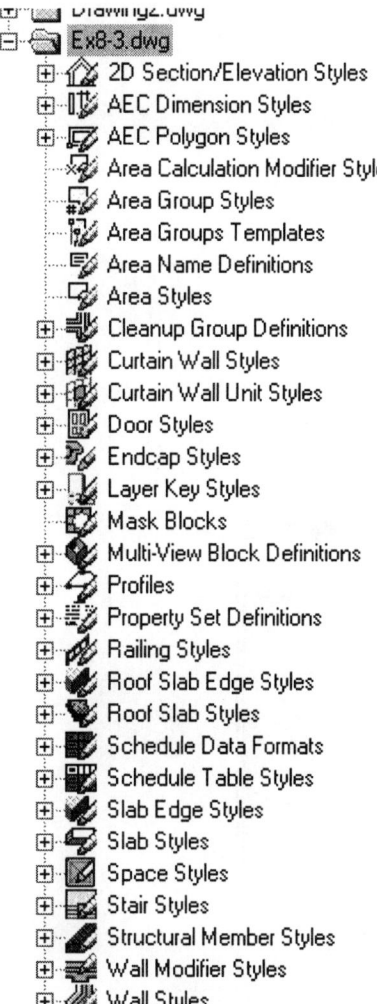

Most of the objects in Autodesk Architectural Desktop have a default Standard style. In addition, Autodesk Architectural Desktop includes a starter set of styles that you can use with your drawings. The Autodesk Architectural Desktop templates contain some of these styles. Any drawing that you start from one of the templates includes these styles. You can also access the styles for doors, endcaps, spaces, stairs, walls, and windows from drawings located in *c:\Program Files\Autodesk Architectural\Content\Imperial or Metric\Styles*. Property set definitions and schedule tables are located in the *c:\Program Files\Autodesk Architectural\Content\Imperial or Metric\Schedules* folder.

You can access the Style Manager directly by choosing it from the Desktop menu. You can also access it when you choose object styles from other Autodesk Architectural Desktop menus.

Wall Styles and Endcap geometry

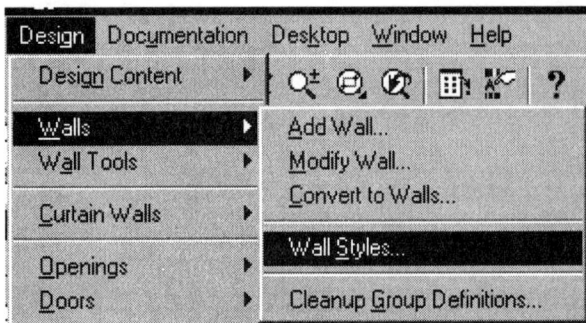

To create or modify wall styles, use the Menu.
Go to Design->Walls->Wall Styles.

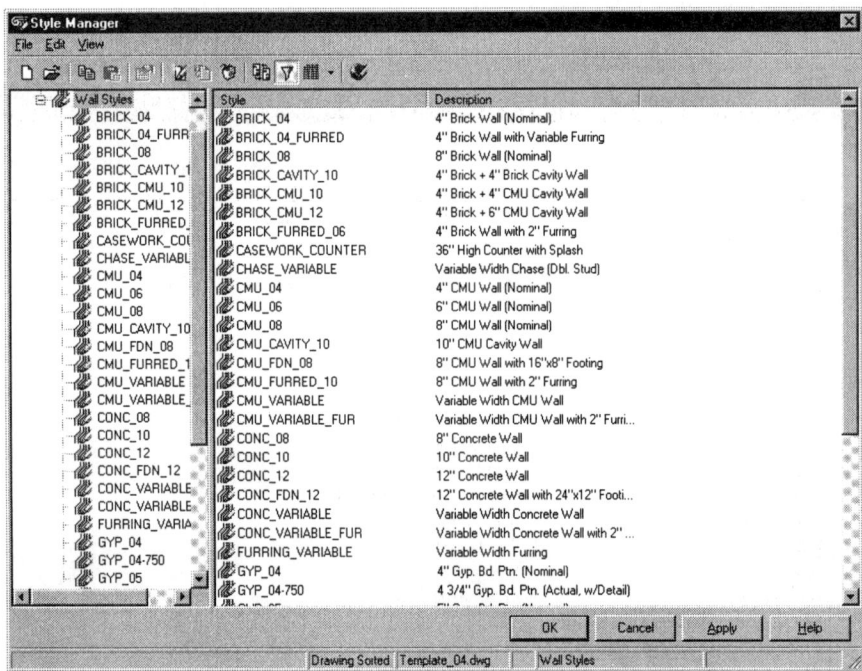

The Wall Style Manager allows the user to modify existing wall styles and create new wall styles.

> **TIP:** I do not recommend modifying ADT's standard styles as this may affect drawings you bring in from outside sources. Instead, copy the existing style to a NEW style and modify it using the desired properties.

Door styles

To create or modify door styles, use the Menu.
Go to Design->Doors->Door Styles.

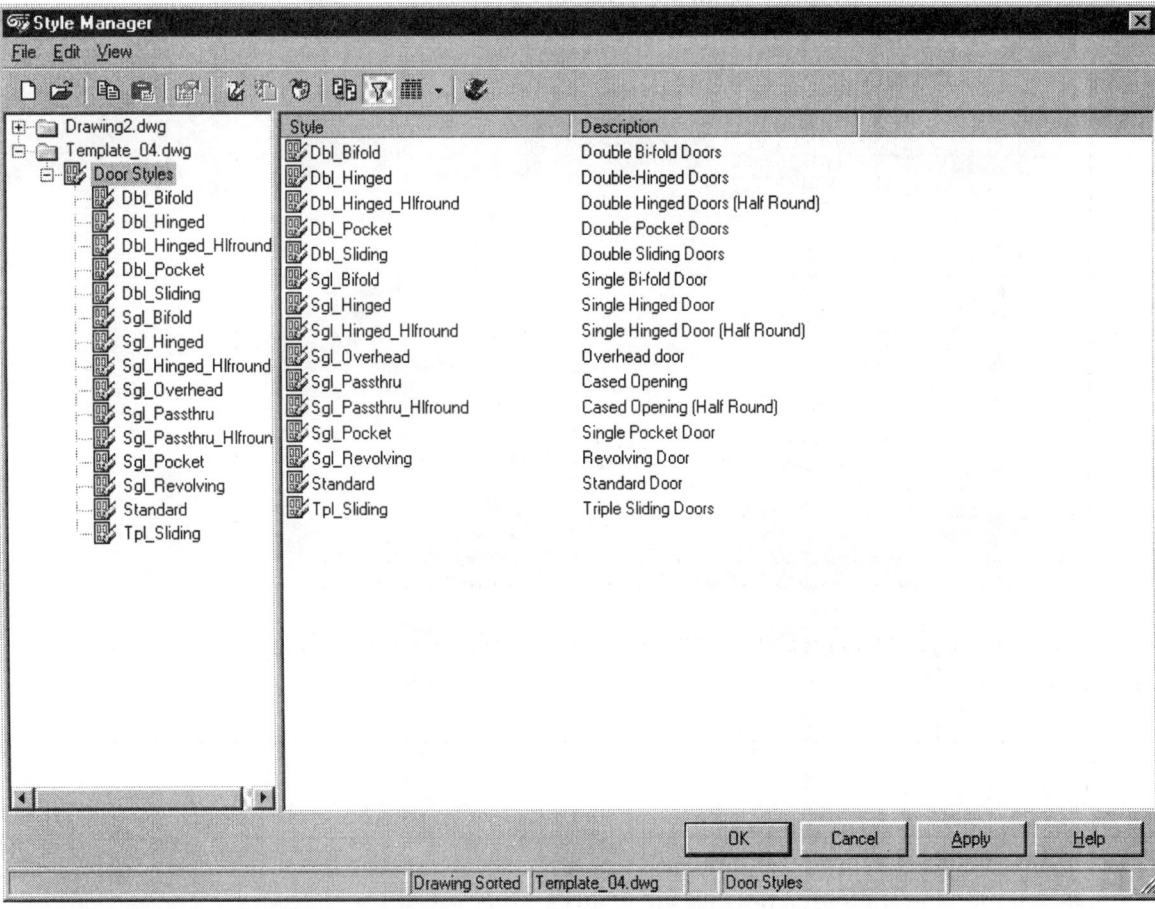

Window styles

To create or modify door styles, use the Menu.
Go to Design->Windows->Window Styles.

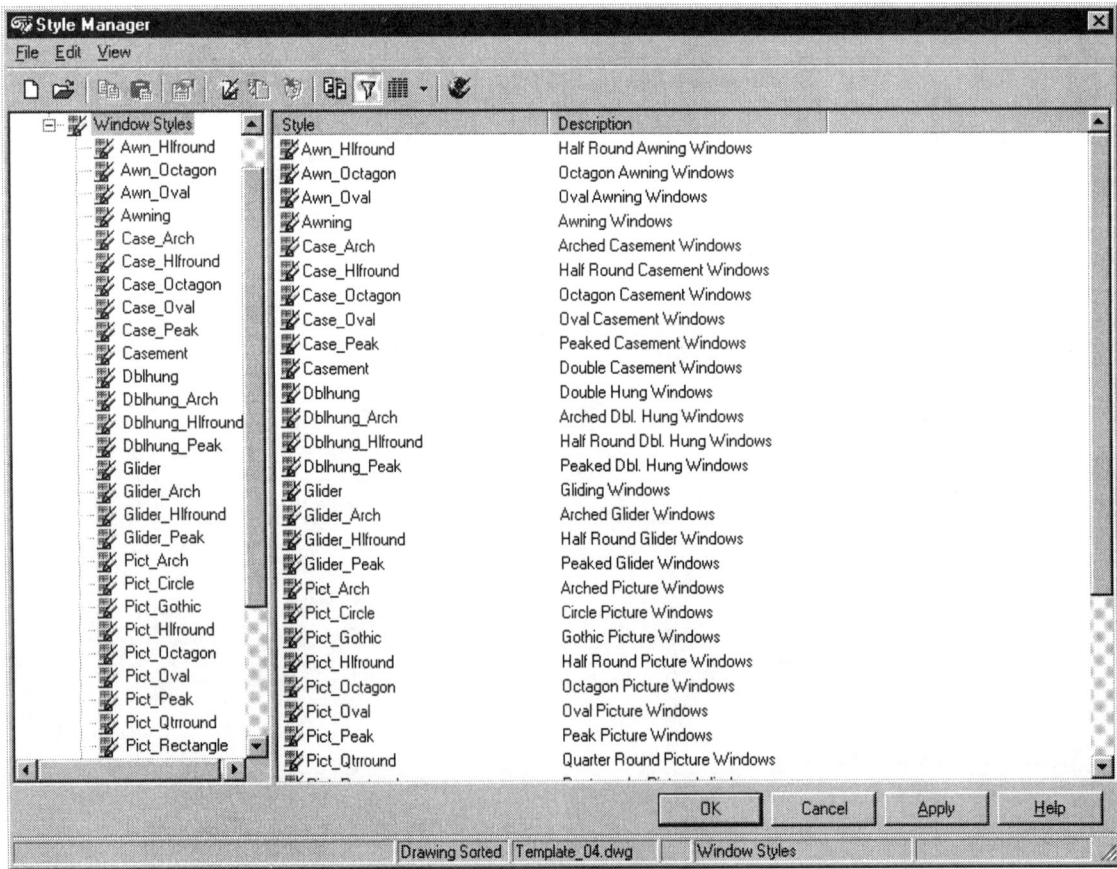

Stair styles

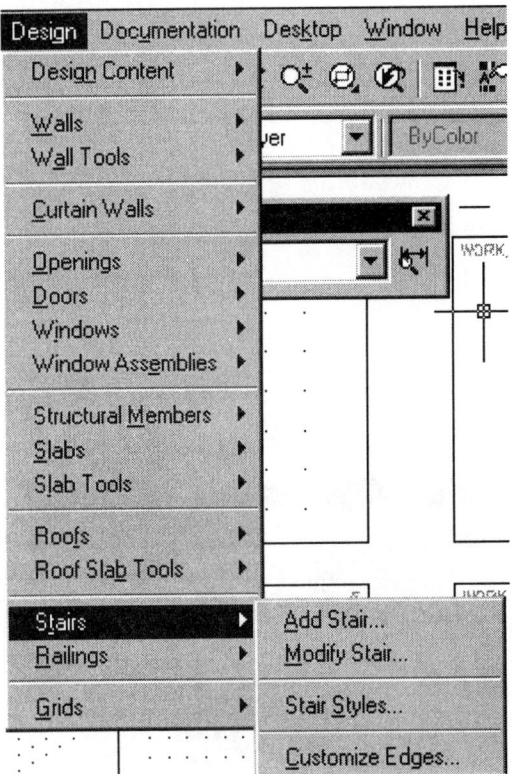

To create or modify door styles, use the Menu.
Go to Design->Stairs->Stair Styles.

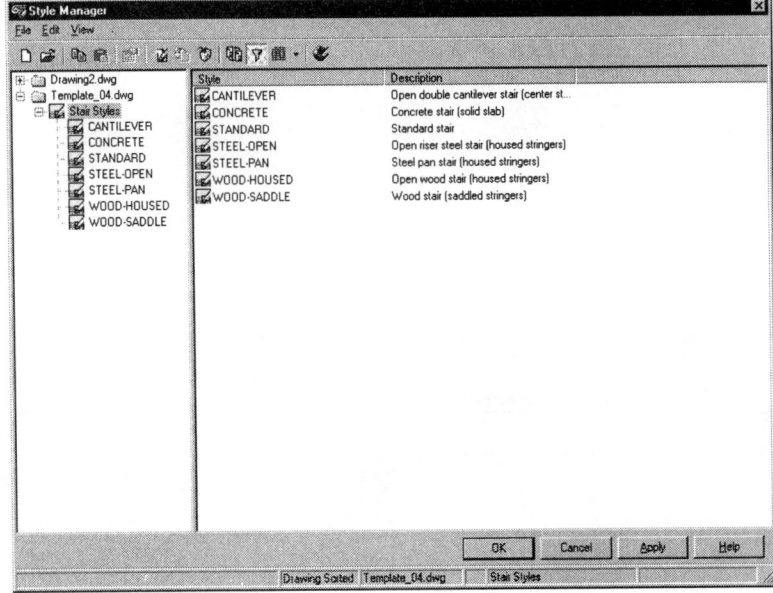

Space styles

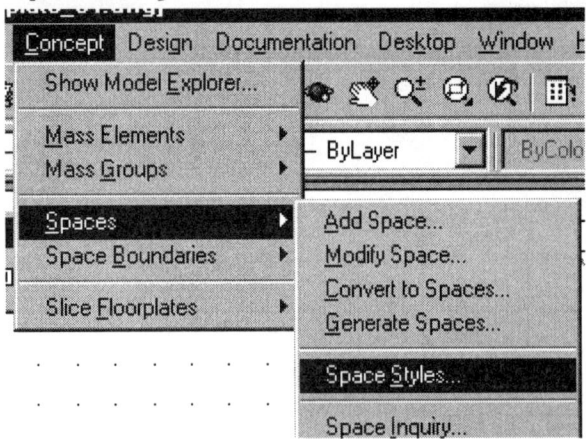

To create or modify space styles, use the Menu.
Go to Concept->Spaces->Space Styles.

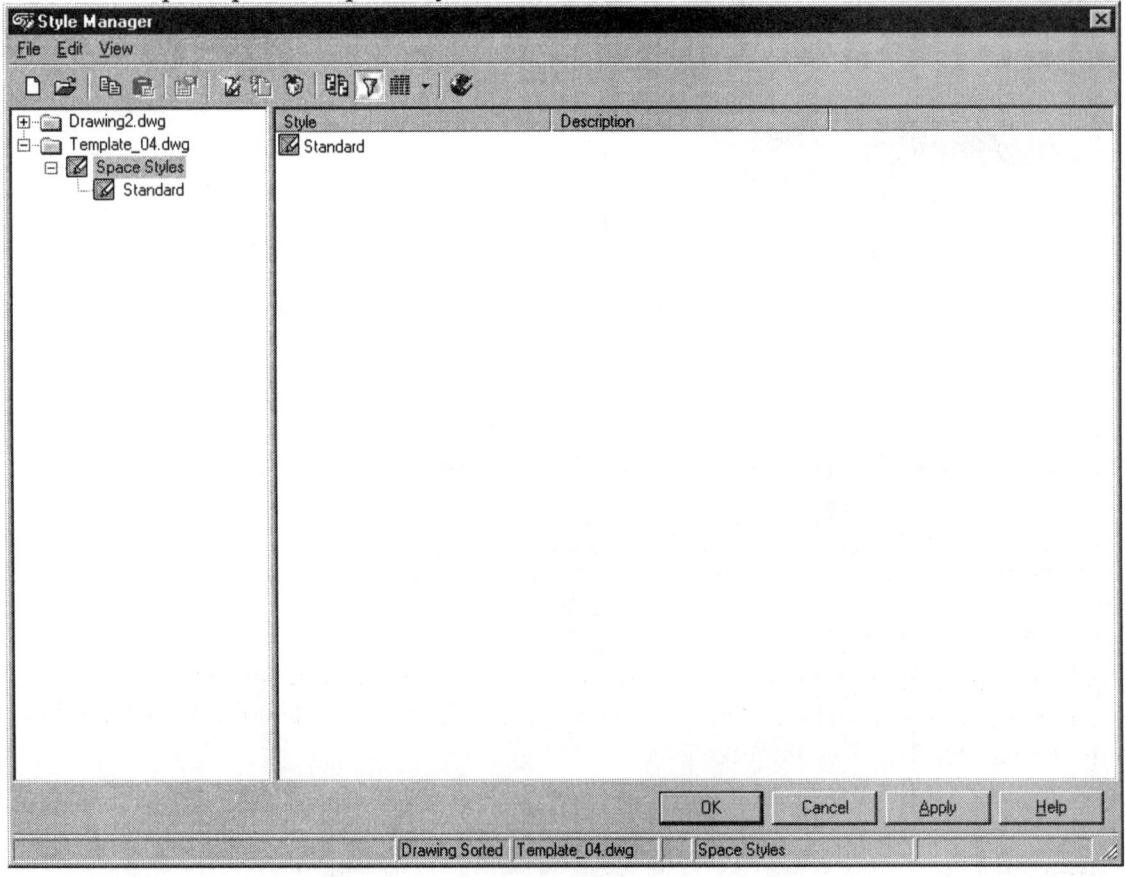

Schedule tables

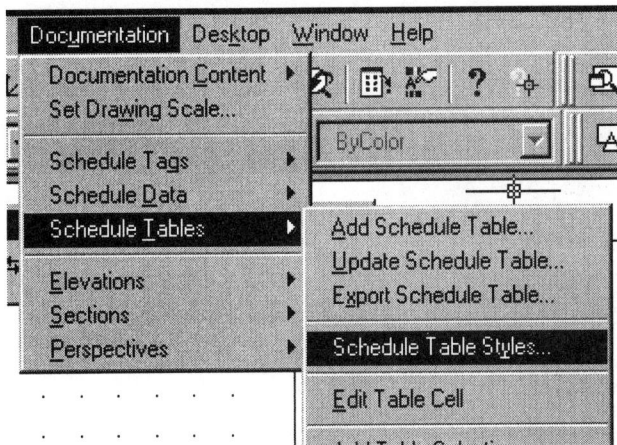

To create or modify Schedule Table styles, use the Menu.
Go to Documentation->Schedule Tables->Schedules Table Styles.

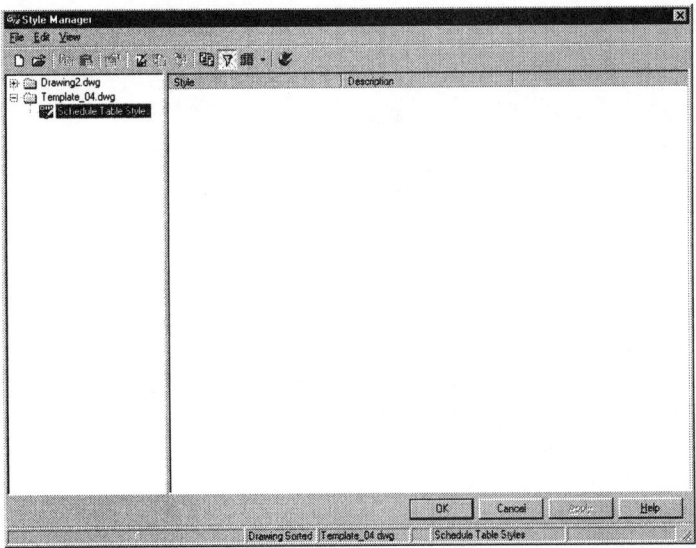

When you open the Style Manager, all the style information from your open drawings and templates is displayed. The Style Manager is split into two resizable panes and has a menu bar, toolbar, and status bar.

The left pane of the Style Manager organizes the styles in your open drawings and templates in a hierarchical tree view that you can navigate by expanding and collapsing the different levels in the tree. You can sort styles in the tree view by the drawing that contains them, or by the object type that they modify. The tree view is always displayed in the left pane, regardless of how you sort the styles.

The right pane of the Style Manager displays different style information, depending on what you select in the tree view in the right pane. You can preview how an object is displayed with a style, view style descriptions, view drawing information, and view the distribution of styles of a selected type across multiple drawings and templates.

The top of the Style Manager includes a menu bar and a toolbar that allow you to quickly access the menu commands. If you position your mouse over a toolbar icon, a tooltip displays with an explanation of the icon.

The status bar at the bottom right of the Style Manager indicates how the styles are sorted in the tree view, and the drawing and style type or style currently selected.

Exercise 2:
Creating a New Geometric Profile

Drawing Name: New using Imperial.dwt
Estimated Time: 15 minutes

This exercise reinforces the following skills:

- Use of AEC Design Content
- Use of Mass Elements in ADT

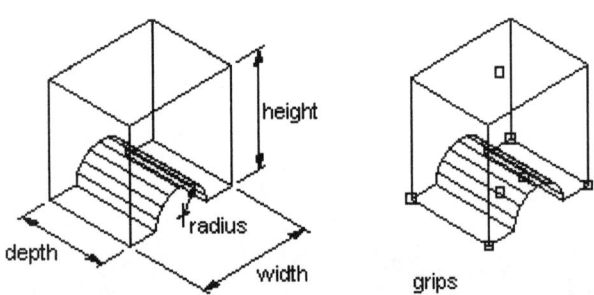

Creating an arch mass element

Start a New Drawing using the Imperial.dwt.

Switch to an isometric view.

Method 1:

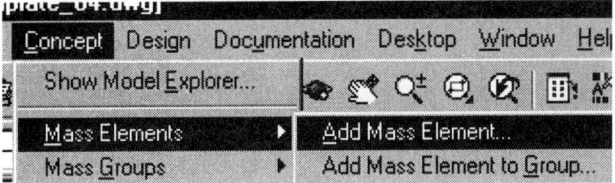

Go to Concept->Mass Elements-Add Mass Element.

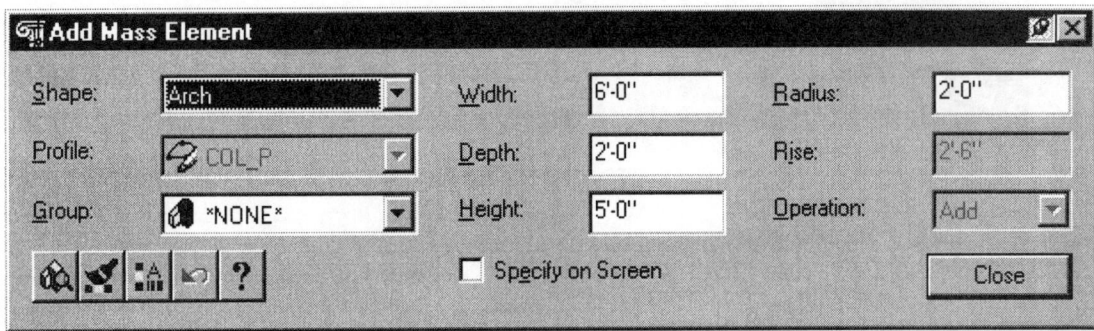

In the Add Mass Element dialog box, from the Shape list, select Arch.

Set Width to 6'.
Set Depth to 2'.
Set Height to 5'.
Set Radius to 2'.

Left mouse click to place the arch anywhere in your drawing.

The insertion point is at the centroid of the bottom face of the arch mass element.

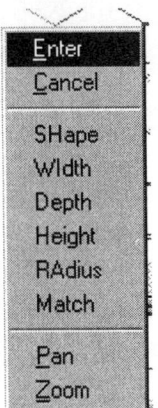

Press ENTER for a rotation angle of zero degrees or right click and select ENTER.

Lesson 1
Desktop Features

Click Close or press ENTER to end the command.

Use the Flat Shaded tool to shade your arch.

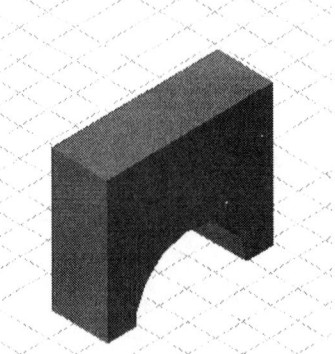

Method 2:

Activate the Mass Elements toolbar.

To activate the Mass Elements toolbar:

Lesson 1
Desktop Features

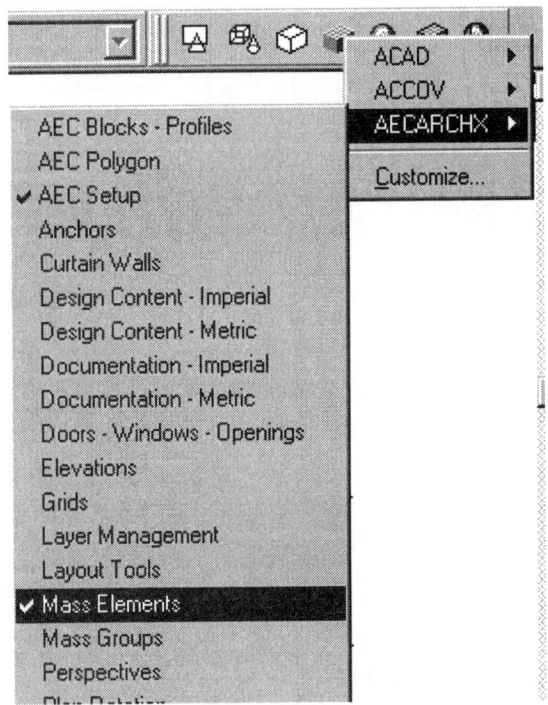

Right click in the gray open area in the menu area (not on any toolbar). Select the AECARCHX header and you will get a toolbar fly out for all the ADT toolbars. Select the Mass Elements Toolbar.

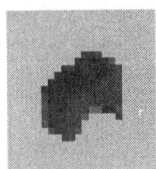

Select the Arch tool from the Mass Elements toolbar.

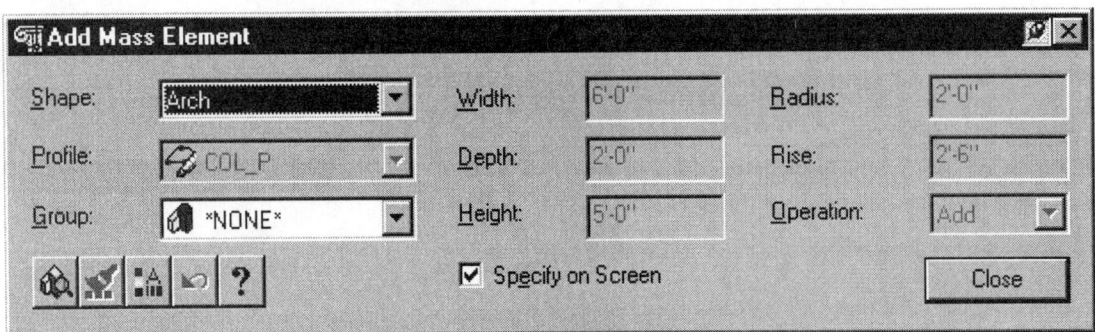

Enable the Specify On Screen option. Notice that this grays out all the values previously entered.

On the command line, you are prompted to select the first corner for your arch.
If you right click, you will get a context sensitive menu.

1-35

Lesson 1
Desktop Features

Select Shape.

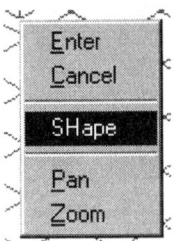

```
First Corner or [SHape]:
Second Corner or [SHape]: SH
Shape [BOx/Arch/BArrel vault/DORic/COne/CYlinder/DOMe/Gable/Isoc triangle/Right triangle/Pyramid/Sphere]:
```

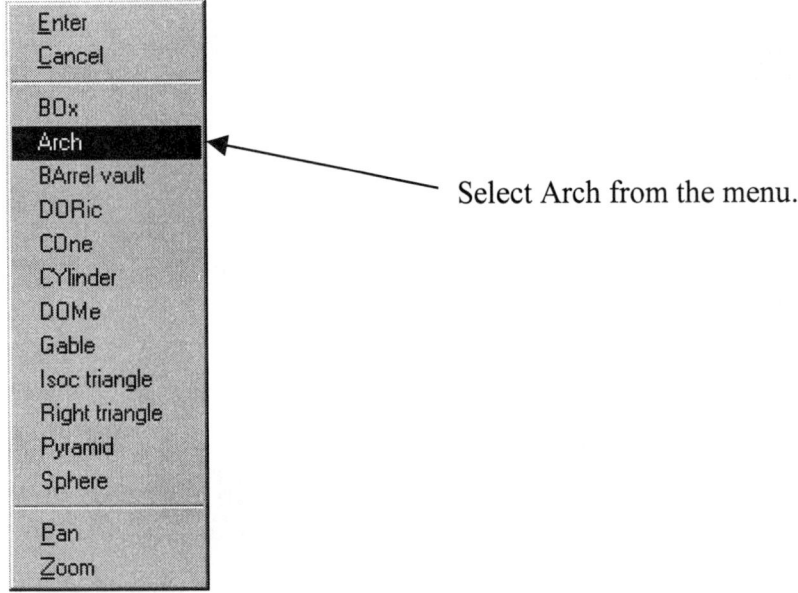

Select Arch from the menu.

On the command line, you see a list of shape options.
If you right click, you get a context sensitive menu with all the shape options.

```
Shape [BOx/Arch/BArrel vault/DORic/COne/CYlinde
triangle/Pyramid/Sphere]: A

First Corner or [SHape]:
Second Corner or [SHape]: @5',2'

Height or [SHape]: 4'

Rotation or [SHape] <0.00>: <ENTER>

First Corner or [SHape/Undo]:   <ENTER>
```

Pick anywhere in your graphics area to select the first corner.
When prompted for the second corner, type @5',2'.
When prompted for the height, type 4'
When prompted for the rotation, press 'ENTER'.

Lesson 1
Desktop Features

When prompted to start a new arch, press 'ENTER'.

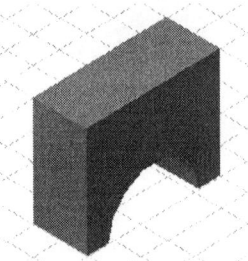

Note that this method used the default value for the arch radius.

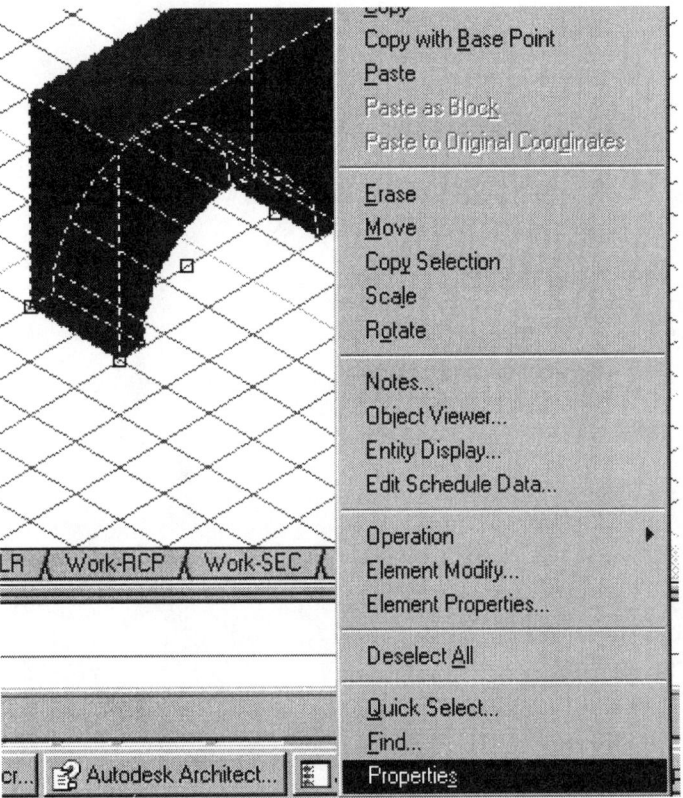

To change the arch properties, select the arch so that it is highlighted.
Right click and select 'Properties'.

1-37

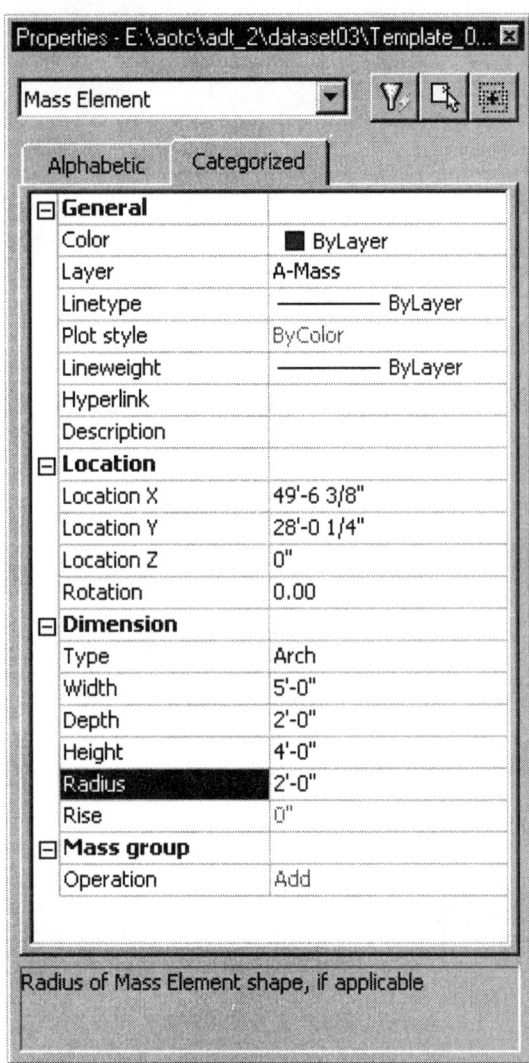

In the Properties table, locate Radius and change it to 2'0".

Note that the arch radius updates immediately.

Close the Properties dialog box.

Close your drawing. You do not need to save it.

Exercise 3:
Creating a New Wall Style

Drawing Name: New using Imperial.dwt
Estimated Time: 15 minutes

This exercise reinforces the following skills:

- Use of AEC Design Content
- Use of Wall Styles

Open a new drawing using the Imperial.dwt.

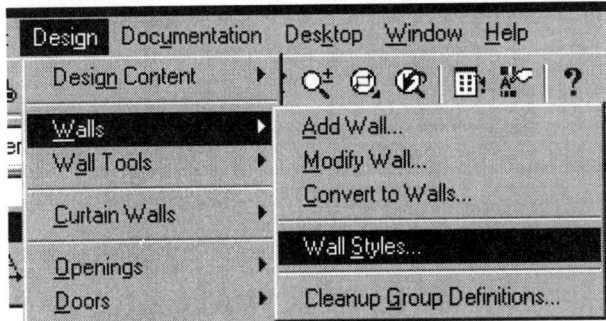

From the Design menu, choose Walls-> Wall Styles, or

From the Walls toolbar, select the Wall Styles tool.

The Style Manager is displayed, with the current drawing expanded in the tree view. The wall styles in the current drawing are displayed under the wall style type. All other style and definition types are filtered out in the tree view.

Select the New Style tool.

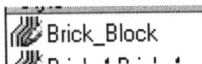

Change the name of the new style by typing 'Brick_Block'.

Highlight the new style you just created.

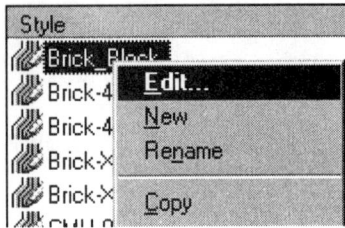

Right click and select 'Edit'.

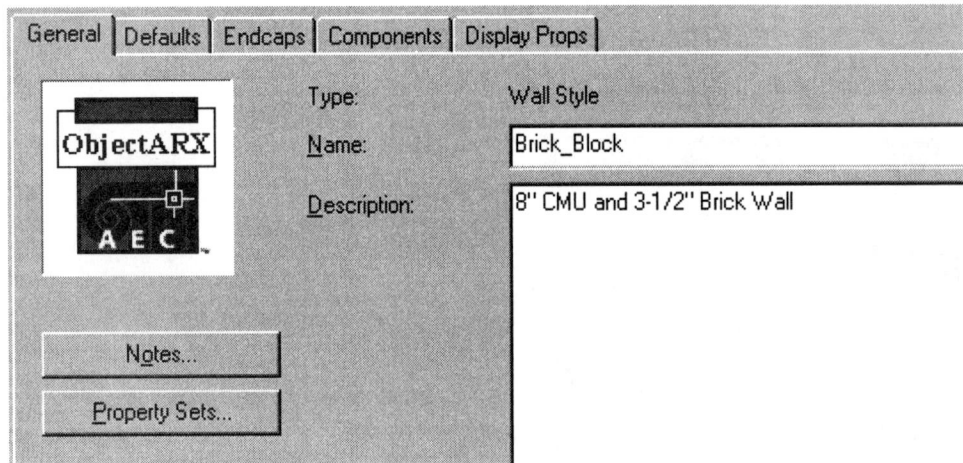

In the General tab, type in a description in the Description field as shown.

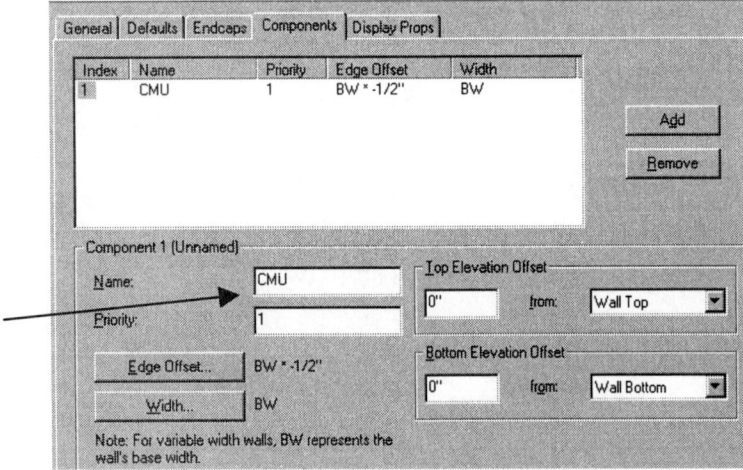

Select the Components tab.

Change the Name to CMU by typing CMU in the Name field indicated.

Press the Edge Offset Button.

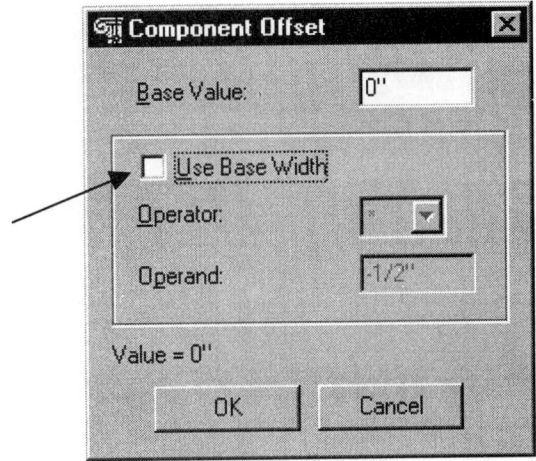

Set the Base Value to 0". Disable the 'Use Base Width' option.

A 0" offset specifies that the outside edge of the CMU is coincident with the wall baseline.

Press 'OK'.

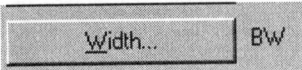

Press the Width button.

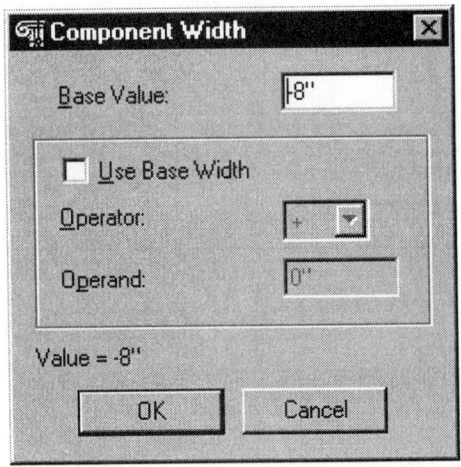

Set the Base Value to –8".

Clear the 'Use Base Width' option.

This specifies that the CMU has a fixed width of 8" in the negative direction from the wall baseline (to the inside).

Press 'OK'.

Lesson 1
Desktop Features

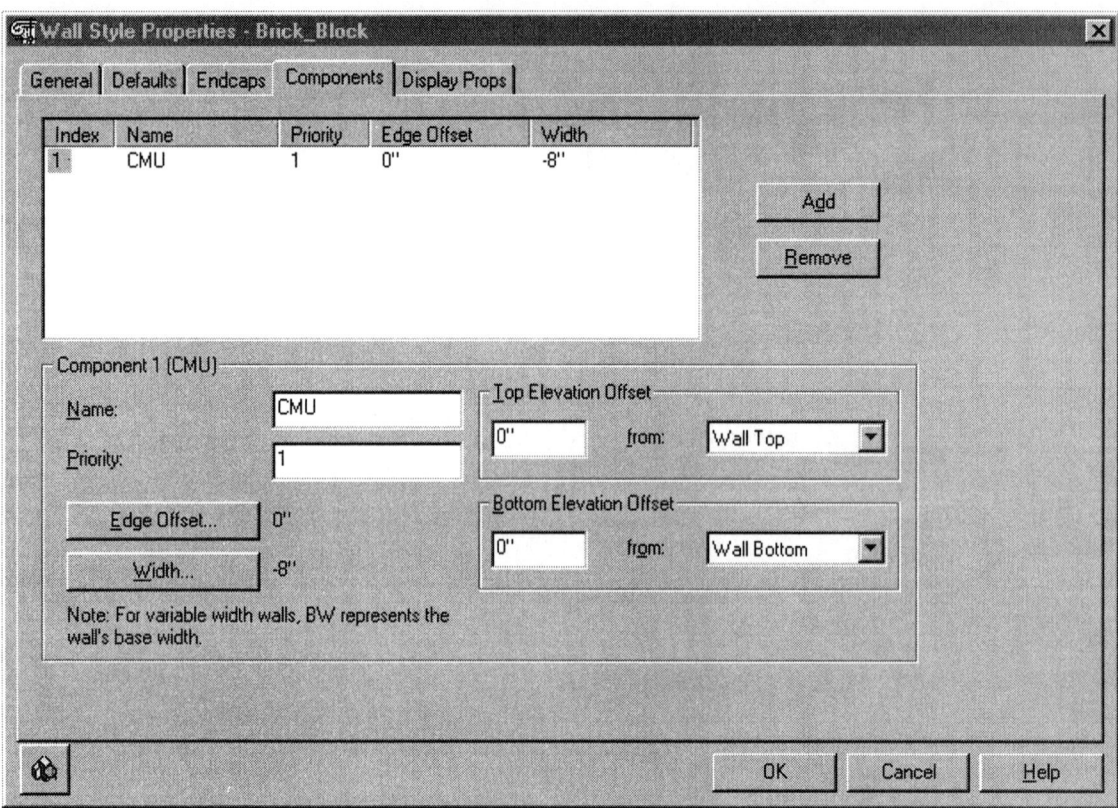

You now have one component defined for your wall named CMU with the properties shown.

Next, we'll add wall insulation.

Press the Add button on the right side of the dialog box.

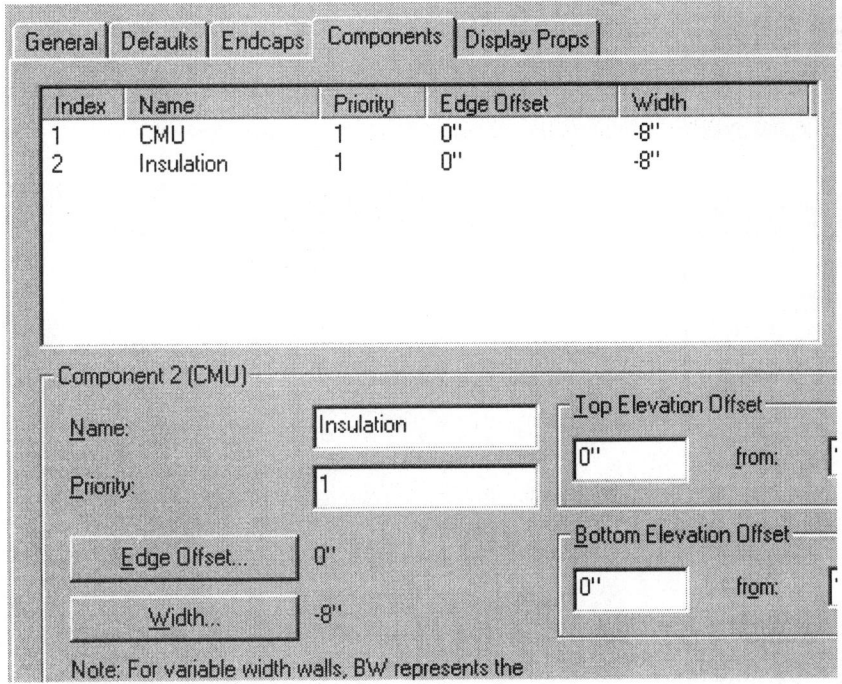

In the Name field, type 'Insulation'.

Set the Priority to 1. (The lower the priority number, the higher the priority when creating intersections.)

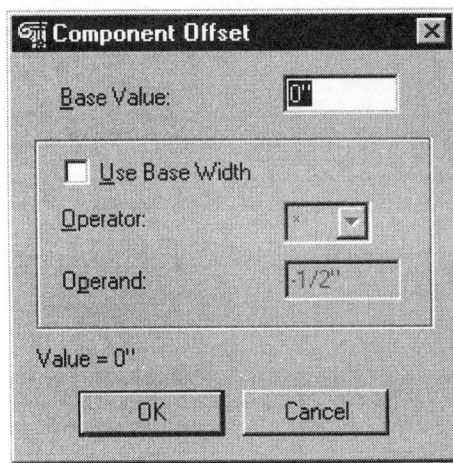

Set the Edge Offset as shown, with a Base Value of 0" and the Use Base Width DISABLED.

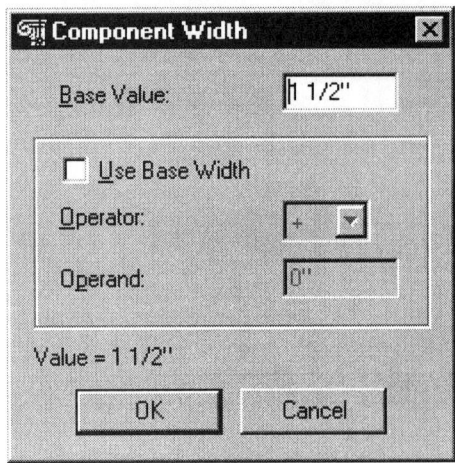

Set the Component Width as shown, with a Base Value of 1.5" and the 'Use Base Width' DISABLED.

This specifies that the insulation has a fixed width of 1.5" offset in a positive direction from the wall baseline (to the outside).

Press 'OK'.

Now, we add an air space component to our wall.

Press the Add button.

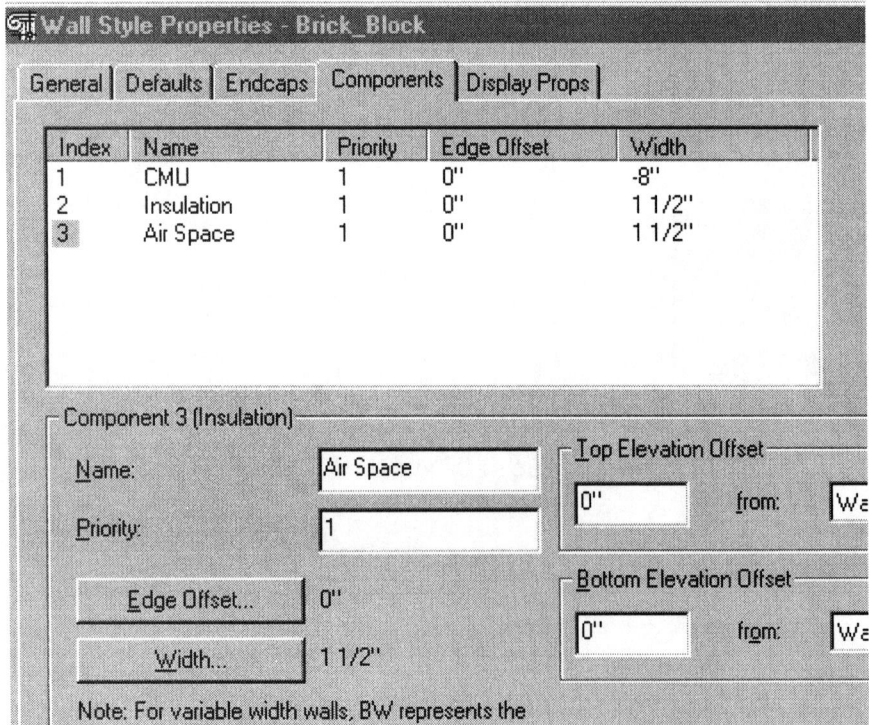

Lesson 1
Desktop Features

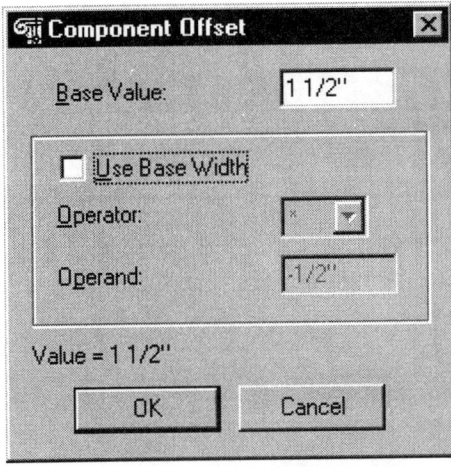

Set the Edge Offset as shown.

This specifies that the inside edge of the air space is coincident with the outside edge of the insulation.

Press 'OK'.

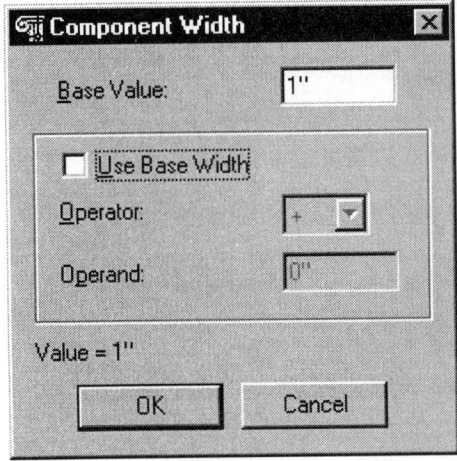

Set the Width as shown.

This specifies that the air gap will have a fixed width of 1" offset in the positive direction from the wall baseline (to the outside).

Lesson 1
Desktop Features

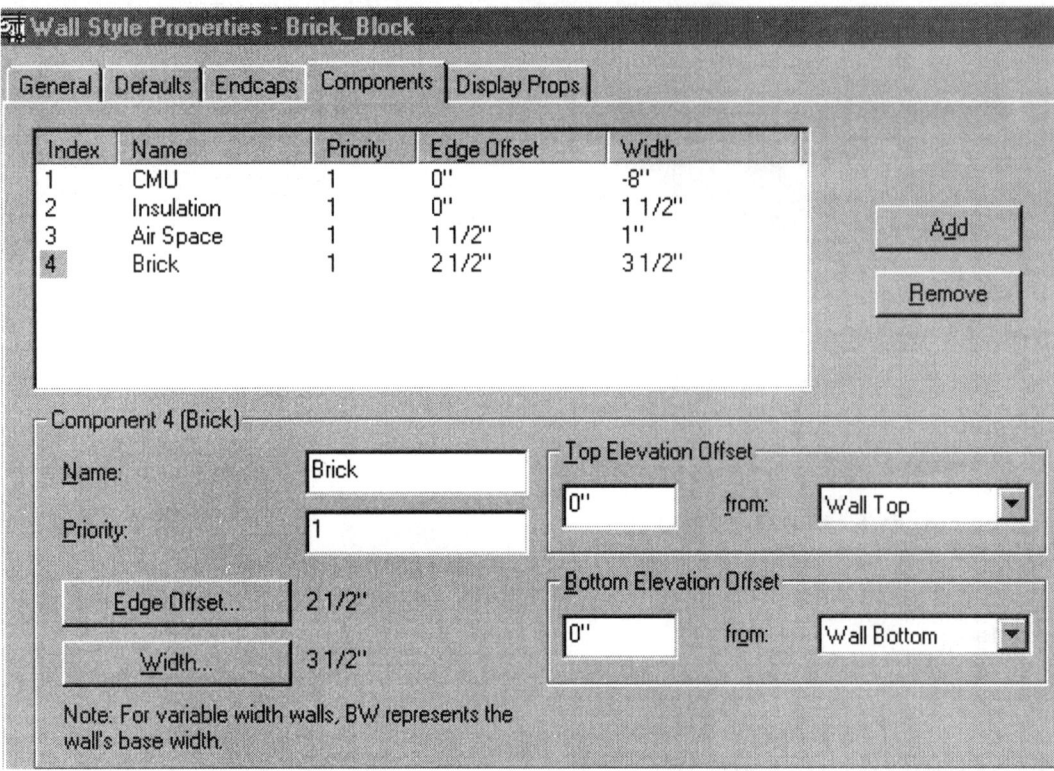

Add a fourth component as shown.

The Name should be Brick.
The Priority set to 1.
The Edge Offset set to 2.5"
The Width set to 3.5"

Press 'OK' twice to save and exit the Wall Styles dialog box.

Place some walls using the 'Add Wall' tool.

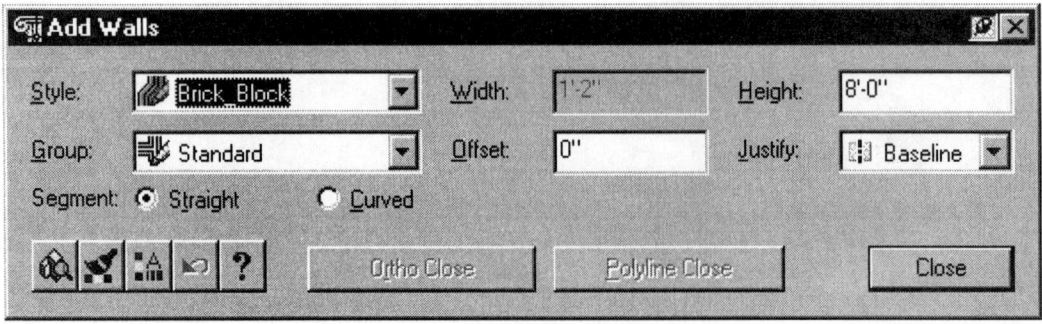

Set the Style to Brick_Block, the style you just created.

1-46

Place some walls in your drawing.

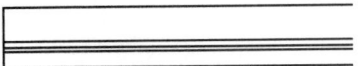

Save your drawing as Styles1.dwg

Exercise 4:
Assigning Wall Properties

Drawing Name: Styles1.dwg
Estimated Time: 15 minutes

This exercise reinforces the following skills:

- Use of AEC Design Content
- Use of Wall Styles

Open the styles1.dwg created in Exercise 3.

 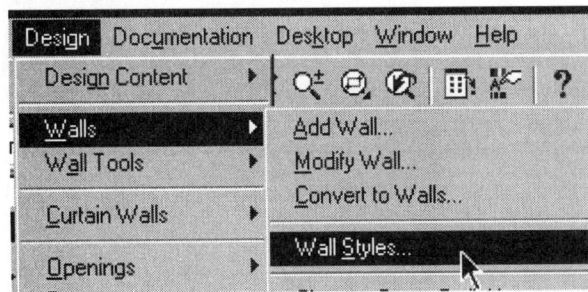

Initiate the Wall Styles dialog box using the Wall Styles toolbar or using Design->Wall->Wall Styles.

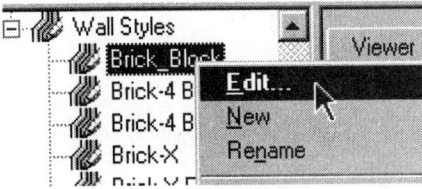

Locate the Brick_Block Wall Style created in Exercise 3. Right click and select 'Edit'.

Lesson 1
Desktop Features

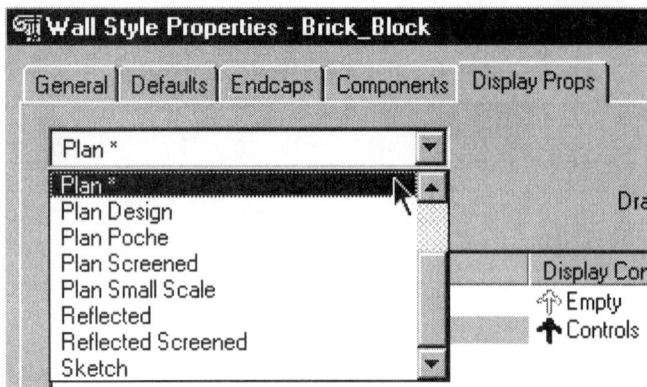

Select the Display Props tab.
Select Plan from the drop down menu.

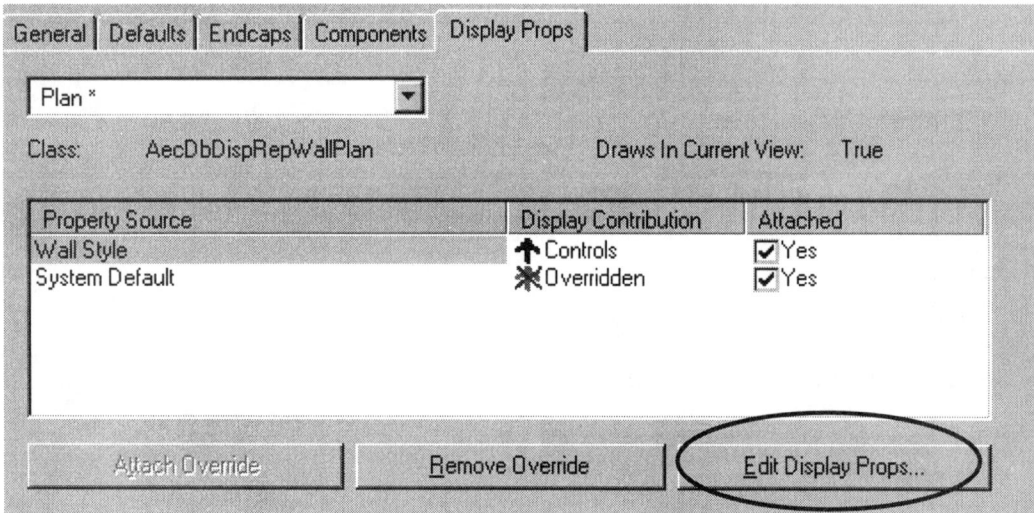

Highlight Wall Style. Left mouse click on Controls under Display Contribution.
The Edit Display Props button will become available and the System Default will change to show Overridden.

Select the Edit Display Props button.

Lesson 1
Desktop Features

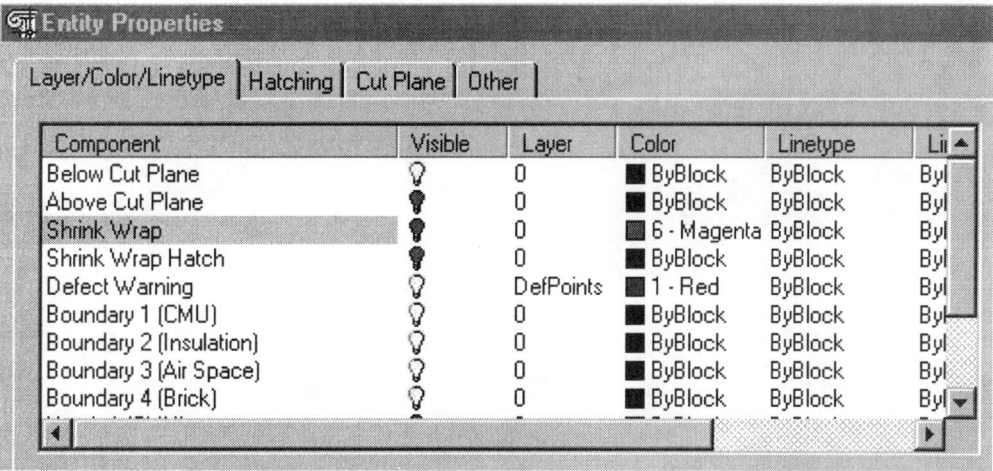

Select the Layer/Color/Linetype tab.
Locate the Component called Shrink Wrap.
Set the color to Magenta.

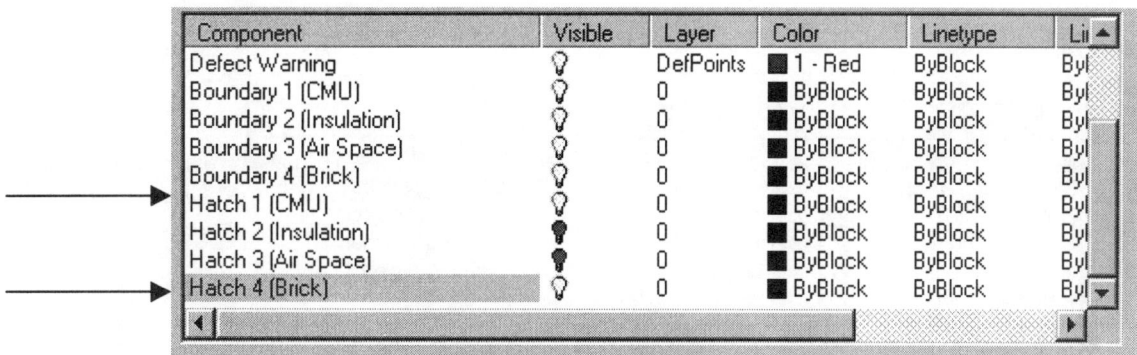

Locate Hatch 1 (CMU) and Hatch 4 (Brick) and turn Visibility to ON.

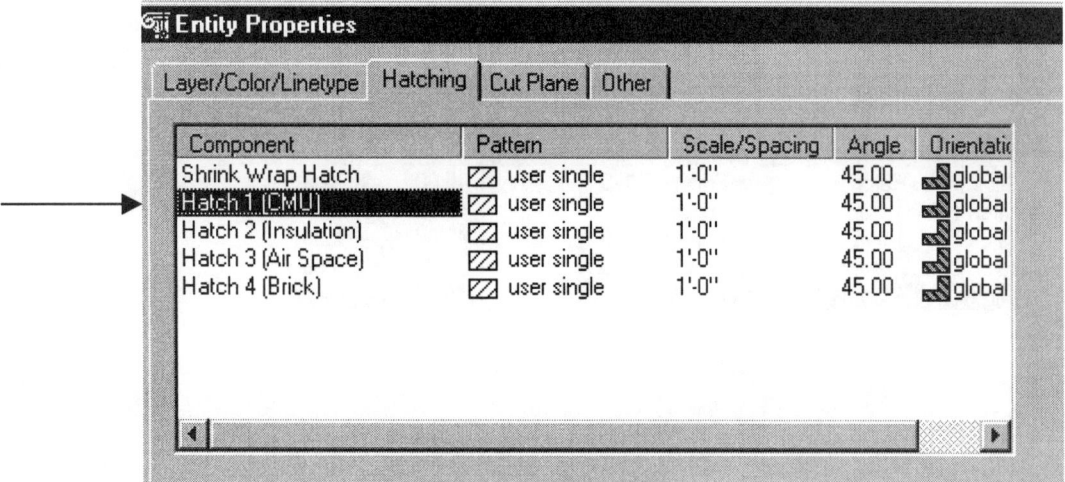

Select the Hatching tab.
Highlight Hatch 1 (CMU).

1-49

Click the Pattern button to display the Hatch dialog.

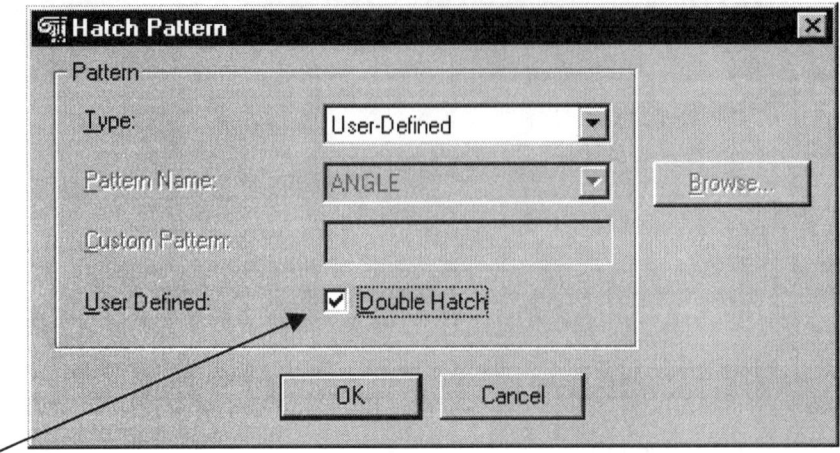

Enable the Double Hatch option.

Press 'OK'.

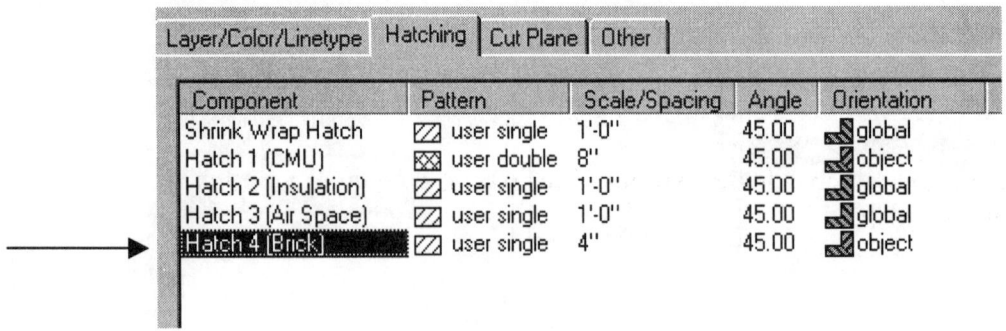

Note that the pattern description has changed to show double.
Change the scale to read 8".
Click on the orientation to switch to object from global.

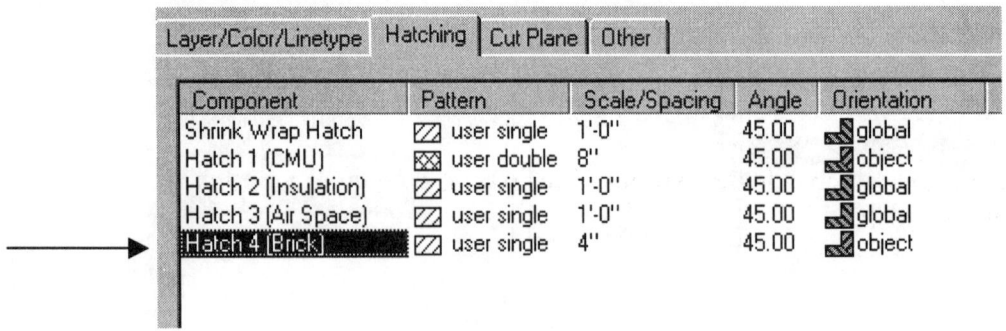

Highlight Hatch 4 (Brick).
Change the scale to 4" and set orientation to object.

Click 'OK' three times to save and exit.

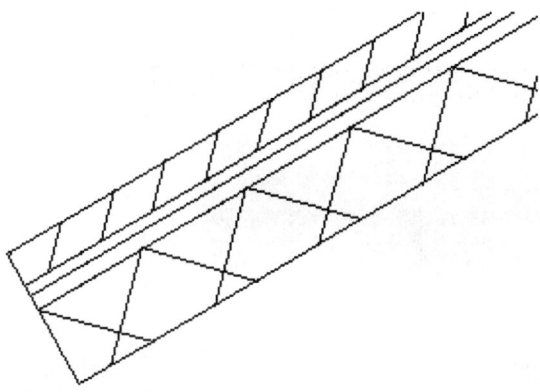

Hatch patterns are now displayed when you create your walls using the Brick_Block wall style.

Autodesk is set up so that the exterior wall is on the outside if you draw clockwise and on the inside if you draw counterclockwise.

Place a wall drawing from left to right.

Place a wall drawing from right to left.

If you need to switch which side the exterior of the wall is on, you can correct it by selecting 'Reverse Wall'.

Lesson 1
Desktop Features

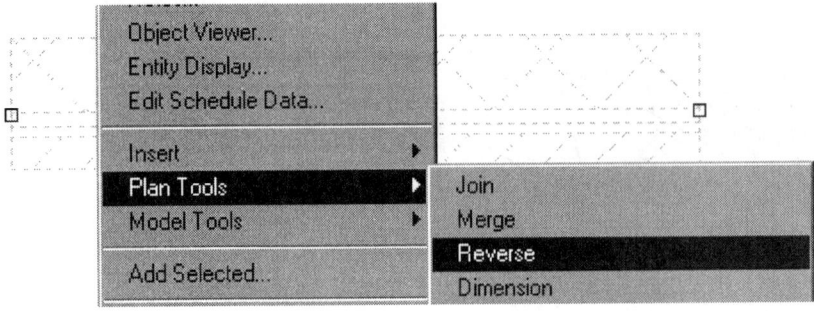

Select the wall drawn from right to left.
Right click and select 'Plan Tools ->Reverse'.

Now both walls have the exterior feature on the outside.

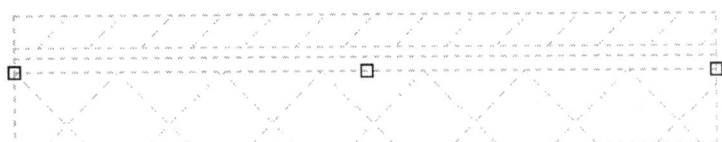

Highlight a wall to activate the GRIPS. The GRIPS are located on the draw line. The location of the draw line is based on the Justification setting.

Lesson 1
Desktop Features

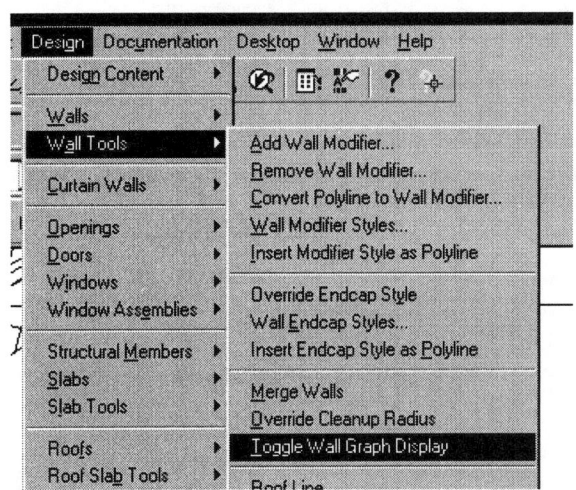

You don't have to activate the GRIPS to see the draw line. You can use the Toggle Wall Graph Display.

Menu	Design->Wall Tools->Toggle Wall Graph Display
Wall Tools Toolbar	
Command Line	WallGraphDisplayToggle

Toggle the Wall Graph Display ON.

A dashed blue line will appear to indicate your draw line.

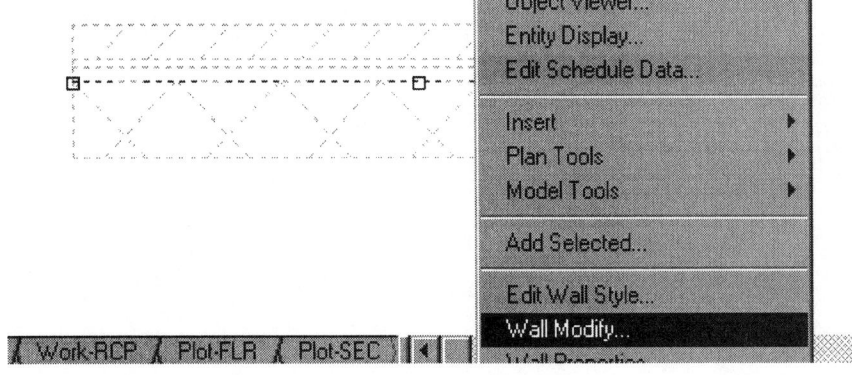

Select a wall. Right click and select 'Wall Modify'.

1-53

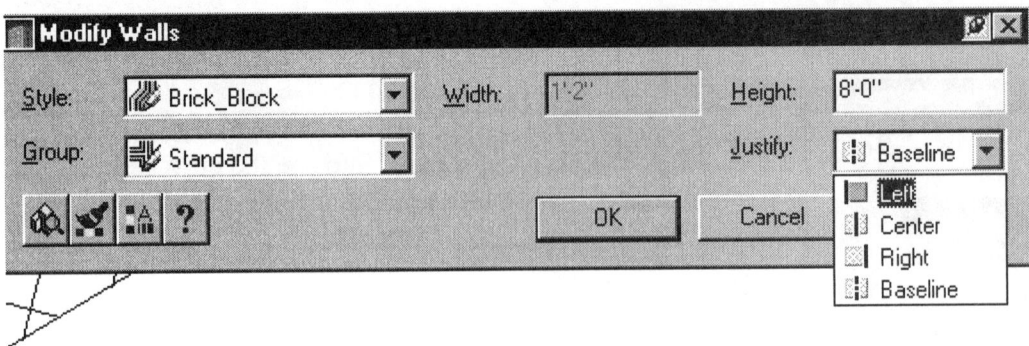

Change the Justification to Left.

Press 'Apply' and 'OK'.

Notice how the draw line has shifted to the top of the wall.

Changing the Justification of the wall can affect the dimensions of your building.

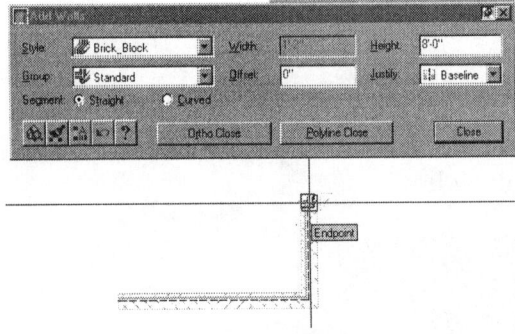

Draw a horizontal wall and a vertical wall.
Set the Style to Brick_Block.
Set the Height to 8'.
Set the Justification to Baseline.
Have the Toggle Wall Graph ON so you can see the draw line.
Move your mouse towards the left.
Press 'ORTHO Close'.

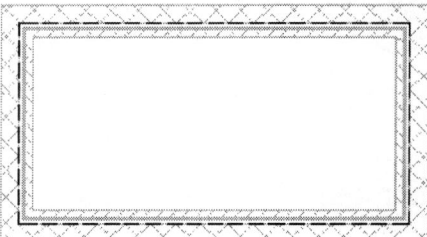

Your walls will automatically complete a rectangle.

Use the Dimension Walls tool located on the Walls Toolbar to add a vertical and horizontal dimension to your room.

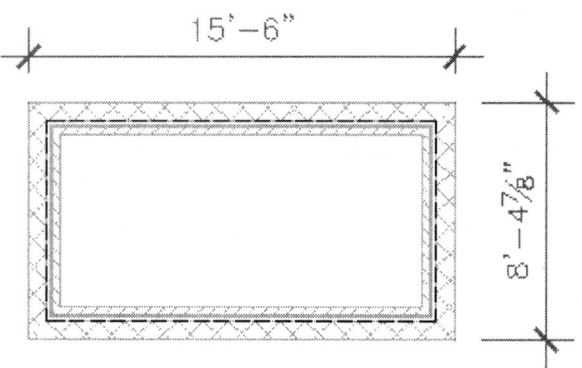

Your dimensions may be different depending on how big you made your walls.

Window around all your walls to select.
Right click and select 'Wall Modify'.

Lesson 1
Desktop Features

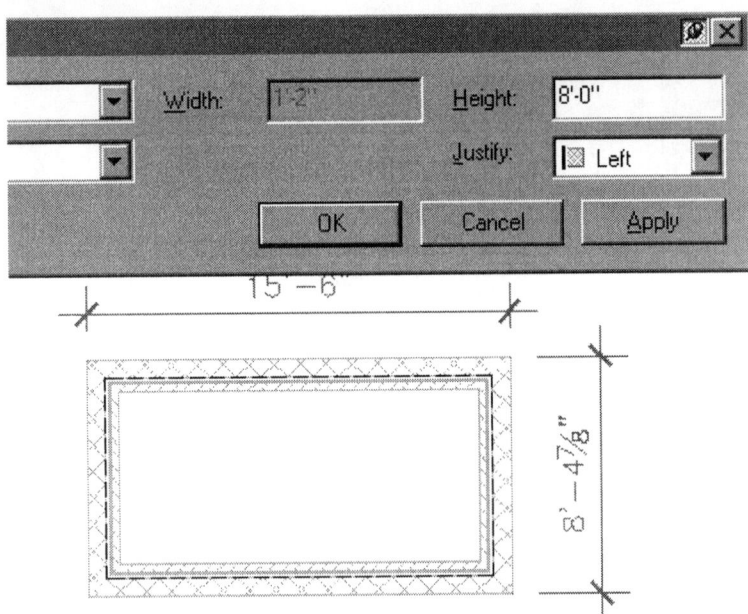

Change the Justification to Left and hit 'Apply'.

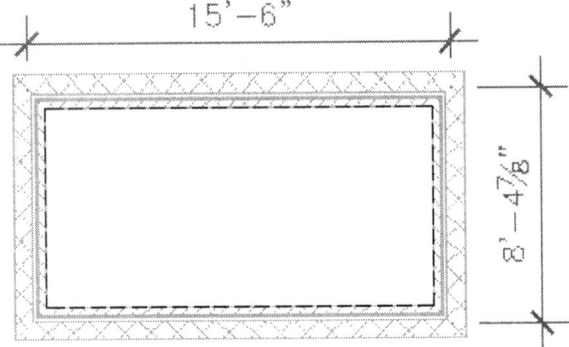

Note how the room changes.

Lesson 1
Desktop Features

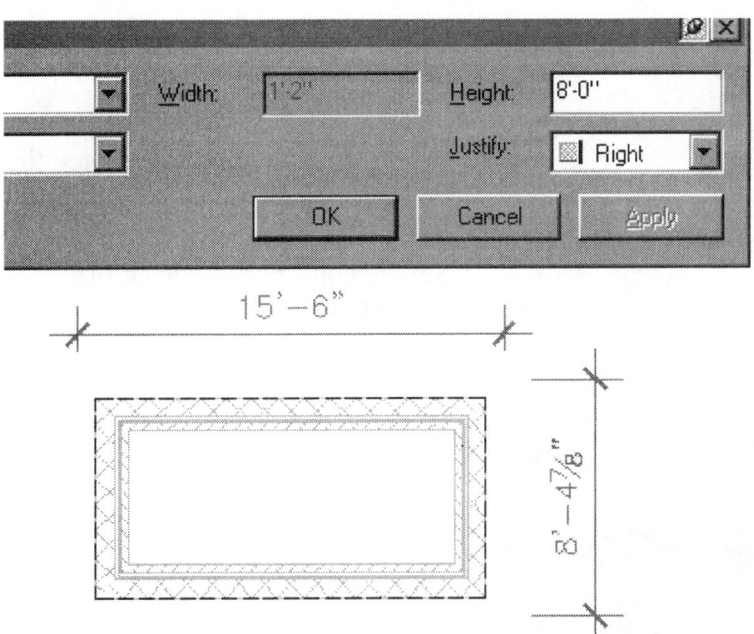

Change the Justification to Right and hit 'Apply'.
Note the change.

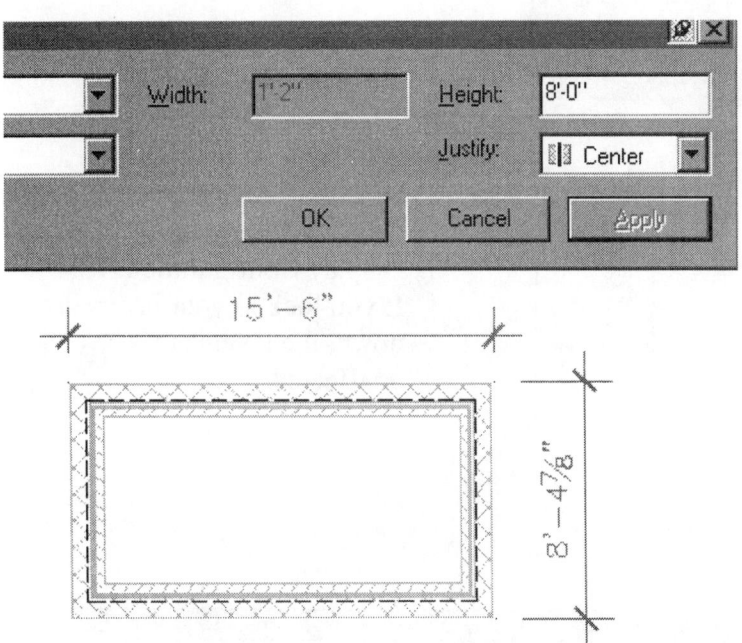

Change the Justification to Center and hit Apply.
Then press 'OK'.

Did you notice that the dimensions did not update? To use Dimensions that will automatically update, you need to apply AEC Dimensions.

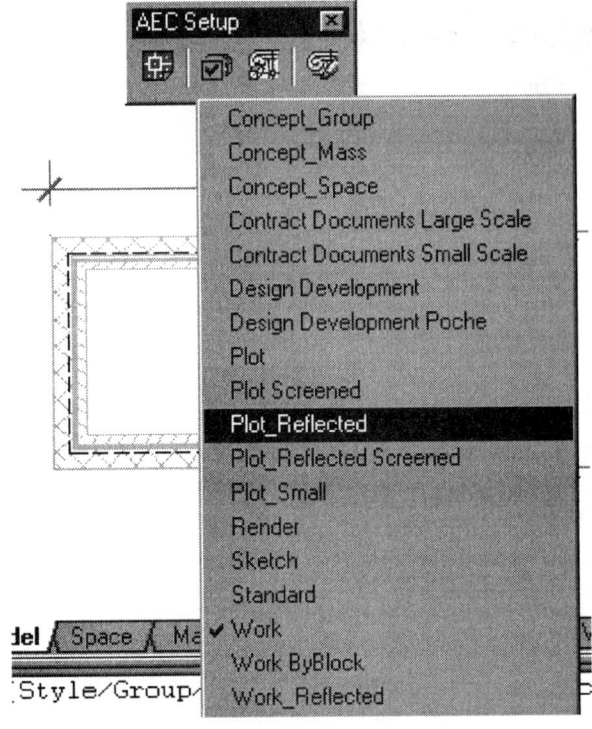

On the AEC Setup toolbar, pick the 'Set Current Display Cofiguration 'tool.
Set the configuration to 'Plot Reflected'.

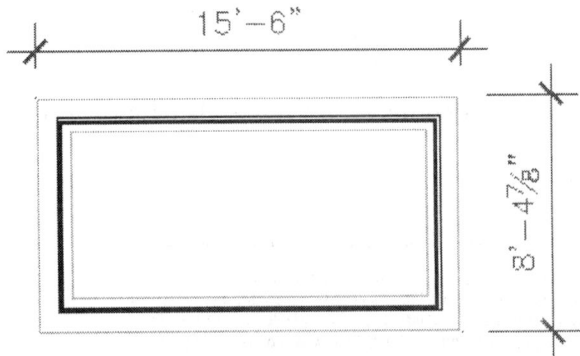

Note that the appearance of the wall changes based on the Display Configuration.
If you look in your layers drop down, the layers are unaffected.

Look at your walls from different perspectives by activating the various layouts in your drawing.

Save your drawing as Styles1.dwg.

Exercise 5:
Downloading Styles from Autodesk's pointA

Drawing Name: Styles1.dwg
Estimated Time: 15 minutes

This exercise reinforces the following skills:

- Use of AEC Design Content
- Use of Wall Styles

NOTE: **This exercise requires access to the Internet.
                    idrop capability should also have been installed.**

Open the styles1.dwg created in Exercise 3.

Open the Wall Styles dialog using the Wall Styles tool on the Walls toolbar.

Select the 'Access Styles on Point A' icon located on the top of the Style Manager dialog.

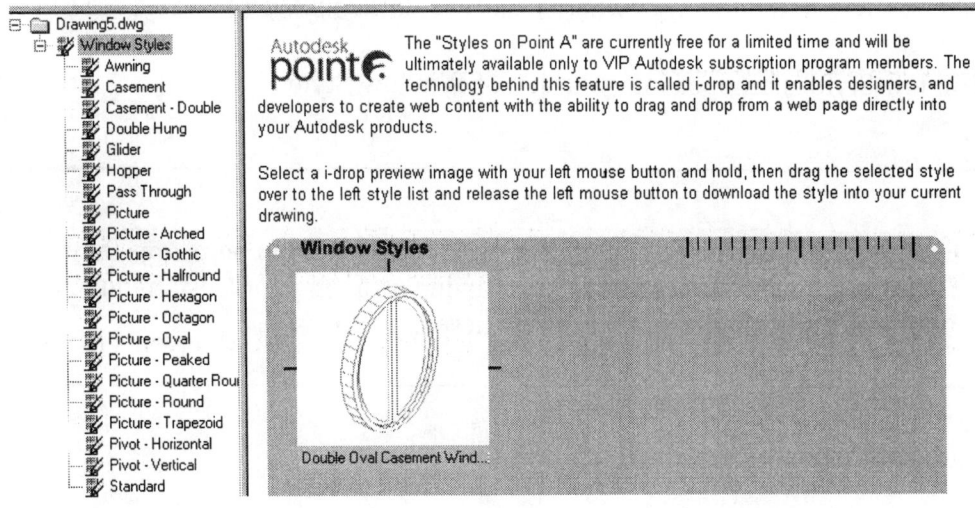

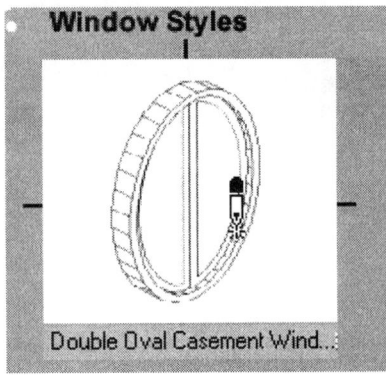

In order to access idrop capabilities, you need to download the idrop software. This is available for free from Autodesk's website.

You will know idrop is properly installed, if you see an eyedropper when you mouse over an idrop-enabled object.

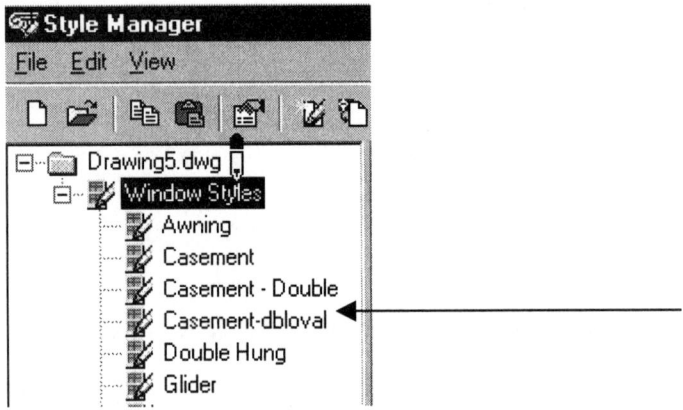

Drag your mouse over to the Explorer where it says Window Styles.

The new Window Style is dropped into your list of Window Styles.

TIP: You may want to download as many styles as possible from pointA as they come available. Store the styles in a separate drawing away in your Custom directory. Autodesk will eventually charge a fee to allow these downloads.

Exercise 6:
Creating a Fireplace/Chimney

Drawing Name: New using Imperial.dwt
Estimated Time: 30 minutes

ADT doesn't come with any chimney or fireplaces as part of the styles or content. Since most residences now have at least one fireplace, it is a good idea to know how to create one. We use the Mass Elements tools included within ADT to assist us in this process.

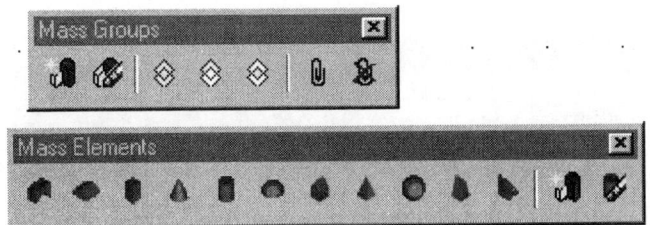

 We can use the Mass-Group layout to create our Mass Elements and Groups or use Model space.

We start by placing a box.

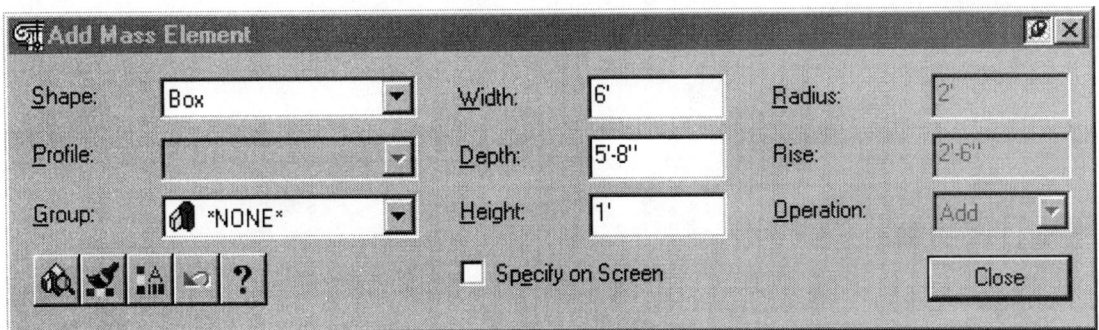

Set the Width to 6'.
Set the Depth to 5'-8".
Set the Height to 1'.

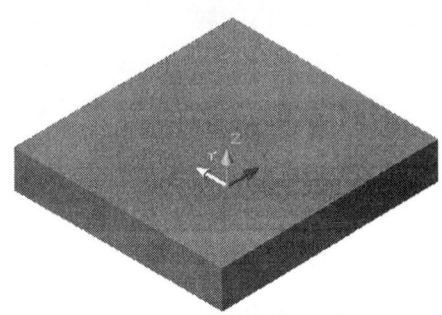

Place at 0,0 with a rotation of 0.

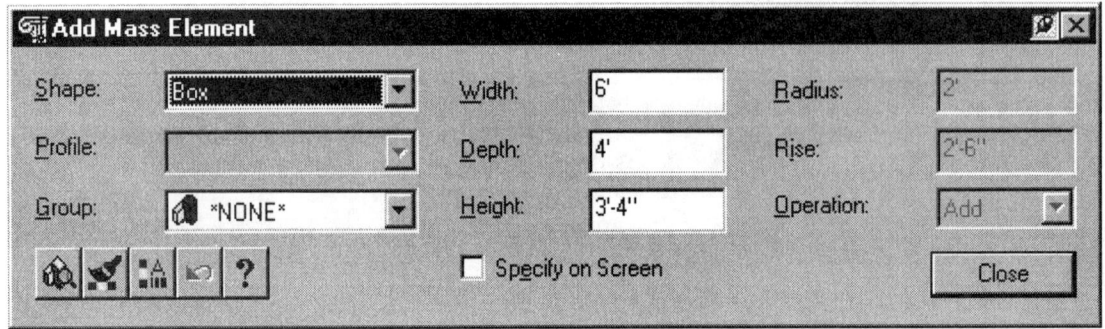

Create a second box.
Set the Width to 6'.
Set the Depth to 4'.
Set the Height to 3'-4".

Place the second box next to the first box.

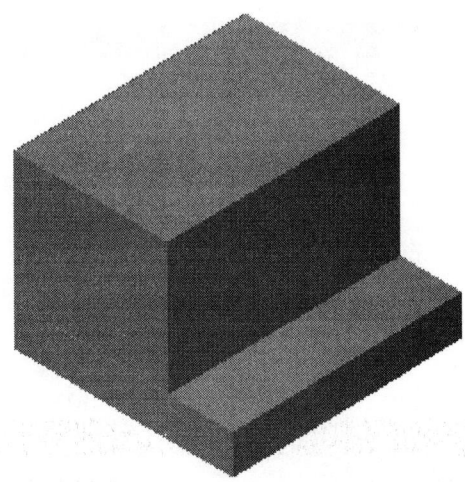

You can then use the MOVE tool to place the second box on top of the first box.

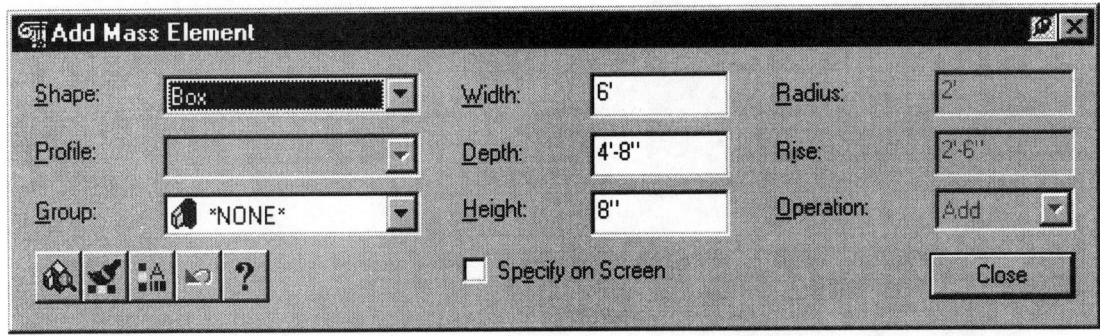

Create a third box. Set Width to 6'. Set Depth to 4'-8". Set Height to 8".

Use the MOVE tool to place the new box section on top as shown.

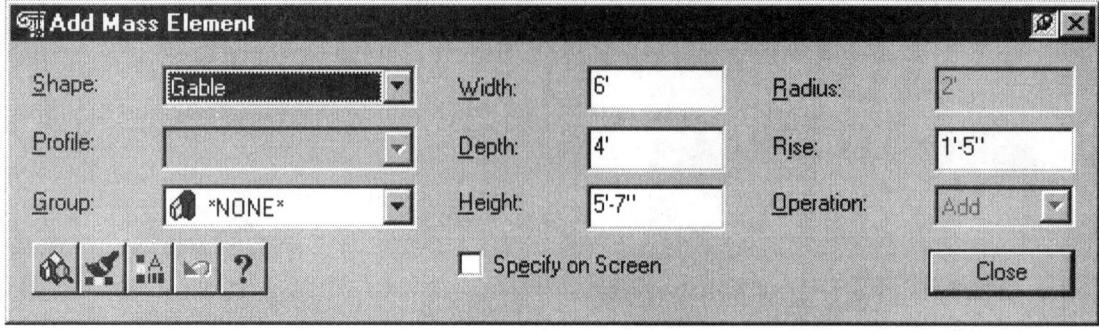

Select Gable for the next Mass Element. Set the Width to 6'. Set the Depth to 4'. Set the Rise to 1'-5". Set the Height to 5'7".

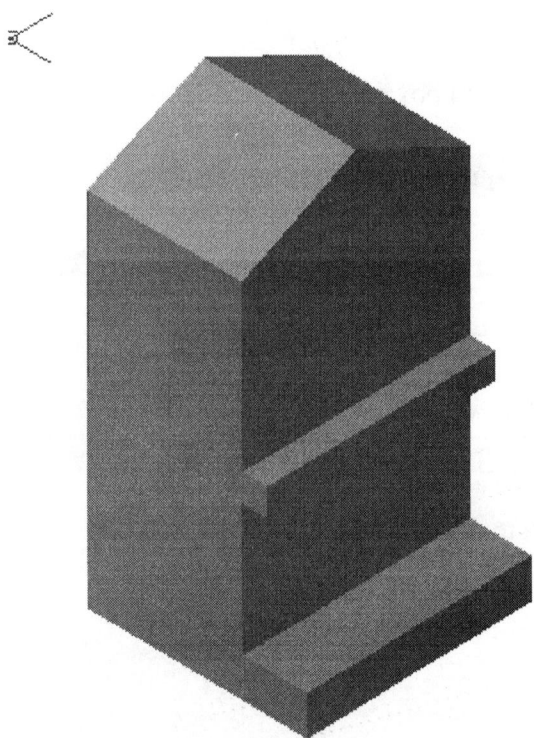

Stack the Gable on top as shown.

TIP: If you see shadows or "ghosts" when creating your mass elements, use REGEN to refresh your screen.

Lesson 1
Desktop Features

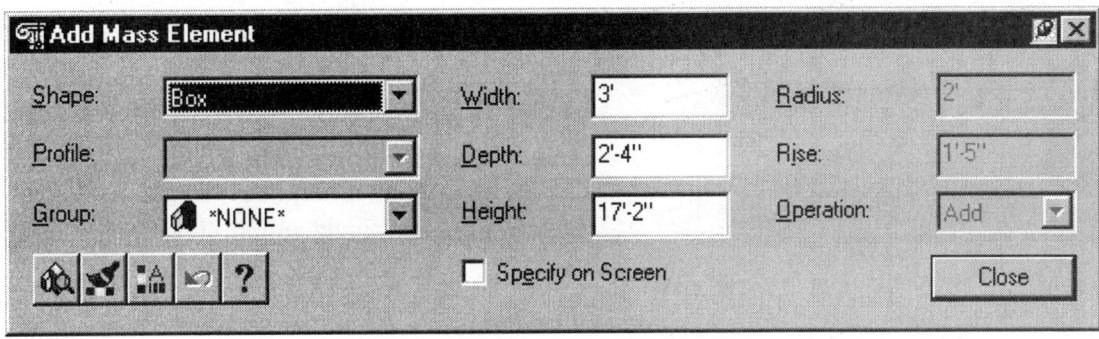

Create a box. Set the Width to 3'. Set the Depth to 2'-4". Set the Height to 17'-2".

Place the tall box as shown.
To locate it properly, use the MOVE tool. Select the midpoint of the rear back wall of the box and then select the rear vertex of the gable.
Then move using the rear vertex of the gable for the first point and @0,0,-2 as the second point to lower the tall box appropriately. This will shift the box down appropriately.

Command: _move
Select objects: 1 found (select tall box)
Select objects:
Specify base point or displacement: _endp of (Select rear vertex)
Specify second point of displacement or <use first point as displacement>: @0,0,-2'

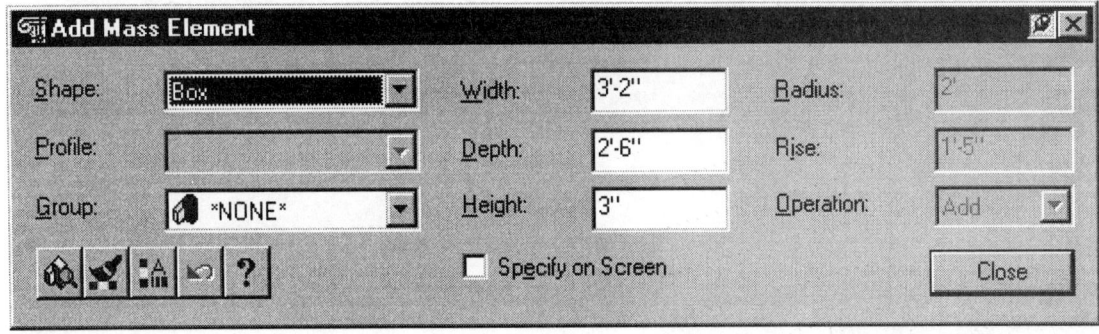

Create another box. Set the Width to 3'-2". Set the Depth to 2'-6". Set the Height to 3".

Lesson 1
Desktop Features

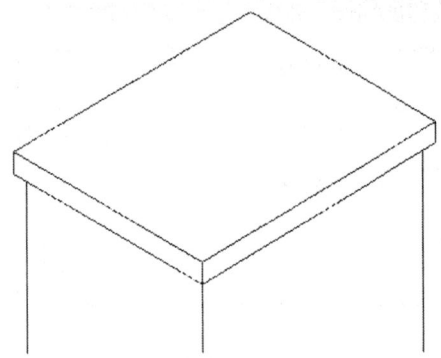

Place the box so there is a 1" overhang on each side. You can do this by using the MOVE command. Place the box so two corners of the boxes are even. Then MOVE again. When prompted for displacement, type –1",-1".

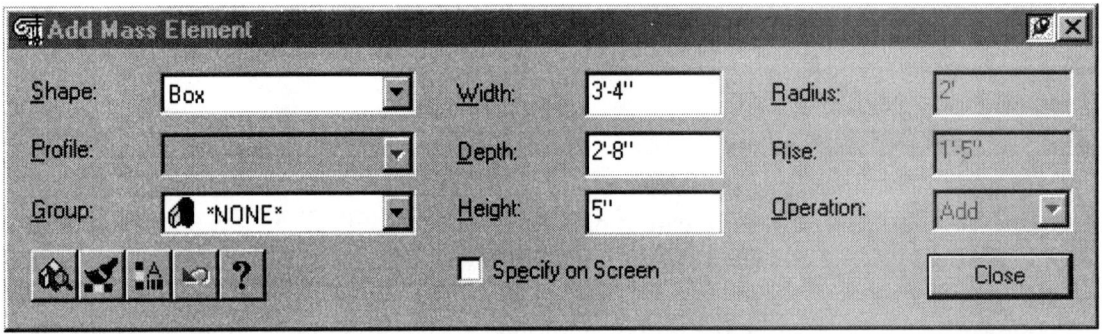

Create another box. Set the Width to 3'-4". Set the Depth to 2'-8". Set the Height to 5".

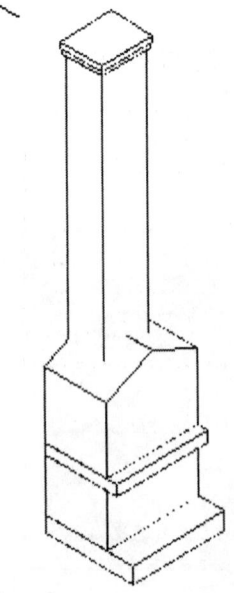

Place this box so it overhangs the previous box 1" on each side.

Create a third box.

Lesson 1
Desktop Features

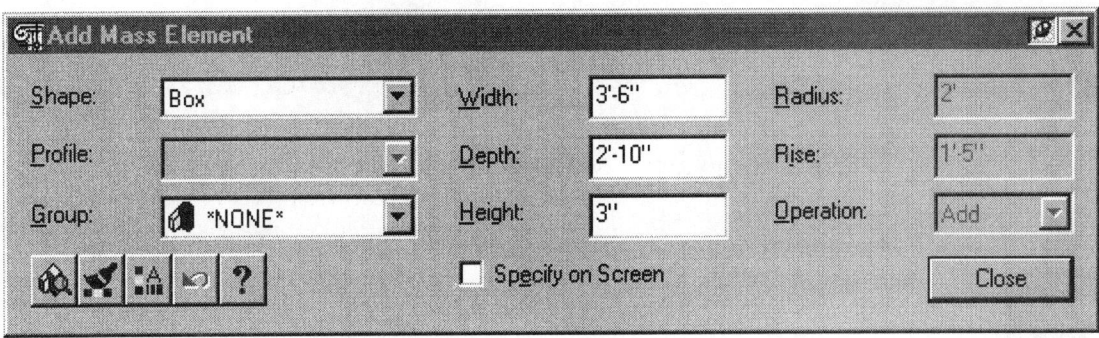

Set the Width to 3'-6". Set the Depth to 2'-10". Set the Height to 3".

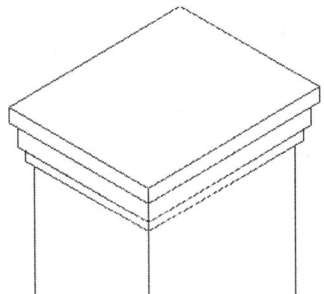

Place this box so it overhangs the previous box 1" on each side.

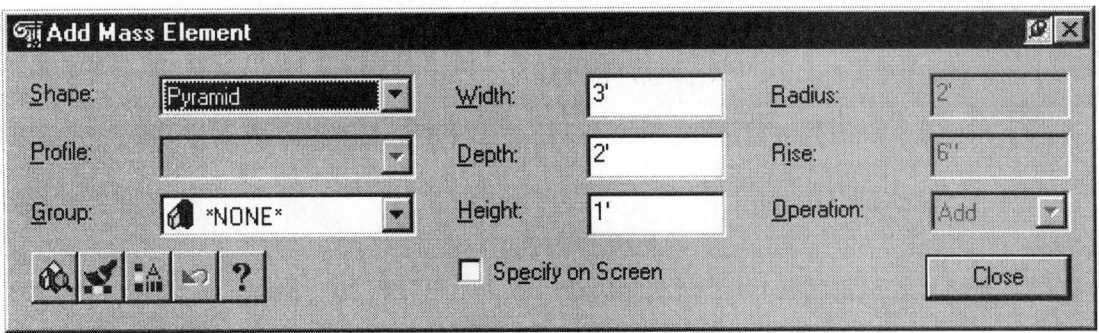

Top your chimney with a pyramid. Set the Width to 3'. Set the Depth to 2'. Set the Height to 1'.

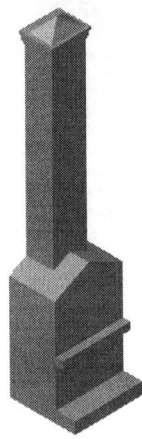

Center the pyramid on top of the chimney.

We've created the base part of our model. We will be adding additional mass elements that will be subtracted from this part to create the openings for the fireplace.

First, we define our Mass Group.

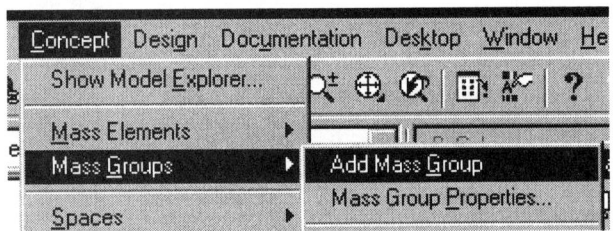

We select the Add Mass Group tool.

```
Command: _AecMassGroupAdd
Location: 0,0,0

Rotation angle <0.00>:
```

When prompted for the location, type 0,0,0.

When prompted for the angle, press ENTER.

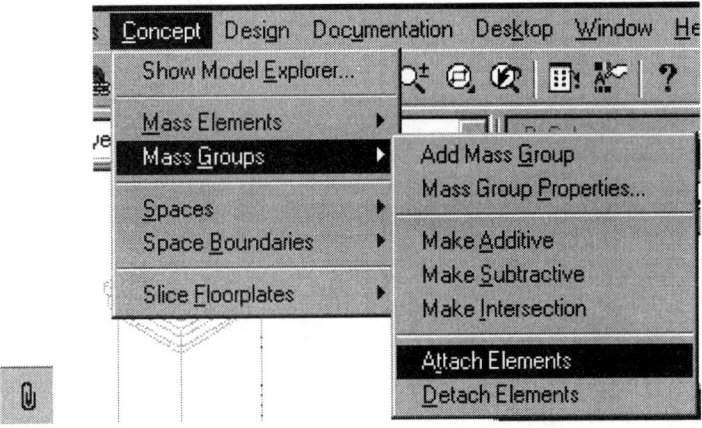

Select Attach Elements.

When prompted for the Mass Element Group, type 'L' for last.

Then select all the elements you just created.

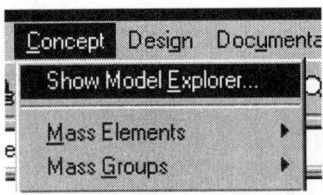 Go to Concept->Show Model Explorer.

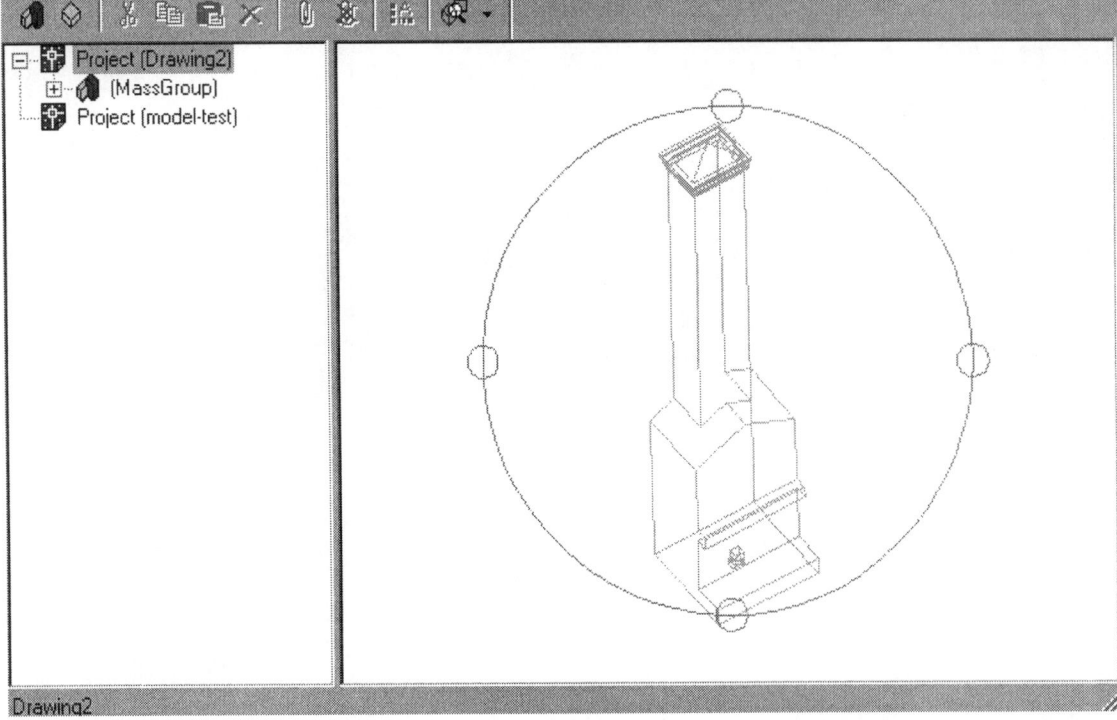

Lesson 1
Desktop Features

A window opens to reveal your Mass Elements group that you have created.

If you right-click in the window, a context sensitive menu will appear providing you with different viewing options.

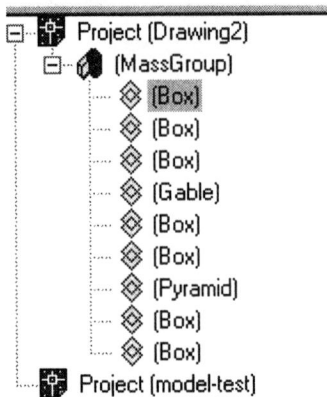

Expanding the Mass Group reveals a feature tree showing all the Mass Elements.

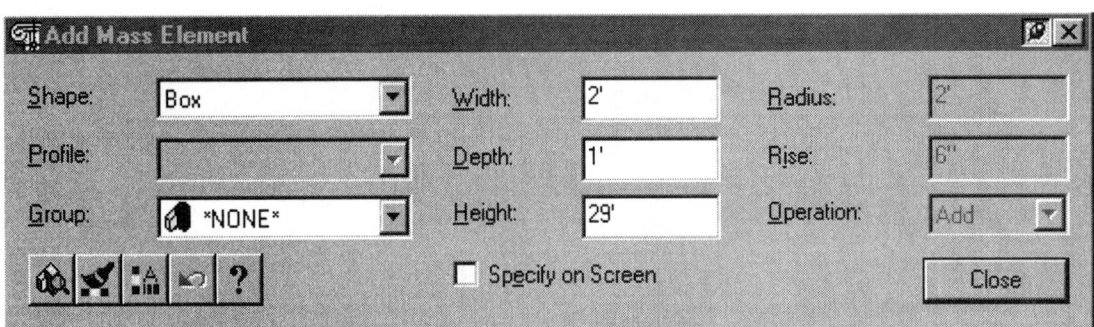

Lesson 1
Desktop Features

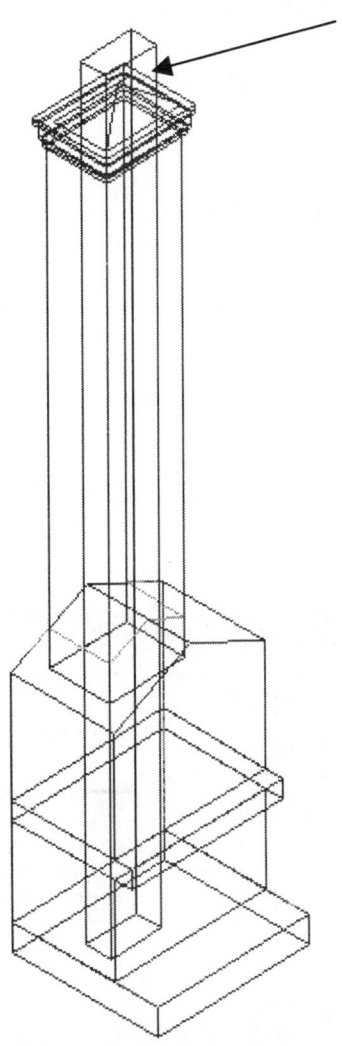

Create a tall box with Width set to 2', Depth set to 1' and Height set to 29'. This will be used to create the flue opening for the chimney. Center the opening inside the chimney.

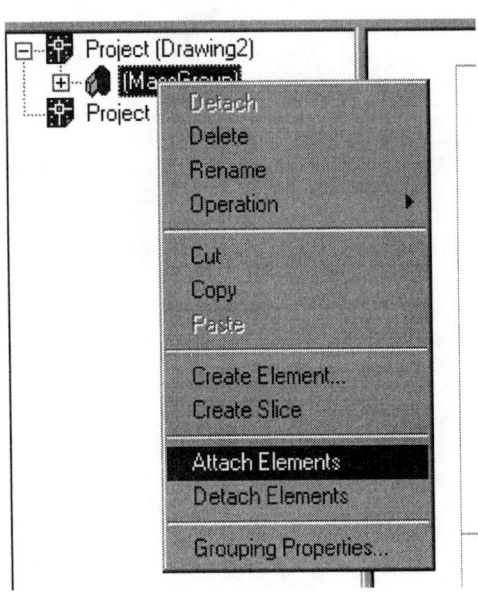

Bring up the Mass Element Explorer. Highlight the Mass Group. Select 'Attach Elements'. Select the tall flue opening we just created.

Lesson 1
Desktop Features

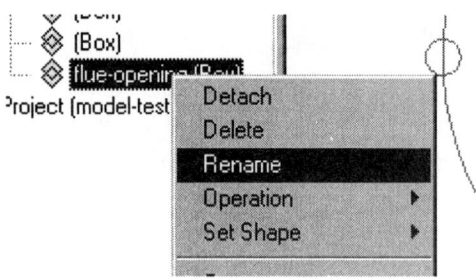

To help us with identifying the various mass elements, we can highlight, right click and select 'Rename' and type in 'flue-opening' as our new name.

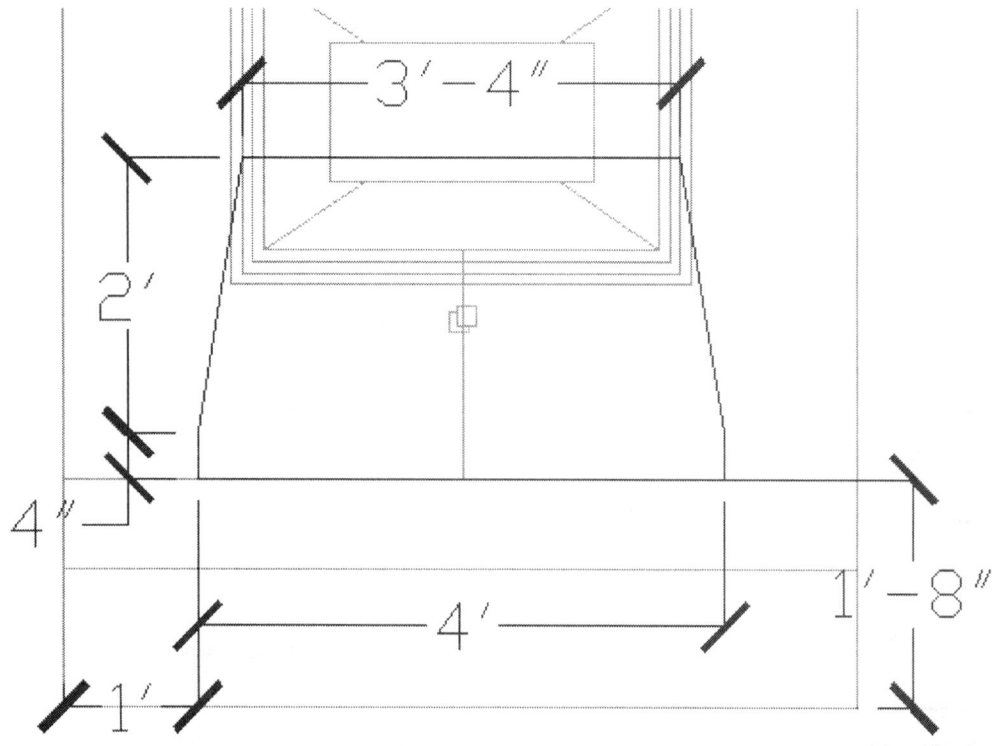

Draw this polyline on the XY plane.

Go to the Model-3D tab for an iso and plan view side by side. Use the plan view to draw your polyline.

Lesson 1
Desktop Features

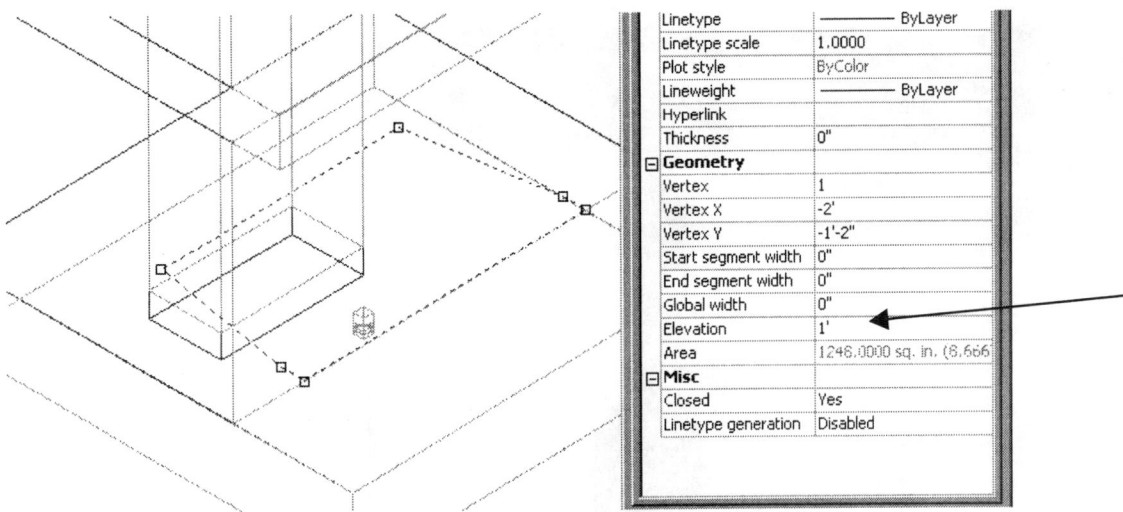

Locate the polyline so it is at 1' elevation as shown. Use Properties to change the elevation.

Extrude the polyline using EXTRUDE 2'-6".

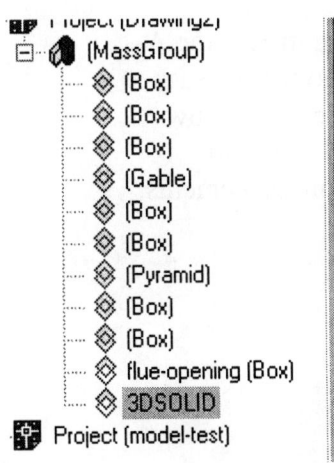

Go back to your Mass Group Explorer.
Attach the 3D Solid you just created using Attach Element.
Highlight the 3D Solid, right click and select 'Operation'->Subtractive.

Lesson 1
Desktop Features

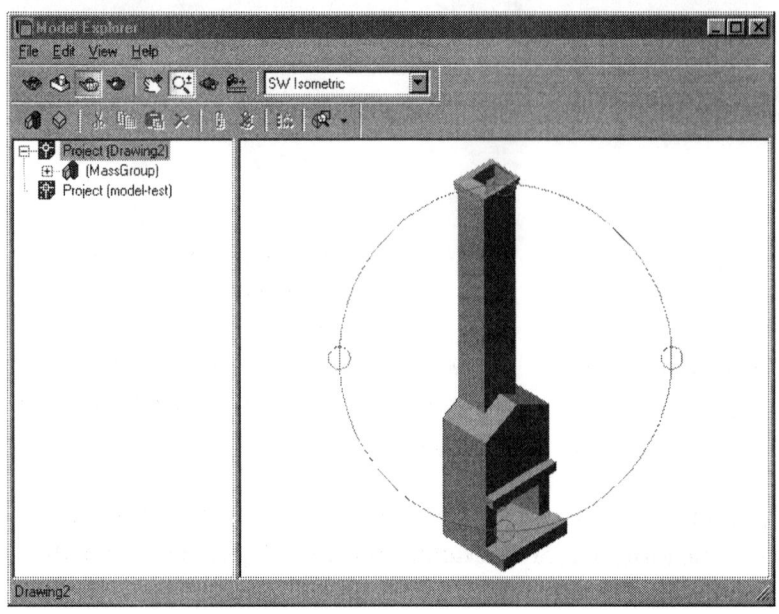

Our completed fireplace with chimney.

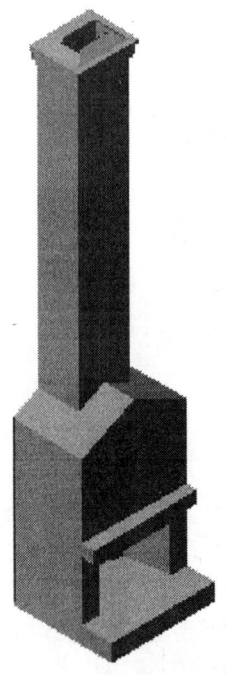

To view the model as shown, freeze the A-Mass Layer. You can also go to the Mass-Group tab, where the left viewport shows mass elements and the right viewport shows mass groups but not mass elements.

Save the file as 'Fireplace with Chimney'. You can use this model in Lesson 5 later.

Layer Manager

Back in the days of vellum and pencil, drafters would use separate sheets to organize their drawings. So one sheet might have the floor plan, one sheet the site plan, etc. The layers of paper would be placed on top of each other and the transparent quality of the vellum would allow the drafter to see the details on the lower sheets. Different colored pencils would be used to make it easier for the drafter to locate and identify elements of a drawing, such as dimensions, electrical outlets, water lines, etc.

When drafting moved to Computer Aided Design, the concept of sheets was transferred to the use of Layers. Drafters could assign a Layer Name, color, linetype, etc. and then place different elements on the appropriate layer.

ADT is unique in that it has a Layer Management system to allow the user to implement AIA standards easily.

Layer Manager

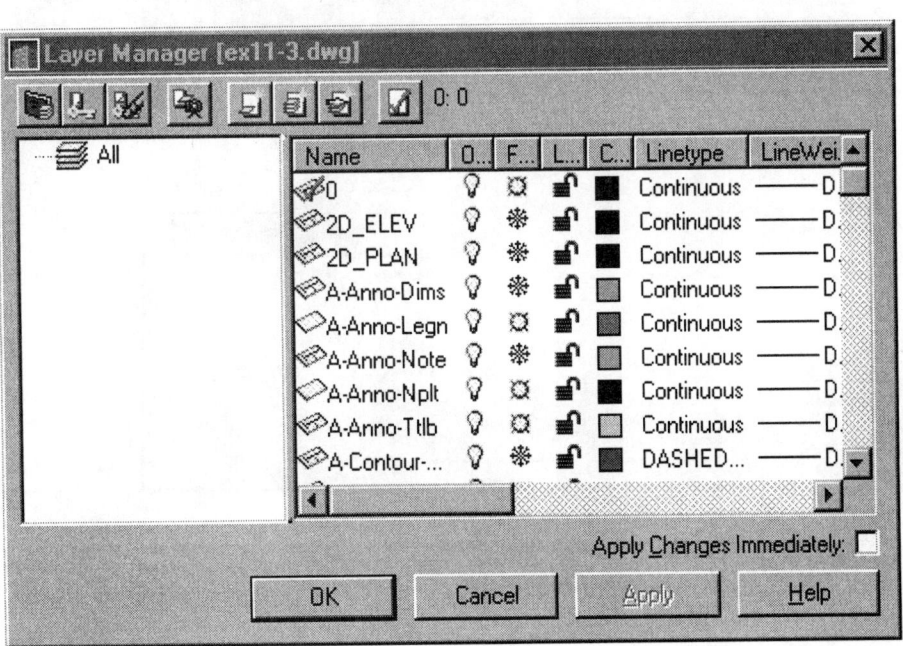

The Layer Manager helps you organize, sort, and group layers, as well as save and coordinate layering schemes. You can also use layering standards with the Layer Manager to better organize the layers in your drawings.

Lesson 1
Desktop Features

When you open the Layer Manager, all the layers in the current drawing are displayed in the right panel. You can work with individual layers to:

- Change layer properties by clicking AutoCAD layer property icons

- Make a layer the current layer

- Create, rename, and delete layers

If you are working with drawings that contain large numbers of layers, you can improve the speed at which the Layer Manager loads layers when you open it by selecting the Layer Manager/Optimize for Speed option in your AEC Editor options.

The Layer Manager has a tool bar as shown.

	Layer Standards You can import Layer Standards from an existing drawing or Export Layer Standards using the current drawing. 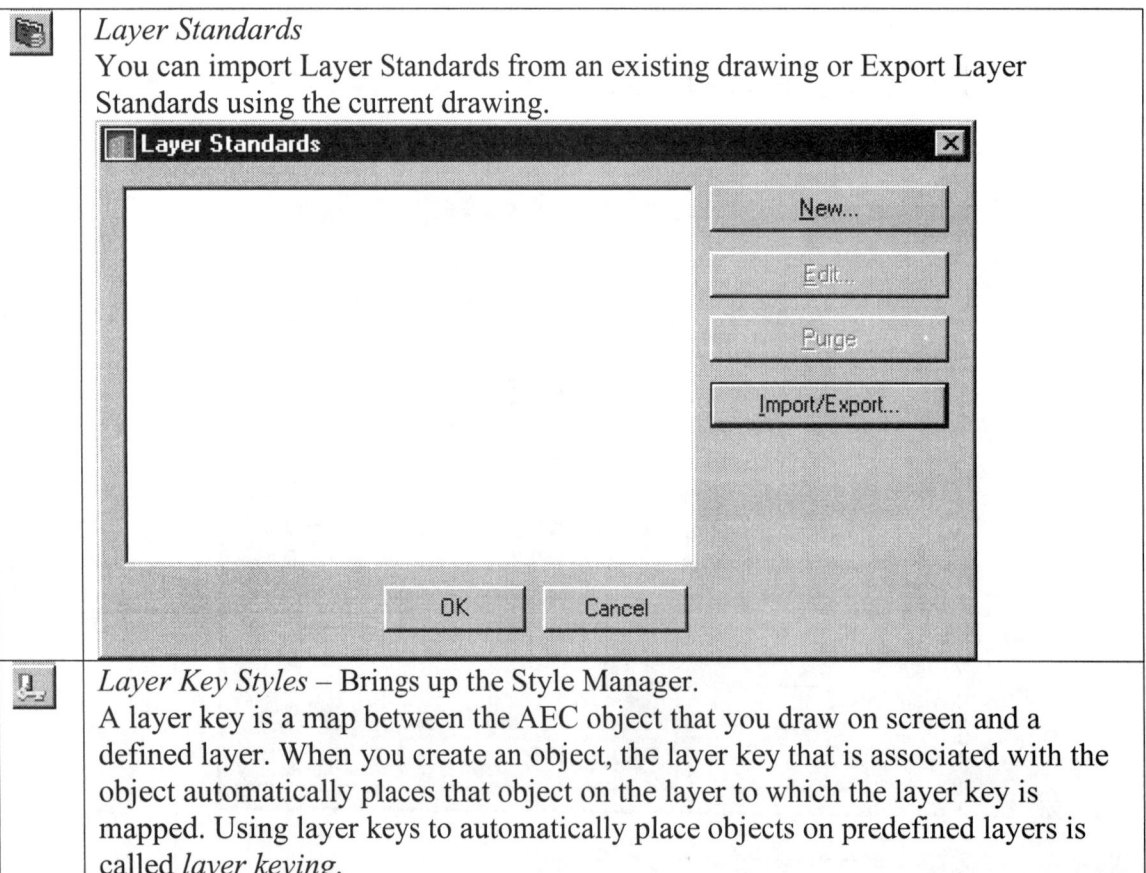
	Layer Key Styles – Brings up the Style Manager. A layer key is a map between the AEC object that you draw on screen and a defined layer. When you create an object, the layer key that is associated with the object automatically places that object on the layer to which the layer key is mapped. Using layer keys to automatically place objects on predefined layers is called *layer keying*.

	Layer Key Overrides You can apply overrides to any layer keys within a layer key style that is based on a layer standard. The structure of the layer name of each layer that each layer key maps an object to is determined by the descriptive fields in the layer standard definition. You can override the information in each field according to the values set in the layer standard definition. You can allow overrides on all the layer keys within a layer key style, or you can select individual layer keys that you want to override. You can also choose to allow all of the descriptive fields that make up the layer name to be overridden, or you can specify which descriptive fields you want to override.
	Snapshots A layer snapshot is a specific set of layers and view information that you can save, edit, and restore in your drawing. By saving layer and view information in a snapshot, you can quickly recall specific layer and view configurations from complex data sets. For example, a facilities manager might create snapshots of individual floor plans and furniture, cable, and HVAC layouts to separate this information from a complete building layout.
	New Layer 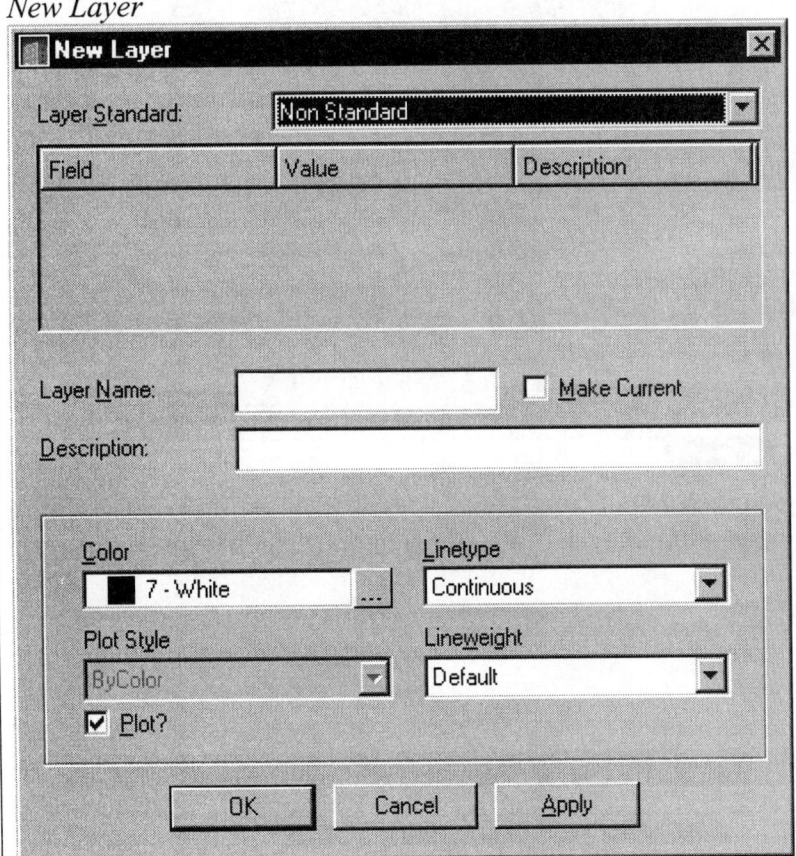

Lesson 1
Desktop Features

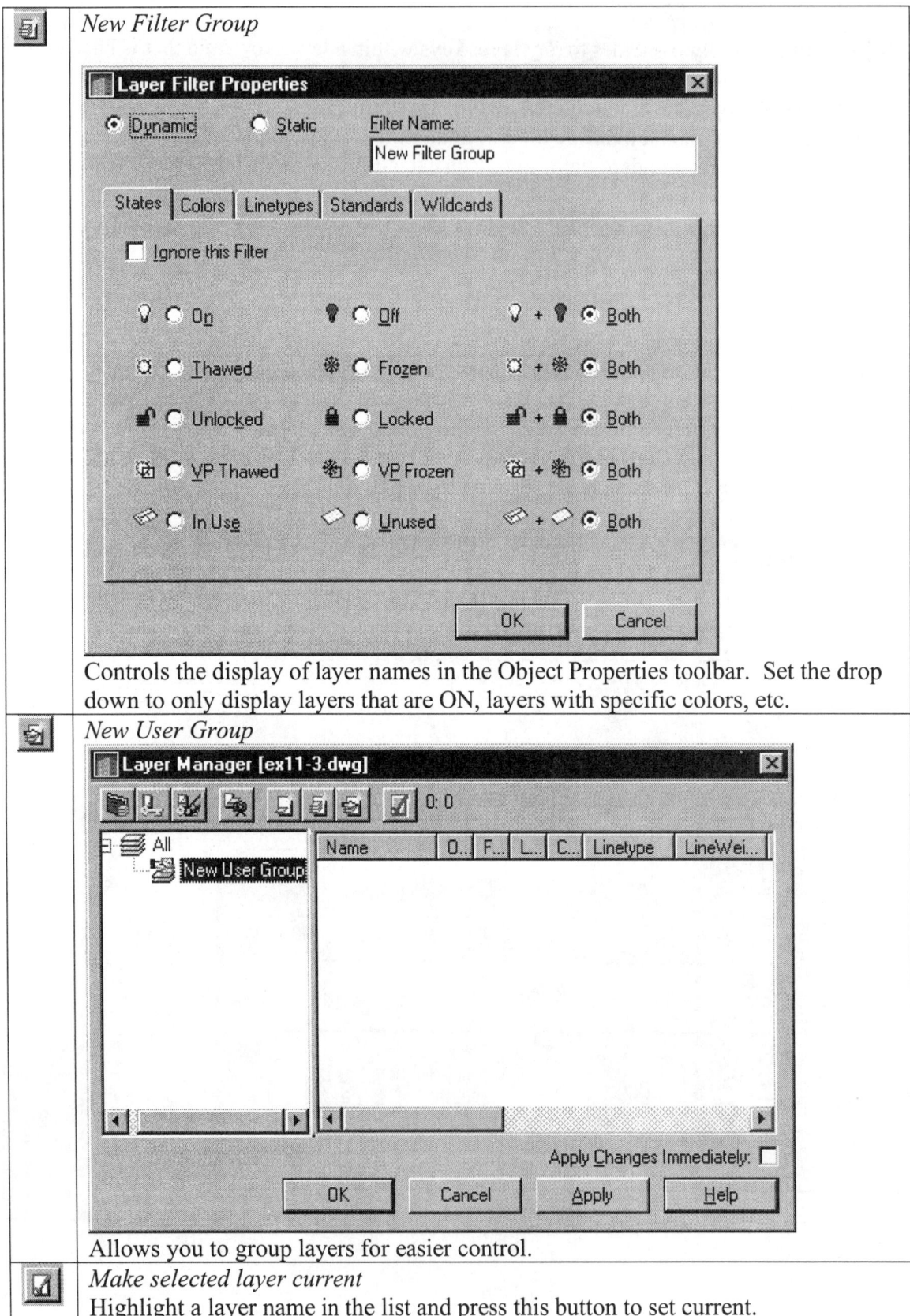

	New Filter Group
	Controls the display of layer names in the Object Properties toolbar. Set the drop down to only display layers that are ON, layers with specific colors, etc.
	New User Group
	Allows you to group layers for easier control.
	Make selected layer current
	Highlight a layer name in the list and press this button to set current.

1-78

	Select Layer Standard	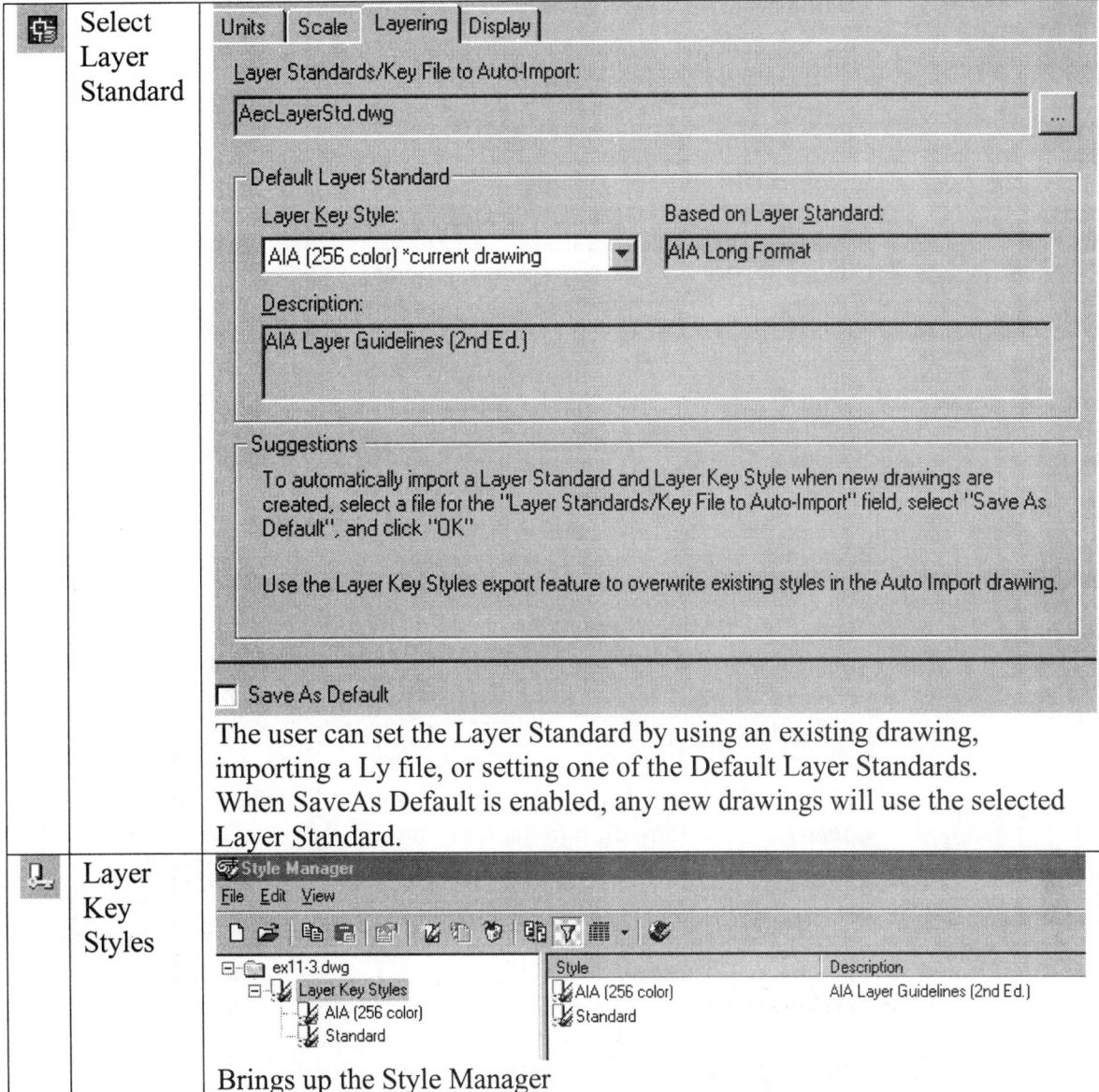 The user can set the Layer Standard by using an existing drawing, importing a Ly file, or setting one of the Default Layer Standards. When SaveAs Default is enabled, any new drawings will use the selected Layer Standard.
	Layer Key Styles	Brings up the Style Manager

	Layer Key Overrides	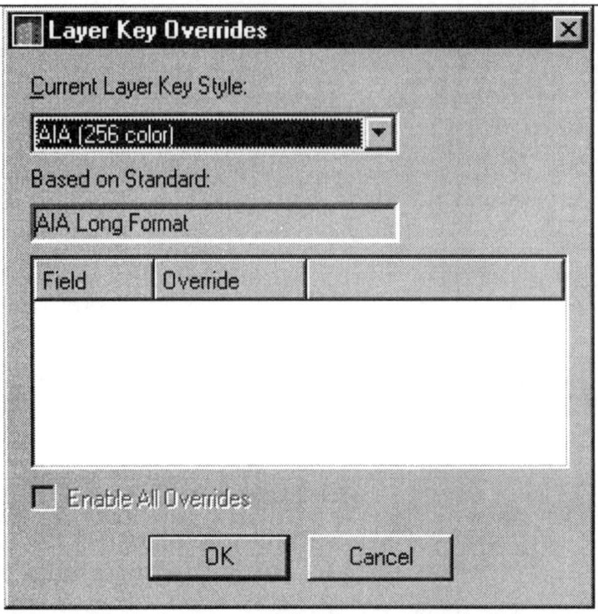 To set layer key overrides, you must have a layer key style that is based on a layer standard. If you base your layer key standard on a nonstandard style, then you cannot set layer key overrides. The layer key overrides that you set apply only to the layer keys that have any or all descriptive field overrides selected in the Layer Key Properties dialog box. In the Field/Override list, type values or click Browse to set values to override the information in the descriptive fields used by the layer standard. The fields in this dialog box depend on what layer standard your layer key style is based on. For example, if you are using the AIA layer standard, then you can enter override values for the Discipline, Major Code, Minor Code, and Status descriptive fields.
	Overrides ON/OFF	Clicking toggles OVERRIDES ON/OFF.
	Remap Object Layers	Changed objects to a Keyed Layer Command is REMAPLAYERS

TIP: The layer key styles in Autodesk Architectural Desktop, Release 3 and above replace the use of LY files. If you have LY files from S8 or Release 1 of AutoCAD Architectural Desktop that you want to use, then you can create a new layer key style from an LY file. On the command line enter **-AecLYImport** to import legacy LY files.

The following are the default layer keys used by Autodesk Architectural Desktop when you create AEC objects.

Default layer keys for creating AEC objects			
Layer Key	Description	Layer Key	Description
ANNDTOBJ	Detail marks	CONTROL	Control systems
ANNELOBJ	Elevation objects	CWLAYOUT	Curtain walls
ANNOBJ	Notes, leaders, etc.	CWUNIT	Curtain wall units
ANNREV	Revisions	DIMLINE	Dimensions
ANNSXOBJ	Section marks	DOOR	Doors
ANNSYMOBJ	Annotation marks	DOORNO	Door tags
APPL	Appliances	DRAINAGE	Drainage
AREA	Areas	ELEC	Electric
AREAGRP	Area groups	ELECNO	Electrical tags
AREAGRPNO	Area group tags	ELEV	Elevations
AREANO	Area tags	ELEVAT	Elevators
CAMERA	Cameras	ELEVHIDE	Elevations (2D)
CASE	Casework	EQUIP	Equipment
CASENO	Casework tags	EQUIPNO	Equipment tags
CEILGRID	Ceiling grids	FINCEIL	Ceiling tags
CEILOBJ	Ceiling objects	FINFLOOR	Finish tags
CHASE	Chases	FIRE	Fire system equip.
COLUMN	Columns	FURN	Furniture
COMMUN	Communication	FURNNO	Furniture tags

Lesson 1
Desktop Features

Layer Key	Description	Layer Key	Description
GRIDBUB	Plan grid bubbles	SECTHIDE	Sections (2D)
GRIDLINE	Column grids	SITE	Site
LAYGRID	Layout grids	SLAB	Slabs
LIGHTCLG	Ceiling lighting	SPACEBDRY	Space boundaries
LIGHTW	Wall lighting	SPACEOBJ	Space objects
MASSELEM	Massing elements	STAIR	Stairs
MASSGRPS	Massing groups	STRUCTBRACE	Structural braces
MASSSLCE	Massing slices	STRUCTBRACEIDEN	Structural brace tags
OPENING	Wall openings	STRUCTCOLS	Structural columns
PEOPLE	People	STRUCTCOLSIDEN	Structural column tags
PFIXT	Plumbing fixtures	SWITCH	Electrical switches
PLANTS	Plants - outdoor	TITTEXT	Border and title block
PLANTSI	Plants - indoor	TOILACC	Arch. specialties
POLYGON	AEC Polygons	TOILNO	Toilet tags
POWER	Electrical power	UTIL	Site utilities
PRK-SYM	Parking symbols	VEHICLES	Vehicles
ROOF	Rooflines	WALL	Walls
ROOFSLAB	Roof slabs	WALLFIRE	Fire wall patterning
ROOMNO	Room tags	WALLNO	Wall tags
SCHEDOBJ	Schedule tables	WIND	Windows
SEATNO	Seating tags	WINDASSEM	Window assemblies
SECT	Miscellaneous sections	WINDNO	Window tags

Set Current Display Configuration

Menu	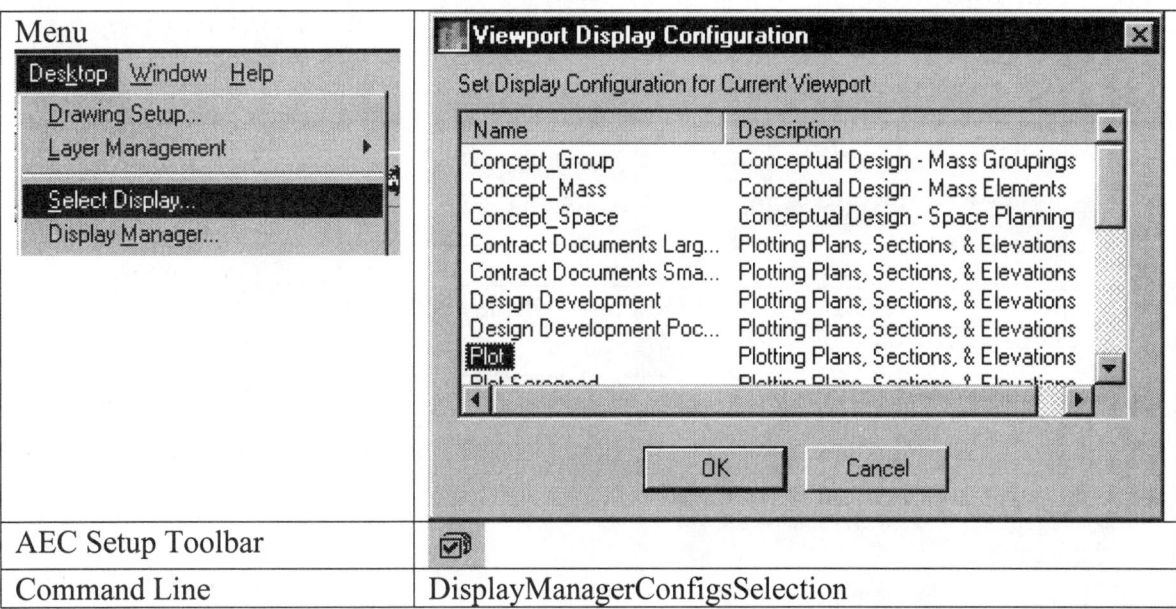
AEC Setup Toolbar	
Command Line	DisplayManagerConfigsSelection

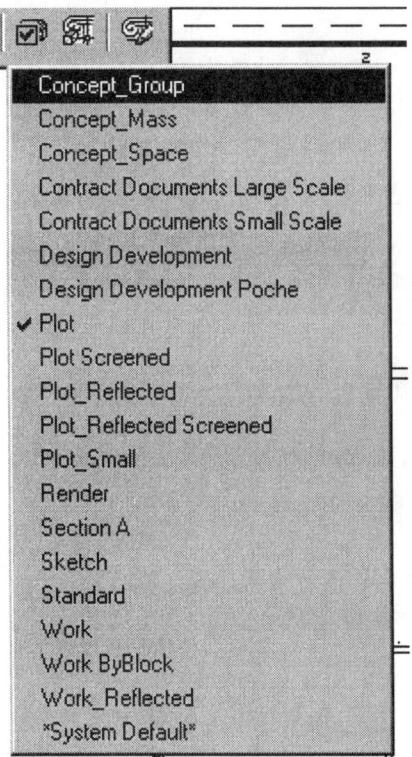

The display system in Autodesk® Architectural Desktop controls how AEC objects are displayed in a designated viewport. By specifying the AEC objects you want to display in a viewport and the direction from which you want to view them, you can produce different architectural displays, such as floor plans, reflected plans, elevations, 3D models, or schematic displays.

Autodesk Architectural Desktop, Release 3 includes templates with predefined display systems applied to viewports. You can use the display systems and viewports supplied by the templates, or you can modify the display system settings to suit your own office standards. If you want to create your own display systems, you can start a drawing from scratch or from a template that does not contain pre-defined display systems.

Display Manager

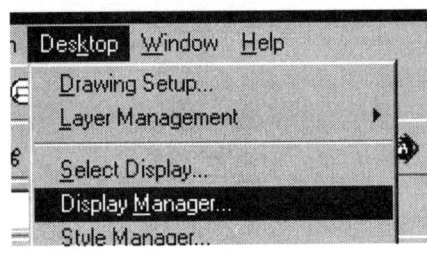 

Menu	Desktop->Display Manager
AEC Setup Toolbar	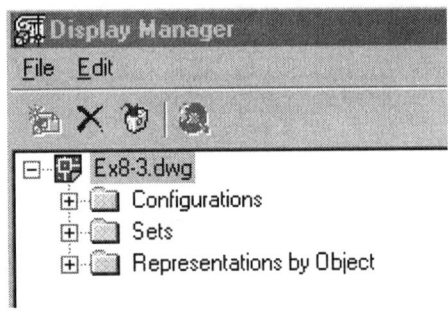
Command Line	

The display system in Autodesk® Architectural Desktop controls how AEC objects are displayed in a designated viewport. By specifying the AEC objects you want to display in a viewport and the direction from which you want to view them, you can produce different architectural displays, such as floor plans, reflected plans, elevations, 3D models, or schematic displays.

To change the display representation of an AEC object in a viewport

1. Set the viewport as current in which you want to make display changes.
2. Make sure that the current viewport is set to the desired view direction with the appropriate display configuration.
3. Open the Display Manager, and expand the Sets folder in the tree view on the left pane of the dialog box. The current display set is in bold text.
4. Click the current display set.
5. Click the Display Control tab on the right pane. This tab shows all the display representations that are active for the current display set.
6. Confirm that the display representations you want to display in your viewport are selected for the current display set. If not, select the appropriate check box(es).
7. Click Apply for the Display Manager to accept your changes.
8. Click OK to close the Display Manager.

The current viewport and all viewports that have the same display configuration assigned to it are drawn with the new display representations.

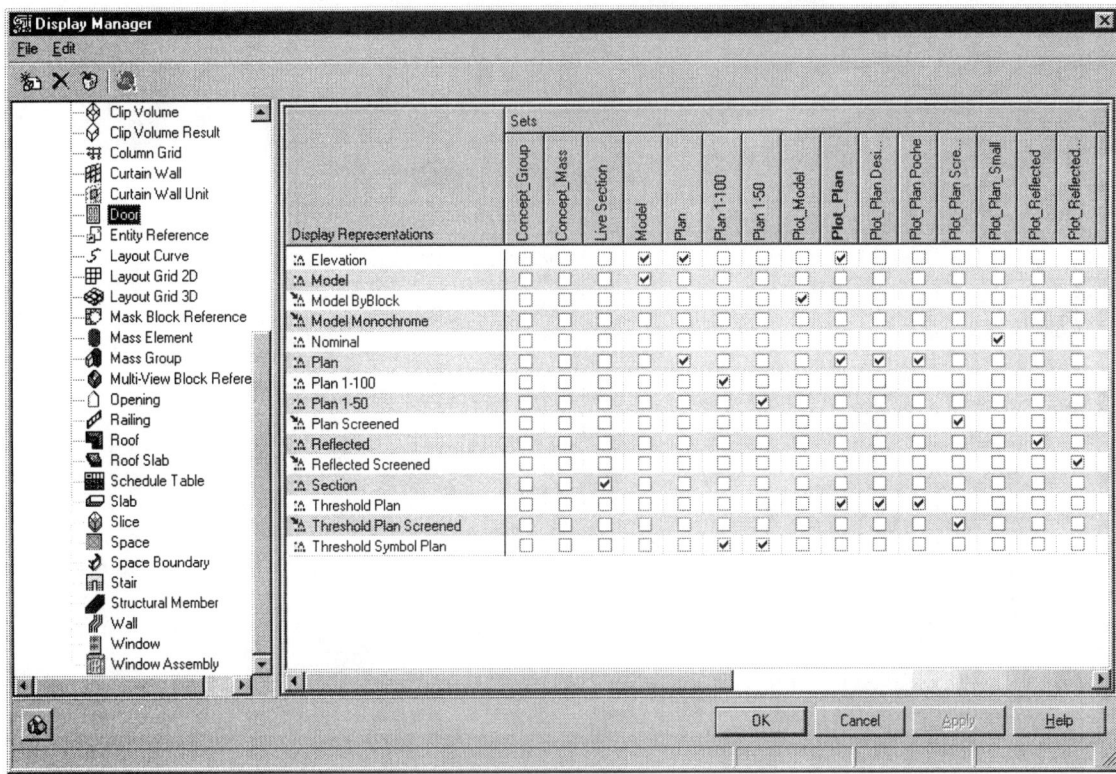

Locate the entity that is not appearing properly and check to see if the visibility is turned on for the display configuration you are in.

NOTES:

Lesson 2:
Site Plans

Most architectural projects start with a site plan. The site plan indicates the property lines, the house location, a North symbol, any streets surrounding the property, topographical features, location of sewer, gas, and/or electrical lines (assuming they are below ground – above ground connections are not shown), and topographical features.

When laying out the floor plan for the house, many architects take into consideration the path of sun (to optimize natural light), street access (to locate the driveway), and any noise factors.

Architectural Desktop allows the user to simulate the natural path of the sun based on the longitude and latitude coordinates of the site, so you can test how natural light will affect various house orientations.

A plot plan must include the following features:

- Length and bearing of each property line
- Location, outline, and size of buildings on the site
- Contour of the land
- Elevation of property corners and contour lines
- North symbol
- Trees, shrubs, streams, and other topological items
- Streets, sidewalks, driveways, and patios
- Location of utilities
- Easements and drainages (if any)
- Well, septic, sewage line, and underground cables
- Fences and retaining walls
- Lot number and/or address of the site
- Scale of the drawing

The plot plan is drawn using information provided by the county/city and/or a licensed surveyor.

It used to be that you would draw to a scale, such as 1/8" = 1', but with Architectural Desktop, you draw full-size and then set up your layout to the proper scale. This ensures that all items you draw will fit together properly.

Lesson 2
Site Plans

Exercise 1:
Creating Custom Line Types

Drawing Name: New
Estimated Time: 15 minutes

This exercise reinforces the following skills:

- Creation of linetypes
- Customization

Architectural drafting requires a variety of custom linetypes in order to display specific features. The standard linetypes provided with Architectural Desktop are insufficient from an architectural point of view. You may find custom linetypes on Autodesk's pointA website, www.cadalog.com, or any number of websites on the Internet. However, the ability to create linetypes as needed is an excellent skill for an architectural drafter.

It's a good idea to store any custom linetypes in a separate file. The standard file for storing linetypes is acad.lin. If you store any custom linetypes in acad.lin file, you will lose them the next time you upgrade your software.

―――――――――   ―――   ―――   ―――――――――

Draw a property line.
This property line was created as follows:
Set ORTHO ON.
Draw a horizontal line 12 units long
Use SNAP FROM to start the next line @3,0 distance from the previous line.
The short line is 6 units long.
Use SNAP FROM to start the next line @3,0 distance from the previous line.
The second short line is 6 units long.
Use SNAP FROM to start the next line @3,0 distance from the previous line.
Draw a second horizontal line 12 units long

The Express Tools have a Make Linetype tool.

TIP: If you do not have the Express Tools loaded, you can download them for free from www.mossdesigns.com. Download the extools04.zip file.

Lesson 2
Site Plans

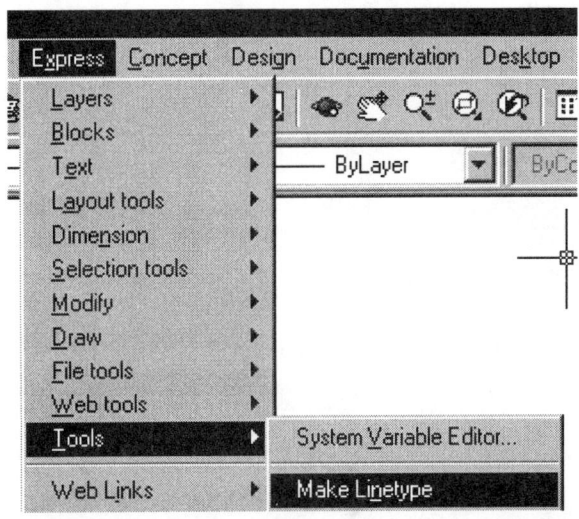

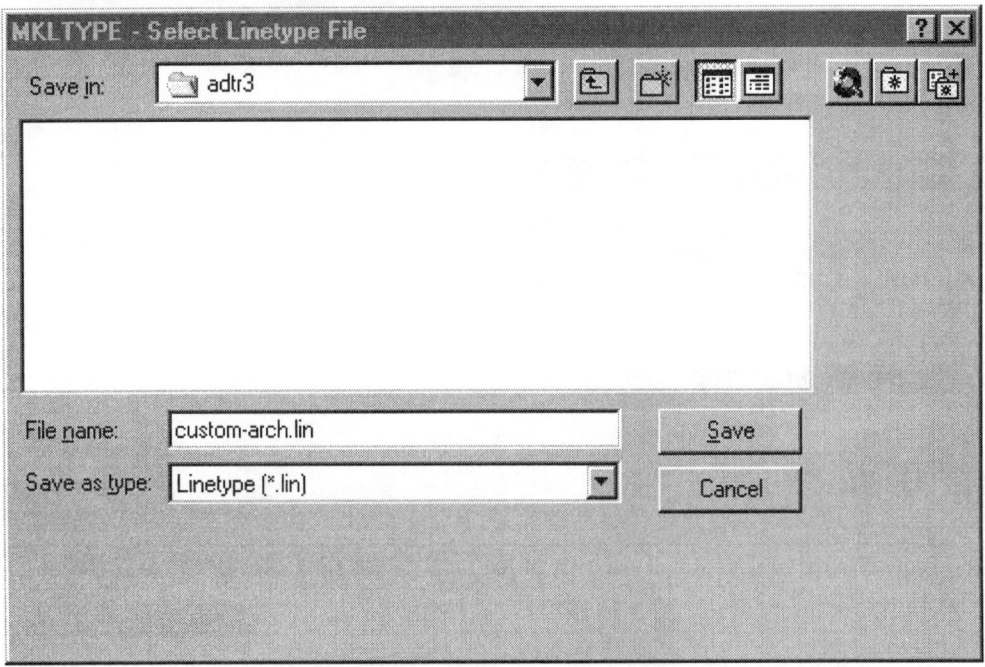

Create a file name called custom-arch.lin and press 'Save'.

When prompted for the linetype name, type: property-line.
When prompted for the linetype description, type: property-line
Specify starting point for line definition; select the far left point of the line type.
Specify ending point for line definition; select the far right point of the line type.
When prompted to select the objects, start at the long line on the left and pick the line segments in order.

Lesson 2
Site Plans

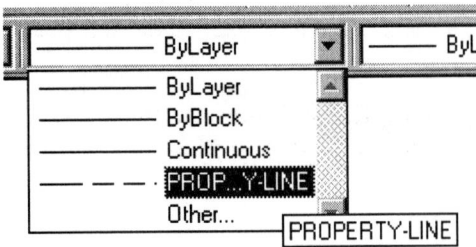

Under the linetype drop-down, locate the property-line just created and make it active.

Draw some lines to see if they look OK. Make sure you make them long enough to see the dashes.

Locate the custom-arch.lin file you created.

Open it using NotePad.

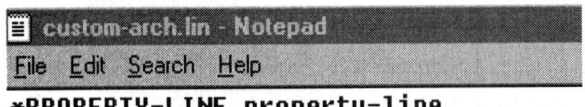

You see a description of the property-line.

Alternate Method:

You can create your linetypes directly using NotePad, if you don't have the Express Tools available.

Start a New file in NotePad.

On the first line, type:
*PROPERTY-LINE,property-line

On the second line, type:
A,12,-3,6,-3,6,-3,12

Save the file as custom-arch.lin.

Load the linetype.

TIP: Place all your custom linetypes in a single drawing file and then use the AutoCAD Design Center to help you locate and load the desired linetype.

Exercise 2:
Creating a Custom Text Style

Drawing Name: New
Estimated Time: 15 minutes

This exercise reinforces the following skills:

- Text Styles
- Customization

Most architects use custom text styles. You may find custom text styles on Autodesk's pointA website, www.cadalog.com, or any number of websites on the Internet. However, the ability to create text styles as needed is an excellent skill for an architectural drafter.

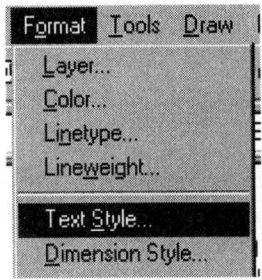

Select Format->Text Style.

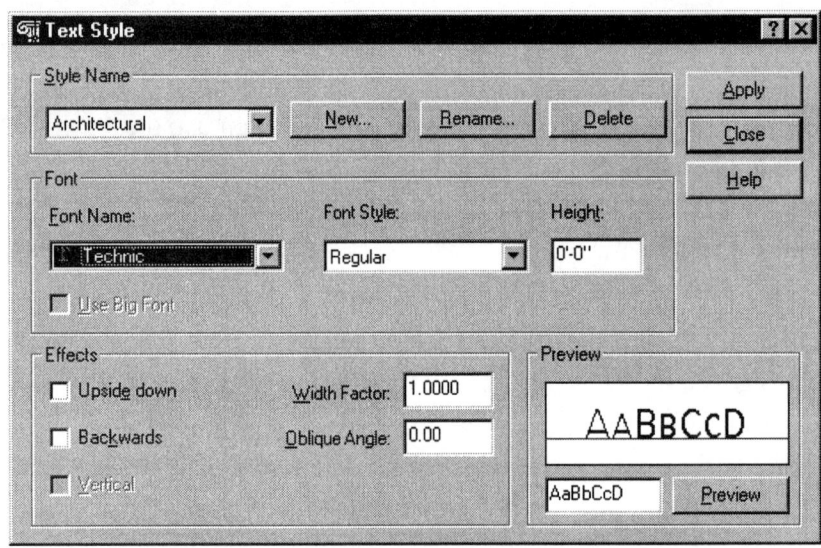

Press the New button. Type 'Architectural' for the new style name. Select 'Technic' for the font name. Technic is a standard font included with Windows.
Press 'Apply'. Then Close.

Save the file as 'fonts.dwg' in your custom subdirectory.

Use this drawing to store all your custom text styles. You can then use the Design Center to load any custom fonts from this drawing.

Exercise 3:
Creating New Layers

Drawing Name: New using the Aec Arch (Imperial) template
Estimated Time: 15 minutes

This exercise reinforces the following skills:

- Use of toolbars
- Layer Manager
- Creating New Layers
- Loading Linetypes

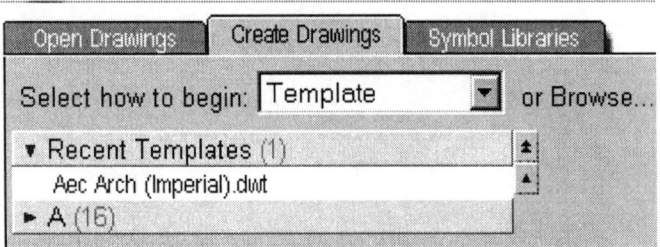

We start by creating a new drawing using the Aec Arch (Imperial) template.

Activate the Layer Management toolbar.

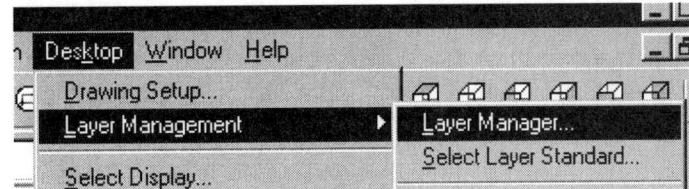

```
Command:
Command: _AecLayerManager
```

Layer Manager

Activate the Layer Manager.

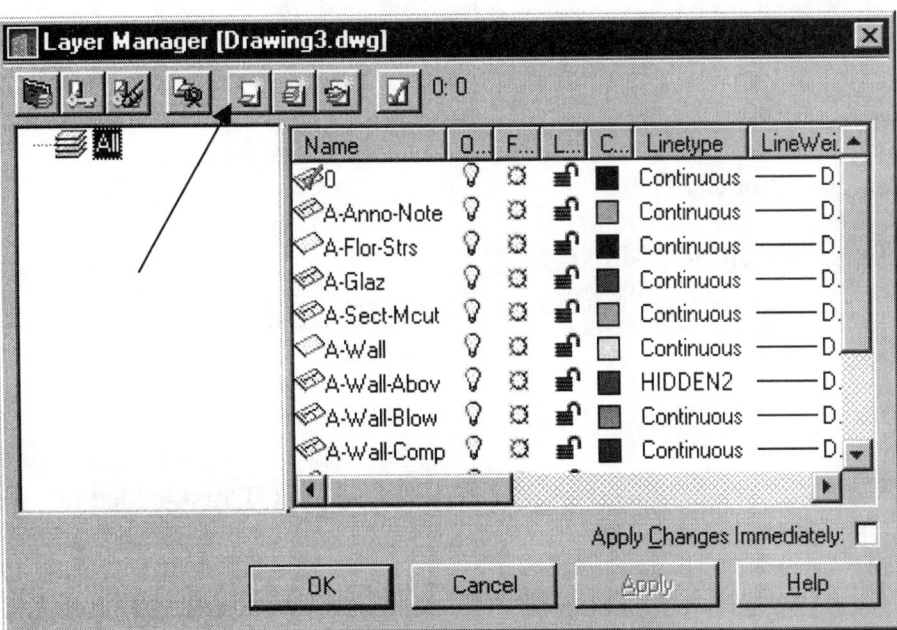

Note that several layers have already been created. The AEC_Arch Imperial template automatically sets your Layer Standards to AIA (256 color).

Select the 'New Layer' tool.

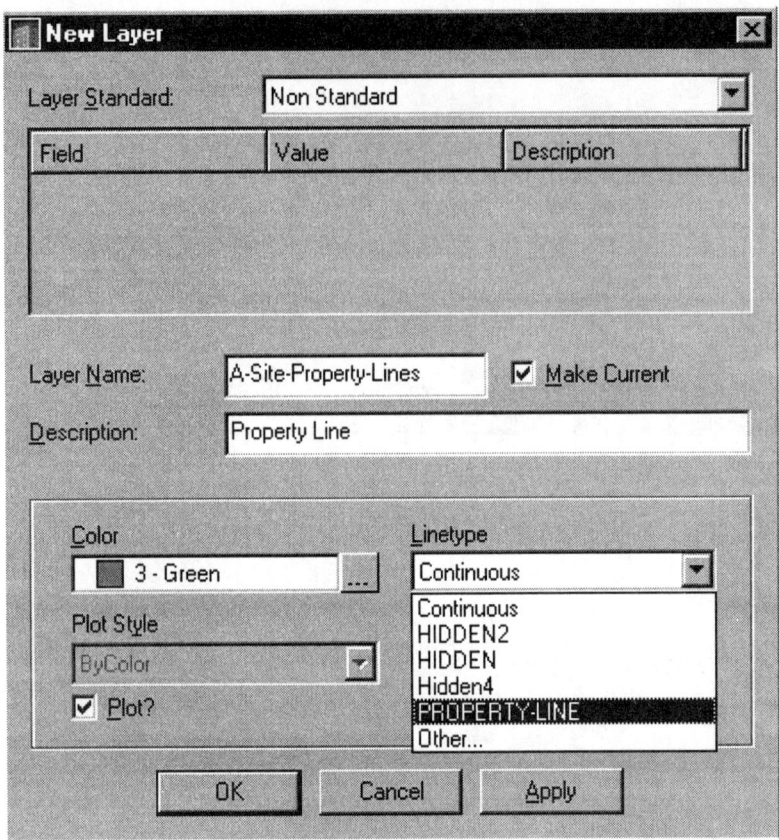

Name the layer 'A-Site-Property-Lines'.
Enable 'Make Current'.
Type in the Description 'Property Line'.
Set the Color to Green.
Under Linetype, select 'PROPERTY-LINE'.

> You will get an error message if the linetype is NOT first loaded in the drawing and it will cause your system to crash. Use Format->Linetype to load.

Highlight PROPERTY-LINE and press 'OK'.

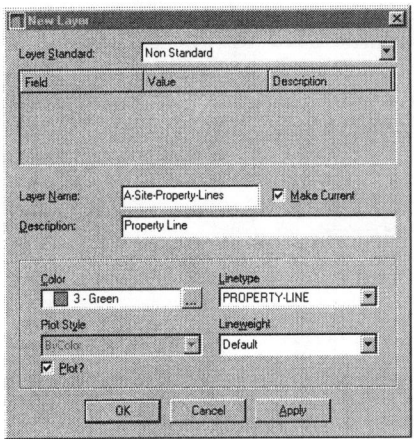

Press 'Apply' and 'OK'.

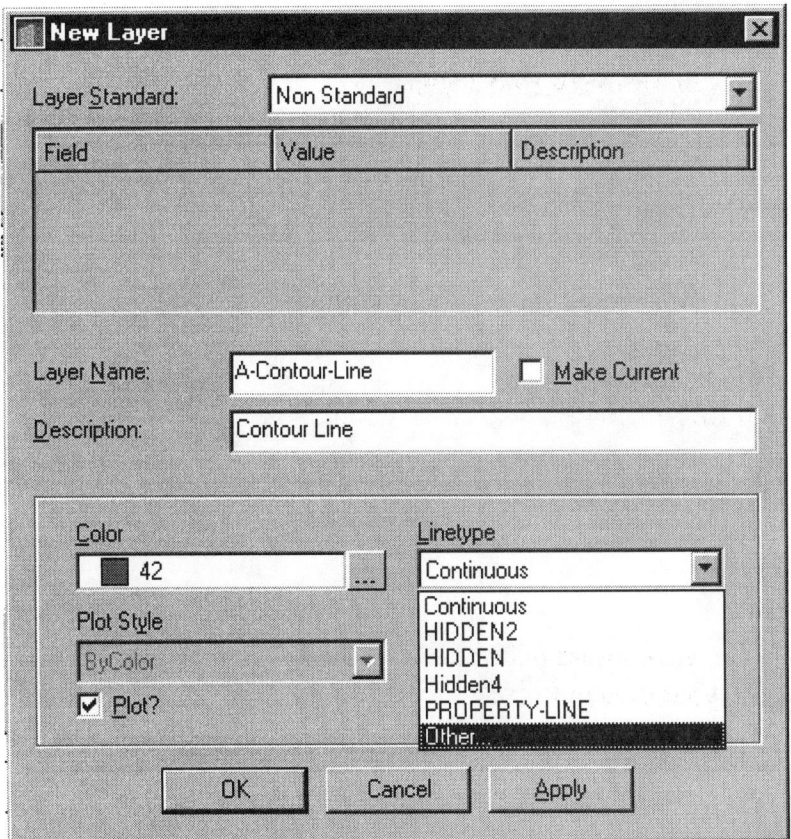

Create a new layer called A-Contour-Line.
Set the color to 42.
Under Linetype, select 'Other'.

Lesson 2
Site Plans

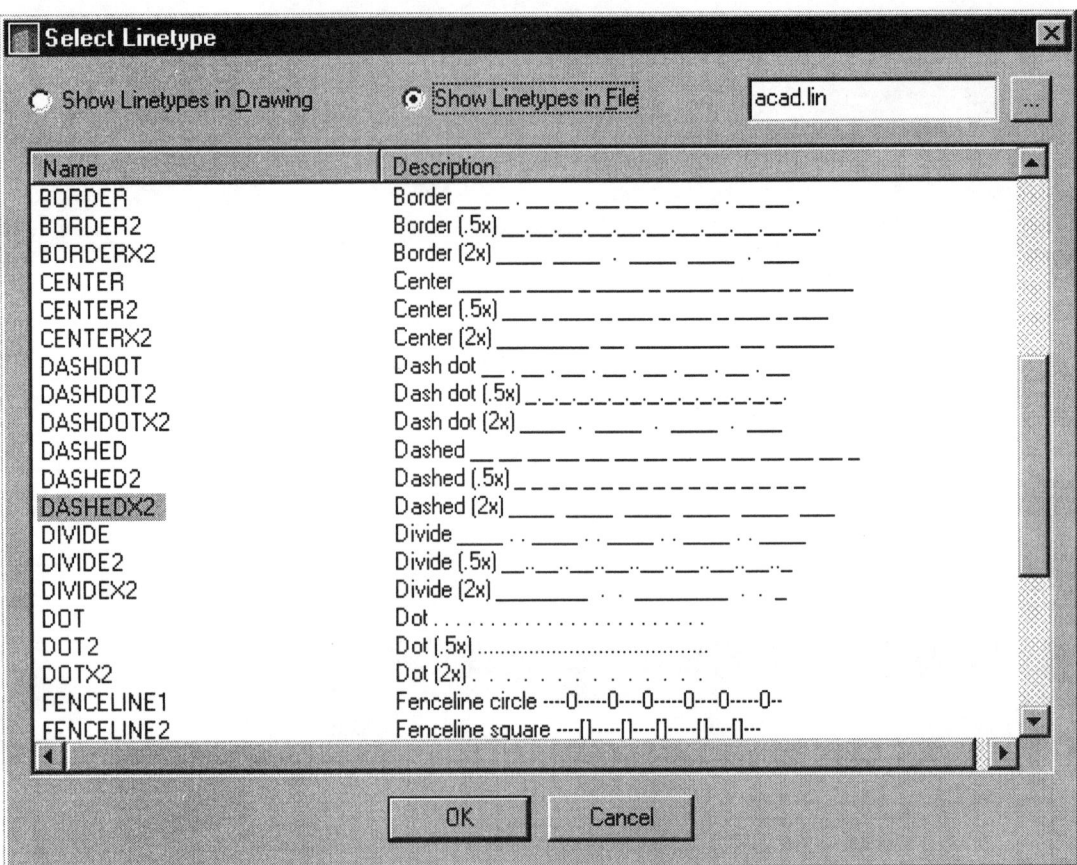

Enable the 'Show Linetypes in File' and set the file to 'acad.lin'.

Locate the Linetype called DASHEDX2.

Press 'OK'.

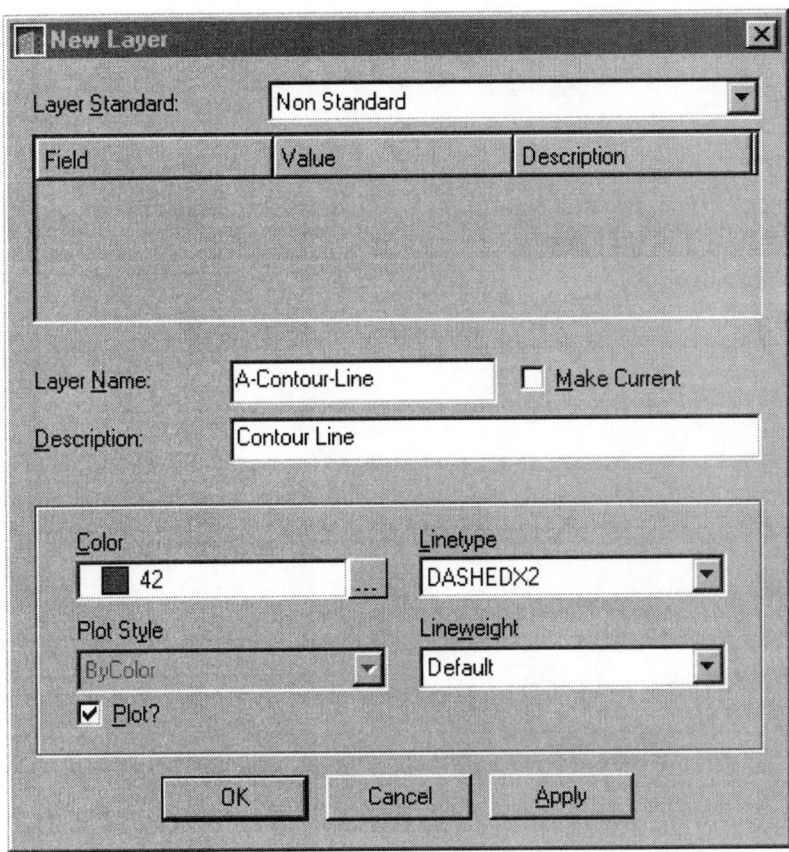

Press 'Apply' and 'OK'.

Save as Ex2-3.dwg

Lesson 2
Site Plans

Exercise 4:
Creating a Site Plan with an AEC Template

Drawing Name: Ex2-3.dwg
Estimated Time: 15 minutes

This exercise reinforces the following skills:

- Use of toolbars
- Documentation Tools
- Elevation Marks
- Surveyor's Angles

Go to Format->Units.

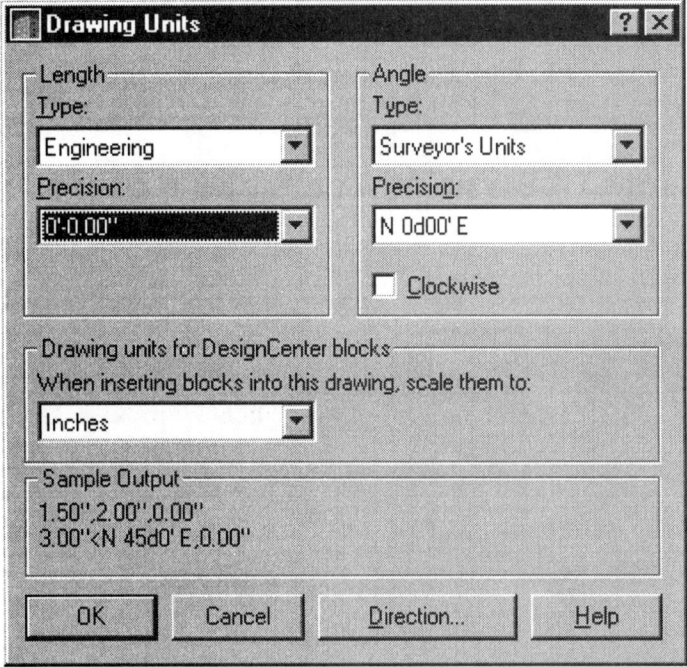

Set up the Length Type to Engineering.
Set the Precision to two decimal places.
Set the Angle Type to Surveyor's Units.
Press 'OK'.

2-12

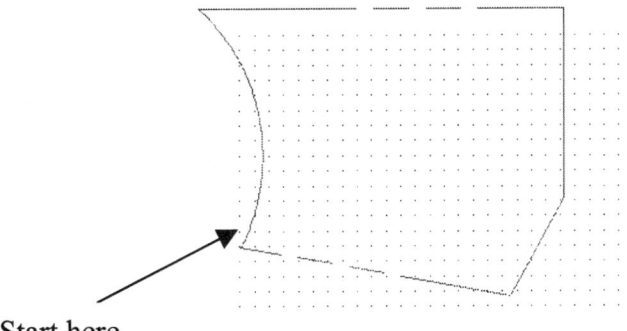

Start here

```
Command: l
LINE Specify first point: 1',50'
Specify next point or [Undo]: @187'<s80de
Specify next point or [Undo]: @75'<n30de
Specify next point or [Close/Undo]: @126'<n
Specify next point or [Close/Undo]: @250'<180
```

This command sequence uses Surveyor's Units.

Draw the property line shown on the A-Site-Property-Lines layer.

Draw an arc using Start, End, Radius.
Select the bottom point as the Start point.
Select the top point at the End point.
The radius is 132'.
(Make sure you use the apostrophe to indicate feet, or your measurements will all be in inches.)

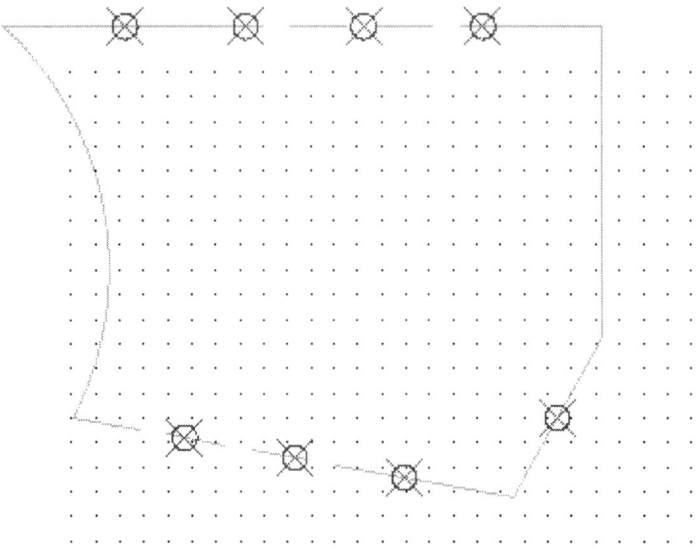

Set the A-Contour-Line layer current.
Use the DIVIDE command to place points as shown. The top line is divided into five equal segments.
The bottom-angled line is divided into four equal segments.
The small angled line is divided into two equal segments.

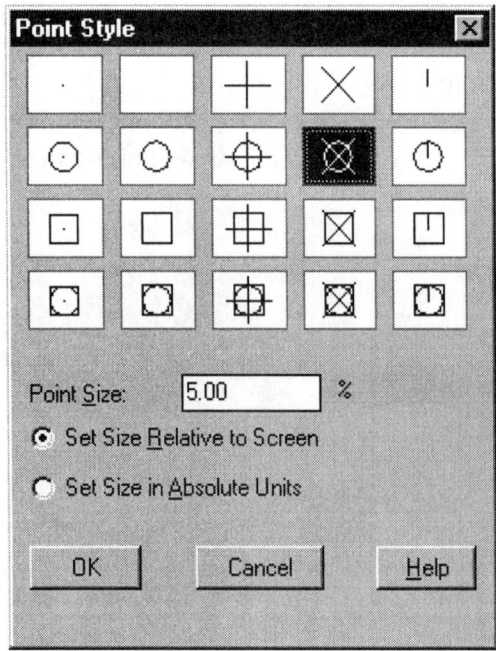

You need to use Format->Point Style to set the points so they are visible.

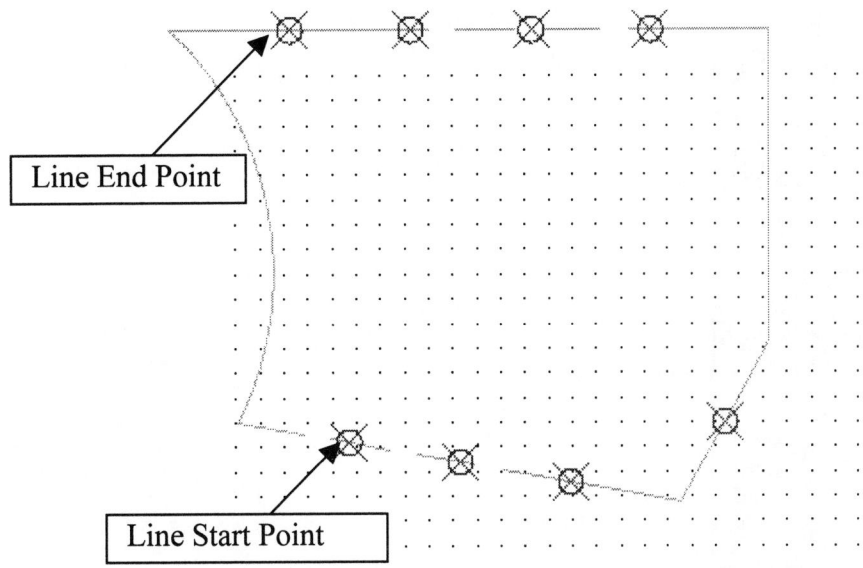

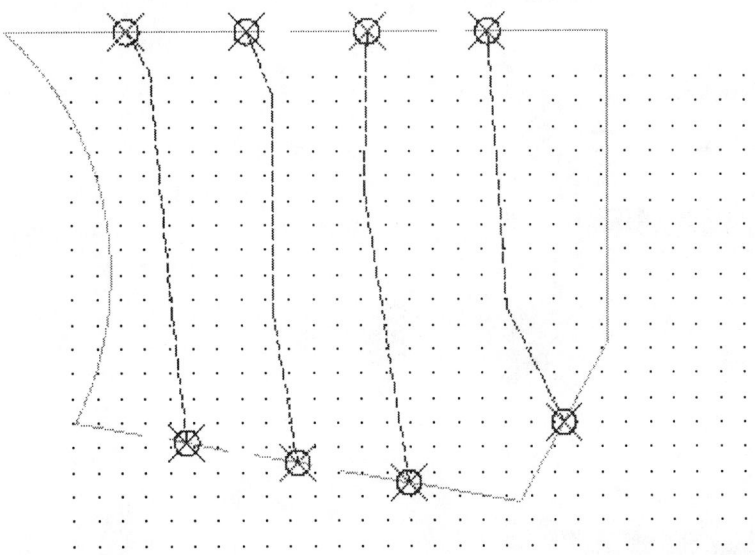

Draw contour lines using Pline and NODE Osnaps as shown.

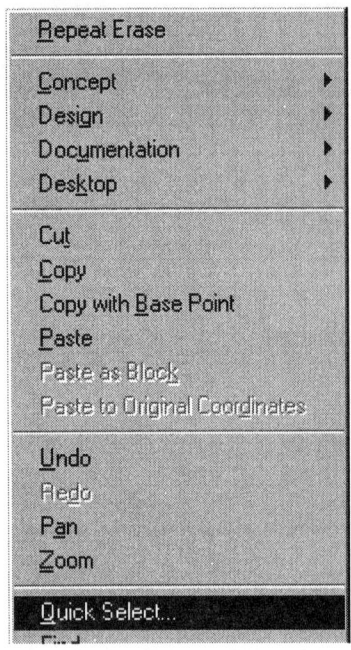

Right click anywhere in the graphics area and select Quick Select.

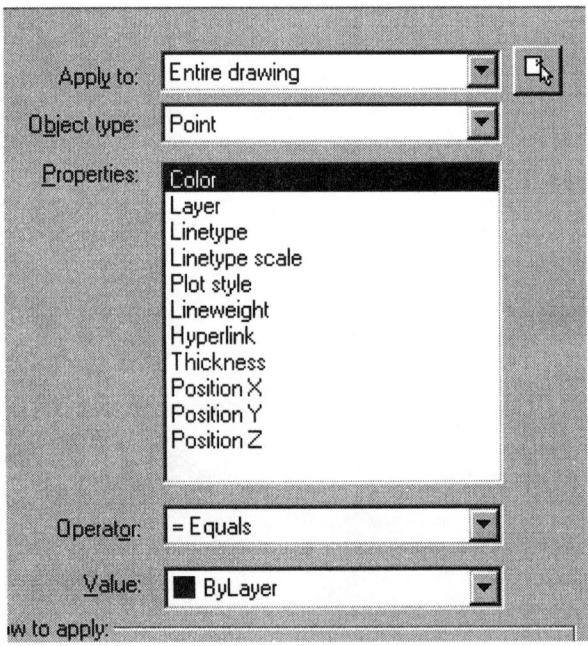

Set the Object Type to Point.
Set the Color Equals ByLayer.

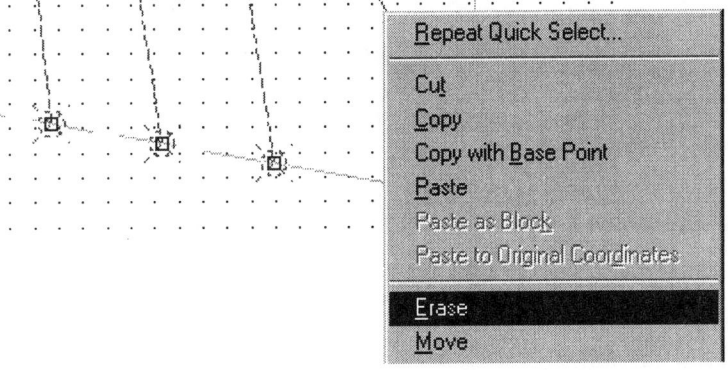

All the points are selected.
Right click and select 'Erase'.

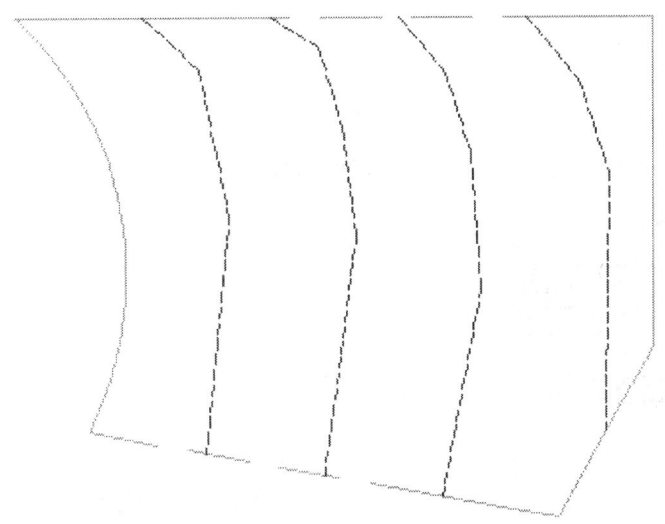

Activate the Documentation-Imperial toolbar.

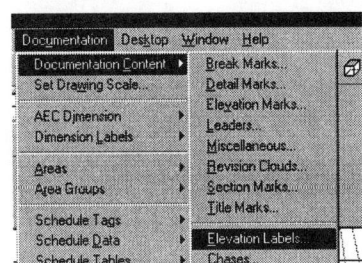

Lesson 2
Site Plans

You can also access Elevation Marks using the Menu.
   Go to Documentation->Documentation Content->Elevation Labels.

Elevation Label

We're going to place an Elevation Label at each vertex of the property line.

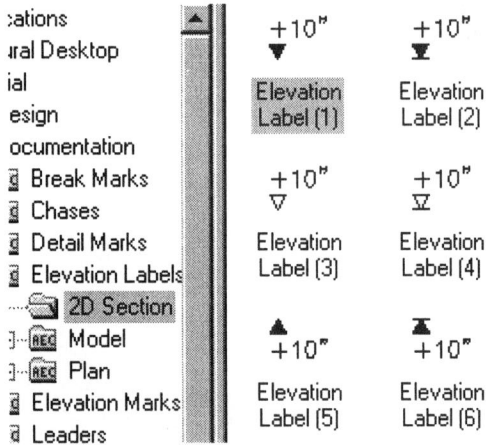

The Design Center will activate.
Select the 2D Section subfolder under Elevation Labels.
We will drag and drop Elevation Label (1) into the drawing.

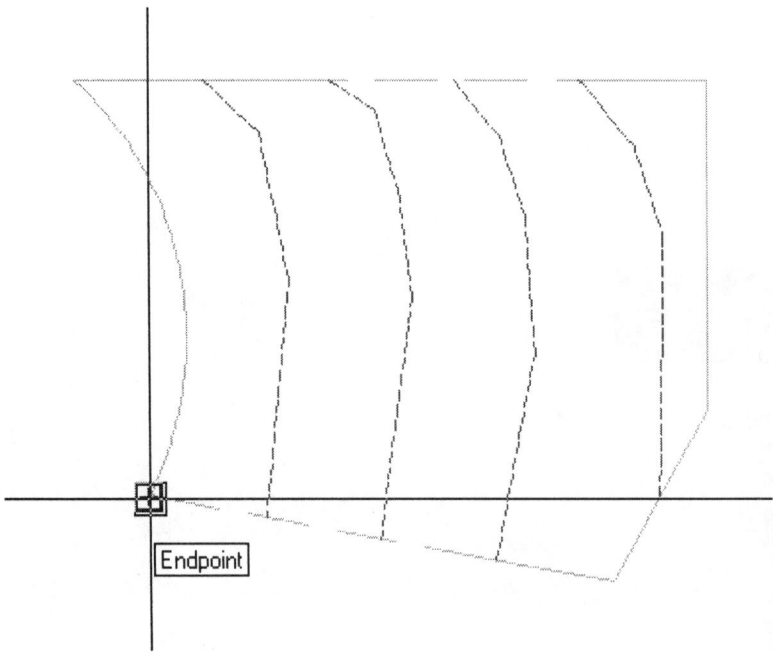

Select the Endpoint shown.

2-18

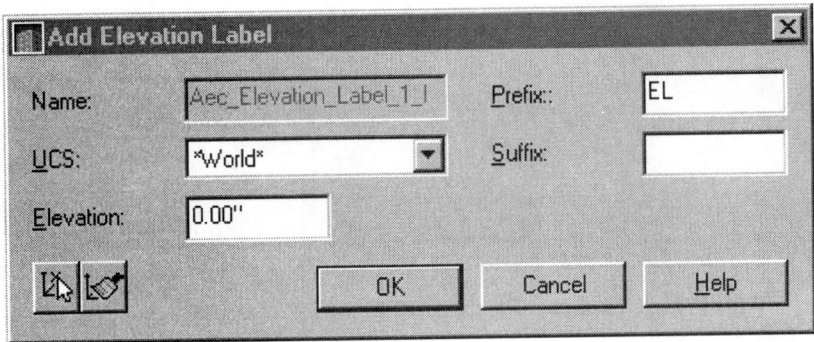

Set the Elevation to 0.00".
Set the Prefix to EL.

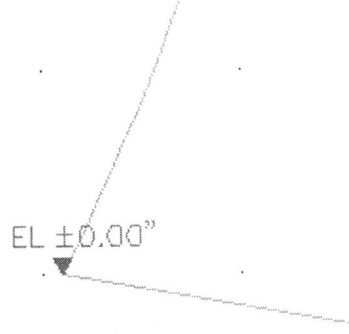

Our elevation label.

 　 💡 　 ✱ 　 🔓 ▬ 221

The Elevation Label is automatically placed on the A-Anno-Dims layer.

EL ±0.00"

Locate an elevation label on the upper left vertex as shown.

Lesson 2
Site Plans

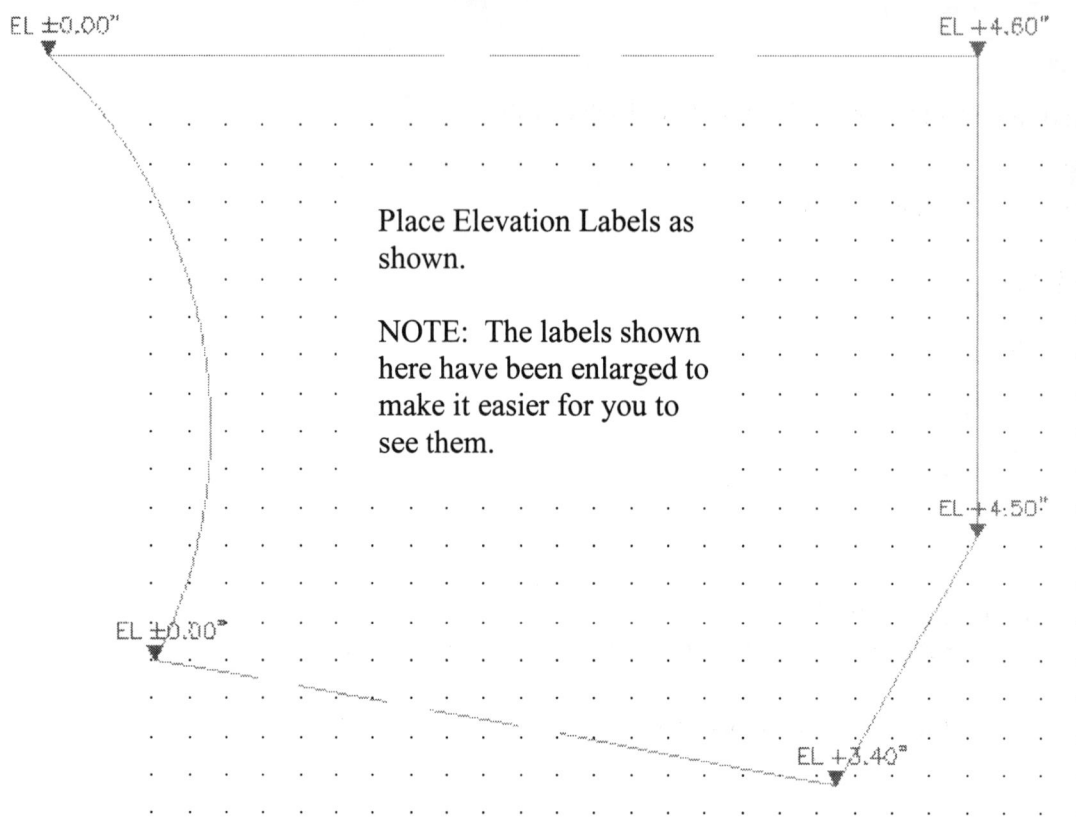

Place Elevation Labels as shown.

NOTE: The labels shown here have been enlarged to make it easier for you to see them.

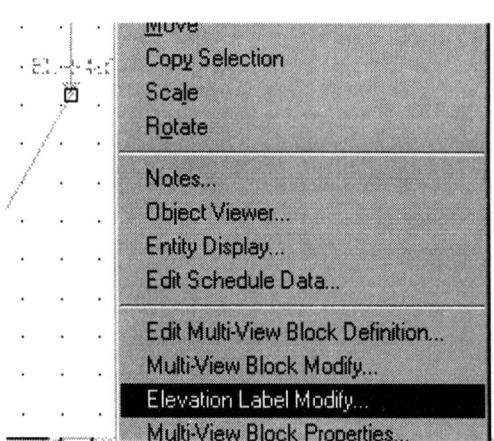

To modify the value of the elevation label, select the label. Right click and select 'Elevation Label Modify'.

Load the custom text style created in Exercise 2-2.

Activate the Design Center.

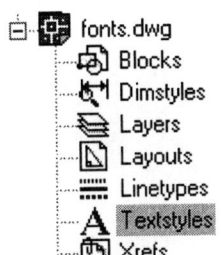

Locate the 'fonts.dwg'. Navigate to the Textstyles subfolder.
Then drag and drop the Architectural text style into the current drawing.

Set the A-Anno-Dims Layer current.

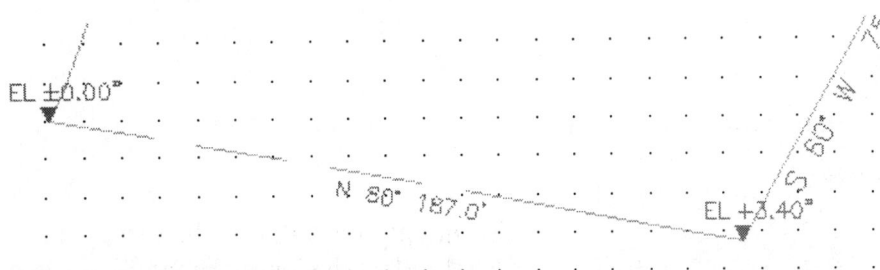

Use the TEXT command to create the text.
Set the text to 5' high so you can see it.
Set the rotation angle to –10 degrees
Use %%d to create the degree symbol.

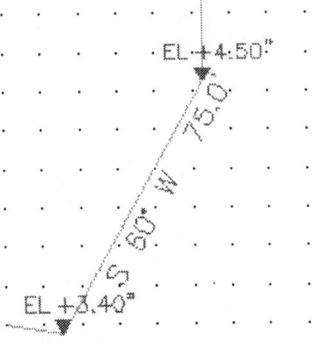

To create the S60... note, use the TEXT command.
Height is 5', rotation angle is 60 degrees.

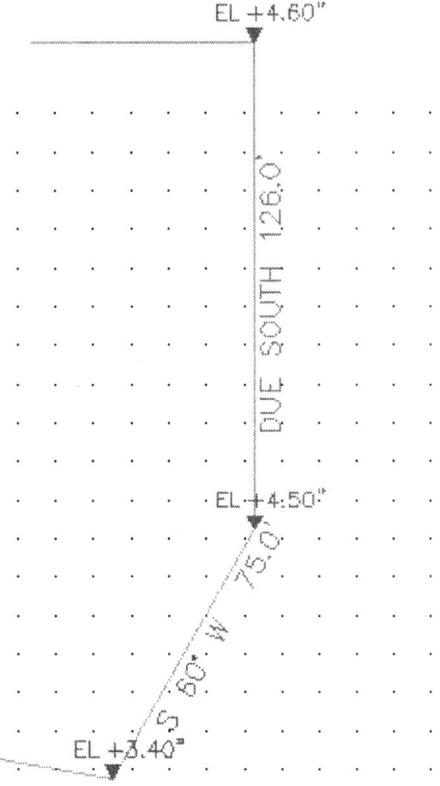

Create the Due South note using height of 5', rotation angle of 90 degrees.

Add the text shown.

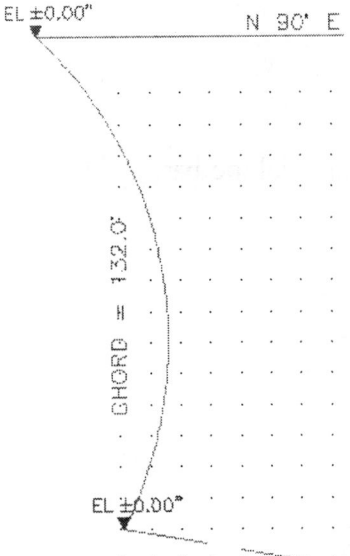

Add the Chord note shown.
Rotation angle is 90 degrees.

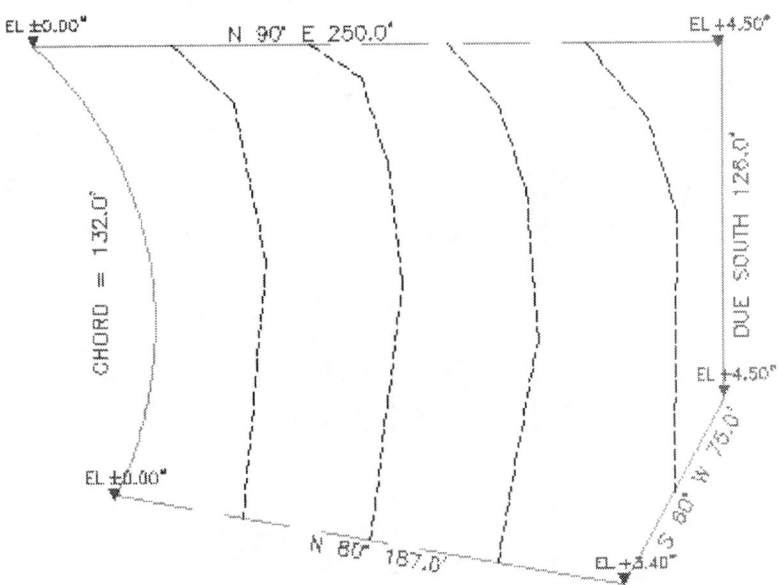

Our site plan so far.

Miscellaneous Tools

Select the Miscellaneous Tools from your Documentation-Imperial toolbar. The Design Center will activate.

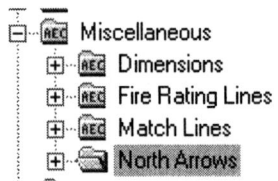

Go to the folder called 'North Arrows'.

Select the North Arrow E. Drag and drop it into your drawing as shown.

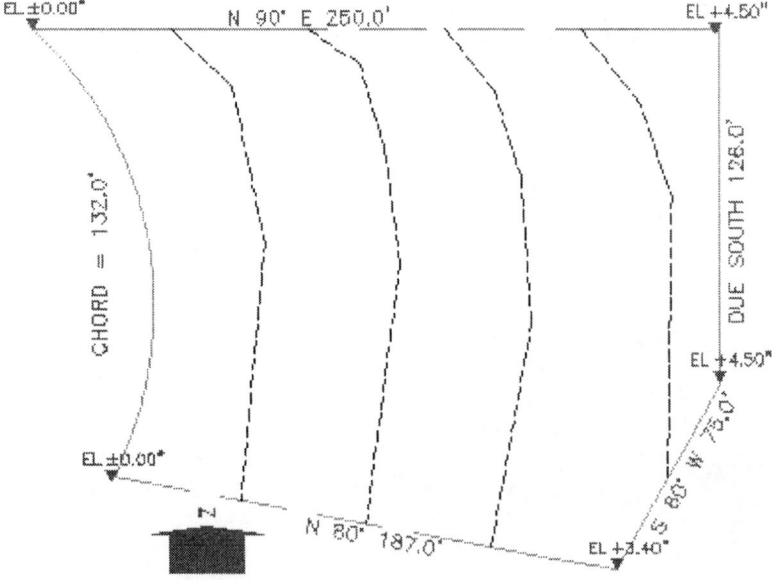

The arrow shown here has been enlarged to make it easier for you to see.

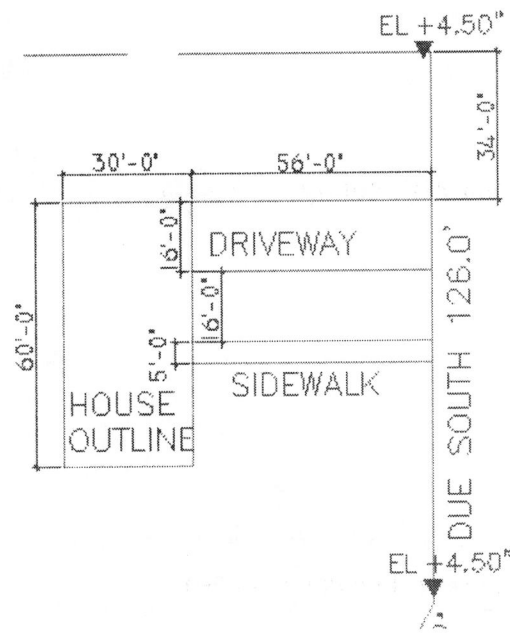

Set A-Site-Property-Lines layer current.
Create the house outline as shown.
Do not create the dimensions or descriptive text. They are there to help you with creating the outlines.

Save the file as Ex2-4.dwg.

Lesson 2
Site Plans

Exercise 5:
Creating a Layer User Group

Drawing Name: Ex 2-4.dwg
Estimated Time: 15 minutes

This exercise reinforces the following skills:

- Use of toolbars
- Layer Manager
- Layer User Group

A Layer User Group allows you to group your layers, so you can quickly freeze/thaw them, turn them ON/OFF, etc.

> **TIP:** The difference between FREEZING a Layer and turning it OFF is that entities on FROZEN Layers are not included in REGENS. Speed up your processing time by FREEZING layers.

The layers we just created are used for our Site Plan. We can create a group of layers that are just used for the site plan and then freeze them when we don't need to see them.

Bring up the Layer Manager.

Select the 'New User Group' tool.

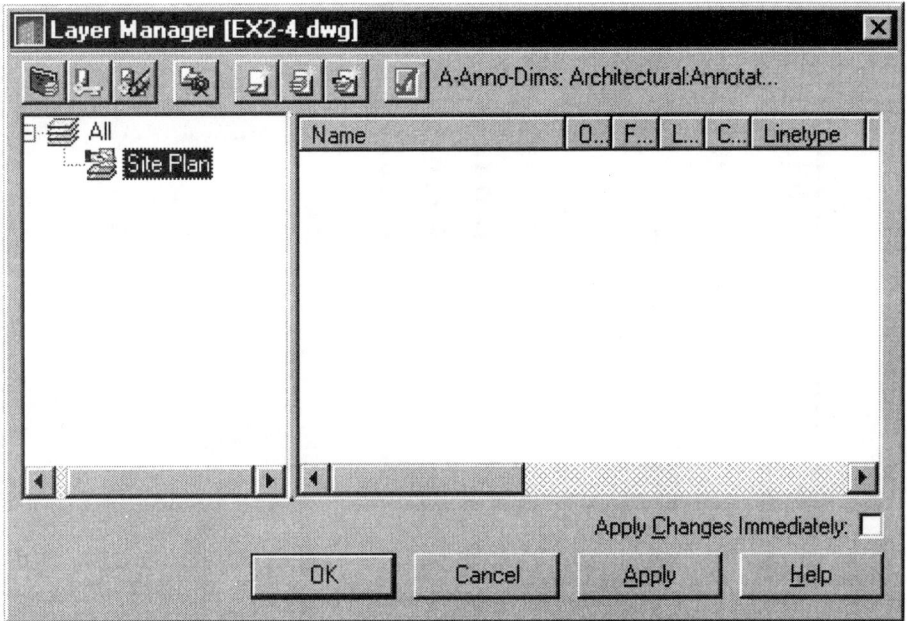

Name our user group as 'Site Plan'.

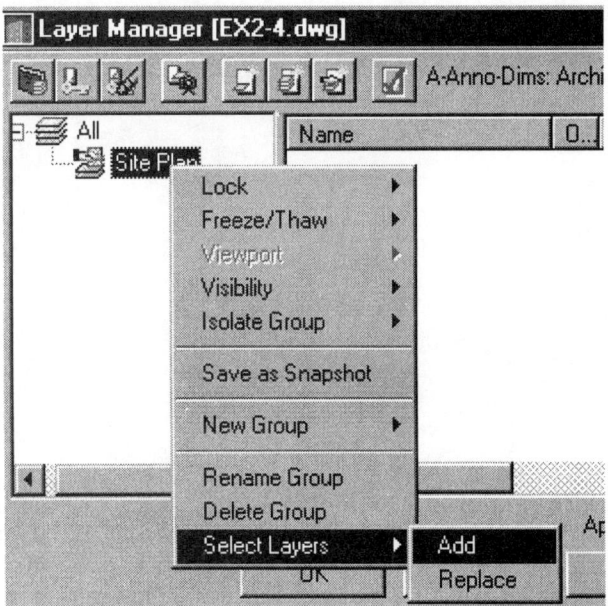

Highlight the group.
Right click and select 'Select Layers->Add'.

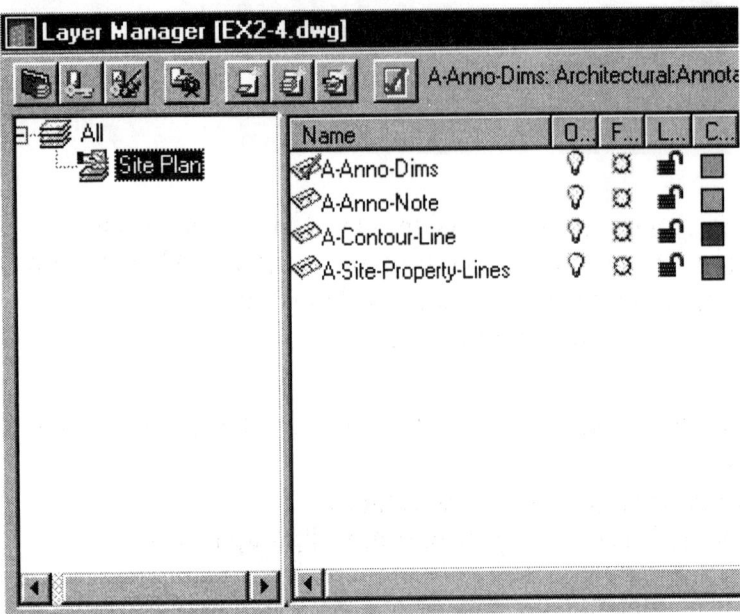

Window around all the items in our drawing.
All the pertinent layers will appear in the window.

We can then use this group to quickly freeze, turn off, etc. those layers.

Save our drawing as 'lesson 2.dwg'.

Quiz 1

True or False

1. Doors, windows, and walls are inserted into the drawing as objects.
2. The Architectural Toolbars are loaded from the AecArchx Menu Group.
3. Templates for beginning an Architectural Desktop drawing are Aec arc (Imperial). Dwt and Aec arch (metric).dwt.

4. If you start a New Drawing using 'New' tool shown above, the Start-Up dialog will appear.
5. Mass elements are used to create Mass Models.
6. The Layer Manager is used to organize, sort, and group layers.
7. Layer Standards are set using the Layer Manager.
8. You must load a custom linetype before you can assign it to a layer.

Multiple Choice
Select the Best Answer.

9. If you start a drawing using one of the standard templates, ADT will automatically create _____ layouts.

    A. Two layouts, plus Model Space
    B. Four layouts, plus Model Space
    C. Ten layouts, plus Model Space
    D. Eleven layouts, plus Model Space

10. ADT Options are set through:

    A. Options
    B. Desktop
    C. User Settings
    D. Template

11. The Desktop Display Manager controls:

    A. How AEC objects are displayed in the graphics window
    B. The number of viewports
    C. The number of layout tabs
    D. Layers

Quiz 1

12. Mass Elements are created under the _____ Menu.

    A. Desktop
    B. Concept
    C. Design
    D. Documentation

13. Wall Styles are created under the _____ Menu.

    A. Desktop
    B. Concept
    C. Design
    D. Documentation

14. To create a New Layer, use:

    A. Layer Manager
    B. Layer Properties Manager
    C. Format->Layer
    D. All of the Above

15. You can change the way AEC objects are displayed by using:

    A. Edit Object Properties
    B. Edit Display Properties
    C. Edit Entity Properties
    D. Edit AEC Properties

ANSWERS:

1) T; 2) T; 3) T; 4) F; 5) T; 6) T; 7) F; 8) T; 9) C; 10) A; 11) A; 12) B; 13) C; 14) D; 15) B

Lesson 3
Foundations

We now are going to use the drawing created in Lesson 2 and add a foundation. We are going to lay down some concrete slab.

Exercise 1:
Convert to Slab

Drawing Name: Lesson 2.dwg
Estimated Time: 15 minutes

This exercise reinforces the following skills:

- Convert to Slab
- Documentation Content – Miscellaneous Tools
- Applying Dimensions from the Design Center
- Layer Manager

Open the drawing created in Lesson 2.

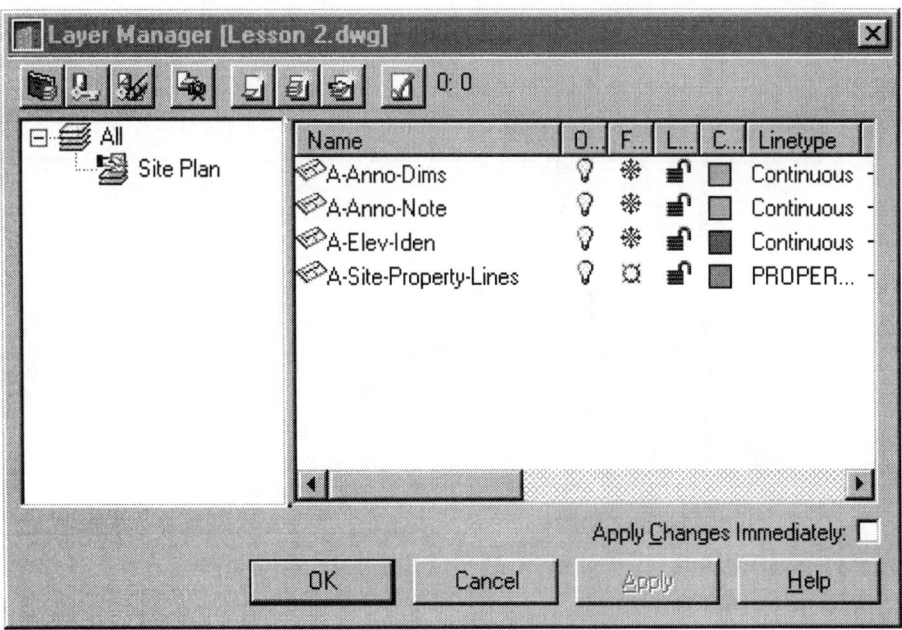

Activate the Layer Manager.
Freeze all the layers under the Site Plan group except the A-Site-Property-Lines layer.

ADT will insert slabs on layer A-Slab.

Lesson 3
Foundations

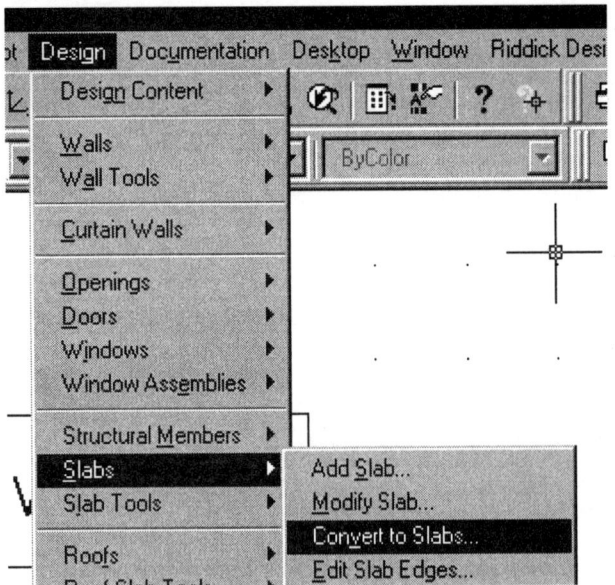

Go to Design->Slabs->Convert to Slabs.

Menu	Design->Slabs->Convert to Slabs.
Command Line	AecSlabConvert
Toolbar (Slabs)	

 TIP: You can only convert a closed polyline into a slab.

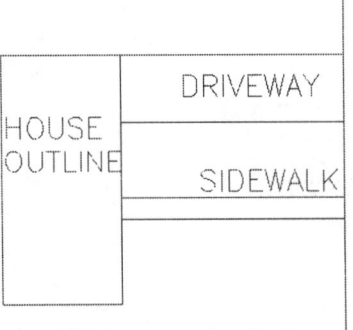

Select the three rectangles created in Lesson 2. If you did not create these as rectangles, use PEDIT to turn the lines into polylines.

```
Select walls or polylines:
Erase layout geometry? [Yes/No] <N>:
```

You're prompted if you wish to erase the layout geometry.

The Convert to Slab command does not actually convert the rectangles. It merely places a slab entity exactly on top of the existing geometry.

Accept the default and retain the existing geometry. (Press 'ENTER' or type 'N').

```
Creation mode [Direct/Projected]<Projected>:
```

Set the Creation Mode to Projected (Press 'ENTER').

```
Specify base height<8'-0.00">: 0"
```

Set the base height to 0" to create a slab starting at a z-height of 0".

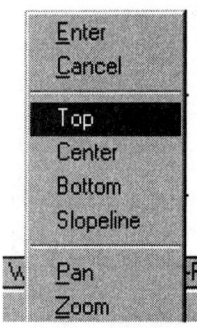

Set the Justification to Top.

This sets the slab so the top is even with the Z = 0 elevation which was the base height previously chosen

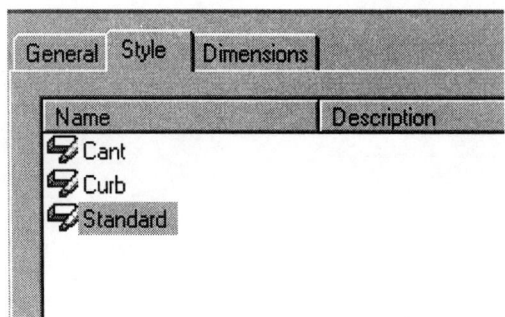

The Slab dialog box pops up. Select 'Standard' Style.

Then select the Dimensions tab.

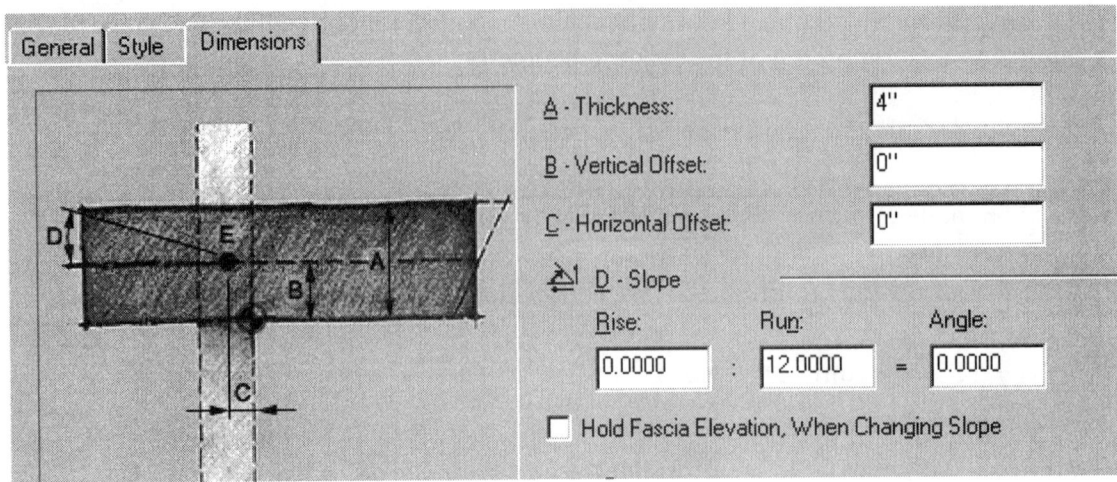

Set the thickness to 4".
Press 'OK'.

Go to tab Work-SEC to see the slabs in iso and side views:

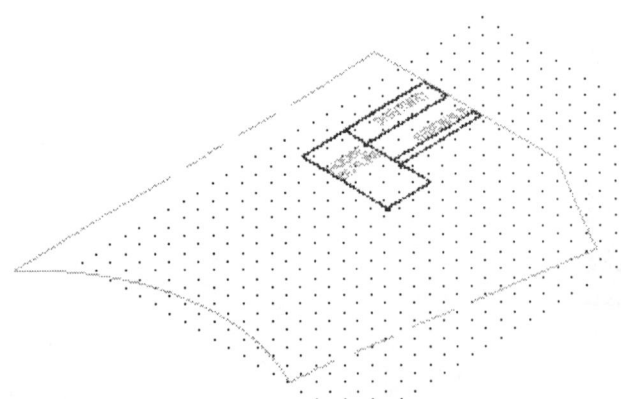

Our slab foundation in isometric view.

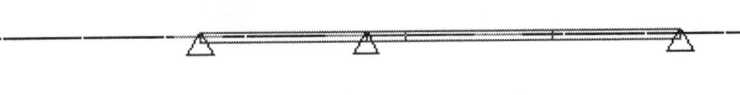

Our slab foundation – front view.
Note how the Justification is when set to Top. Justification affects the location of the pivot point as well as the z-location of the slab relative to the base height.

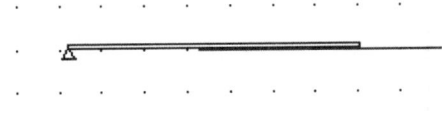

Slab Foundation when Justification is set to Bottom.

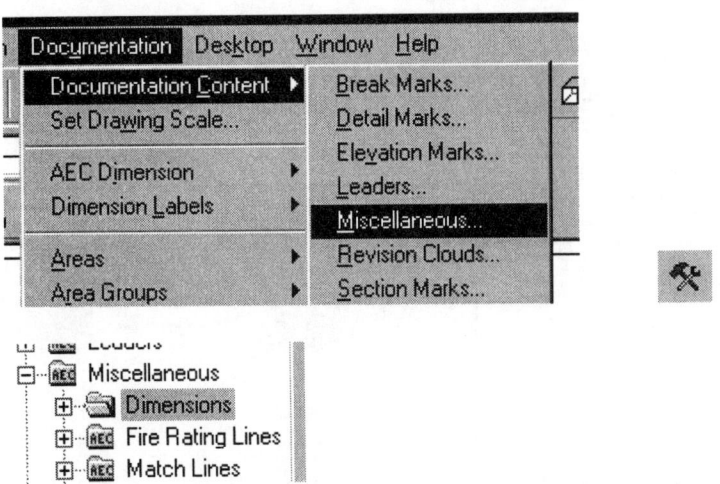

Select the Miscellaneous tools under Documentation Content.
When the Design Center Activates, go to the Dimensions subfolder.

Label and dimension the concrete foundation as shown.

Lesson 3
Foundations

Continue

Linear

Create the dimensions by dragging and dropping Linear
and Continue from the Design Center.

Using the menu, Documentation-> Content->Miscellaneous->Dimensions applies the
usual dimensions using a present dimension style Aec_Arch_1 on layer A-Anno-Dims.
Using the dimension toolbar or dimension menu picks will provide dimensions on the
current layer in the present dimension style Aec_Arch_1 (unless another style has been
set current.)

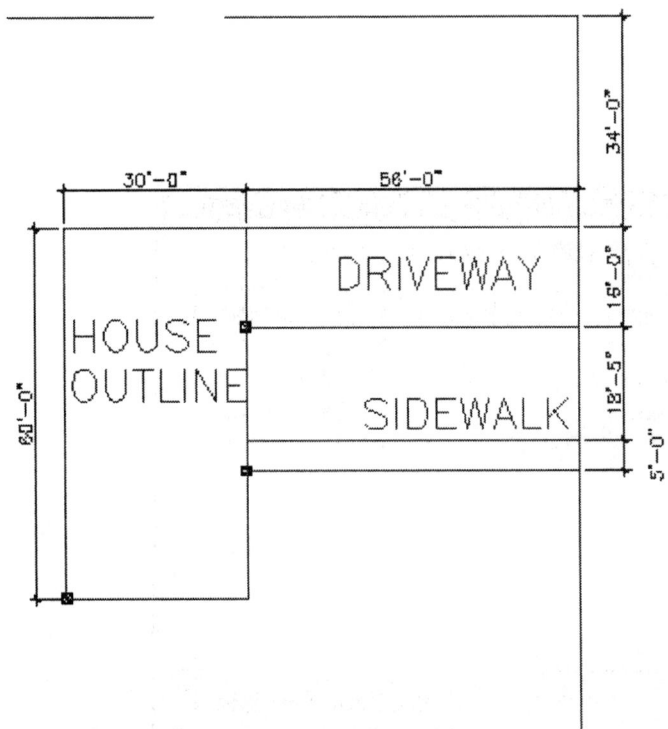

Save the file as Ex3-1.dwg.

Exercise 2:

Layer User Groups

Drawing Name: Ex3-1.dwg
Estimated Time: 15 minutes

This exercise reinforces the following skills:

- Layer User Groups
- Layer Manager

Activate the Layer Manager.

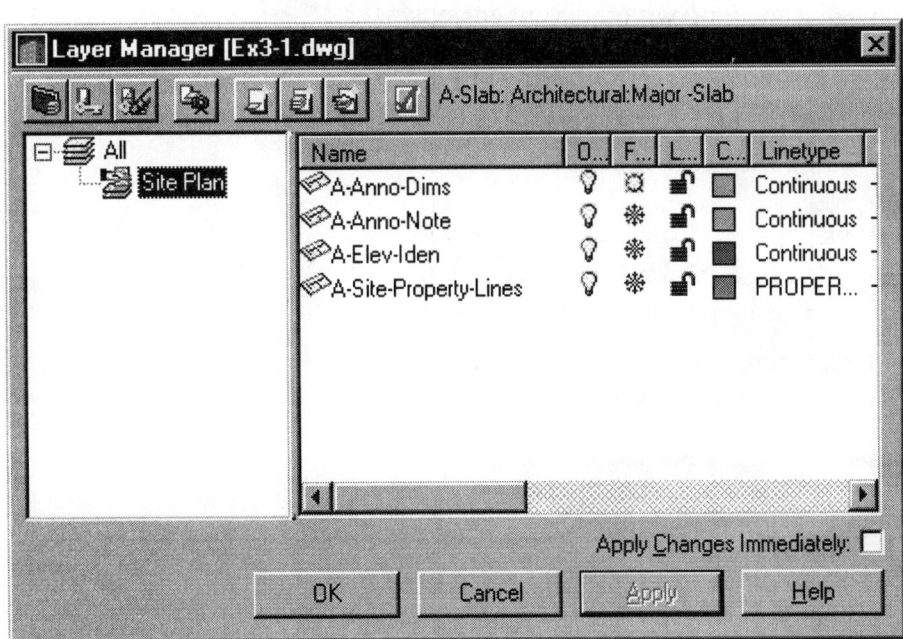

Freeze the A-Site-Property-Lines layer.

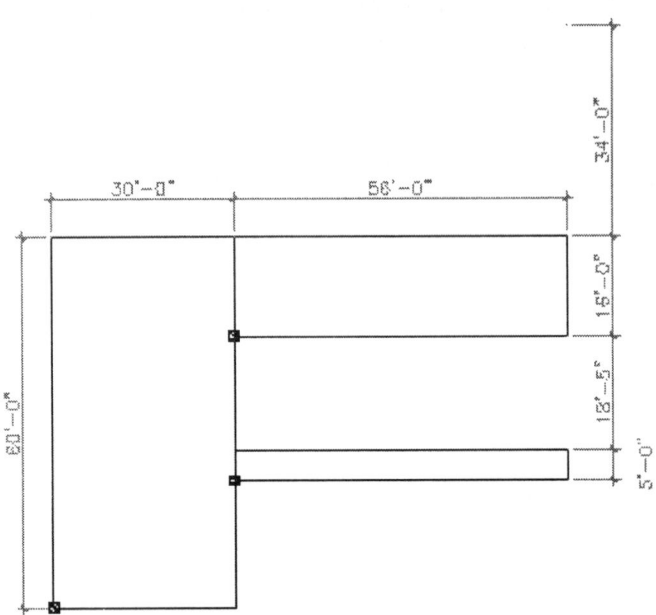

We see the slabs we placed along with their dimensions.

 Create a New Layer User Group.

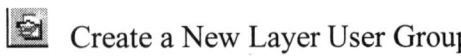

Place your mouse over 'All'.
Pick the New User Group tool.
Rename the user group 'Foundation.'

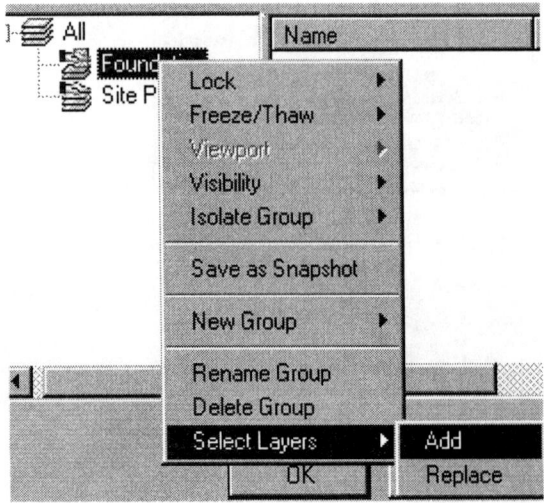

Highlight the Foundation group.
Right click and select 'Select Layers->Add.'

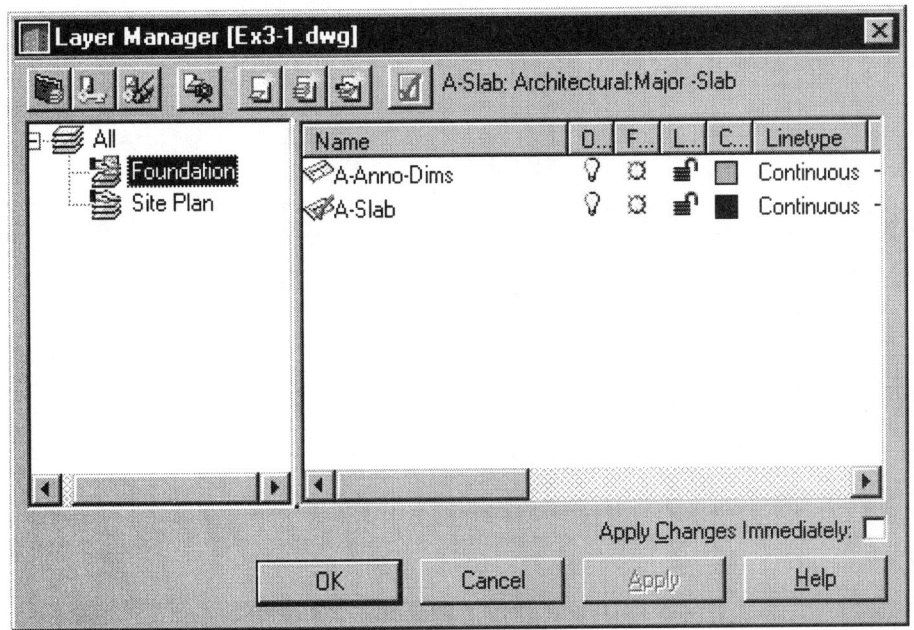

Window around the objects in your drawing.
Press 'Apply'.

Pick 'All' in the Layers tree.

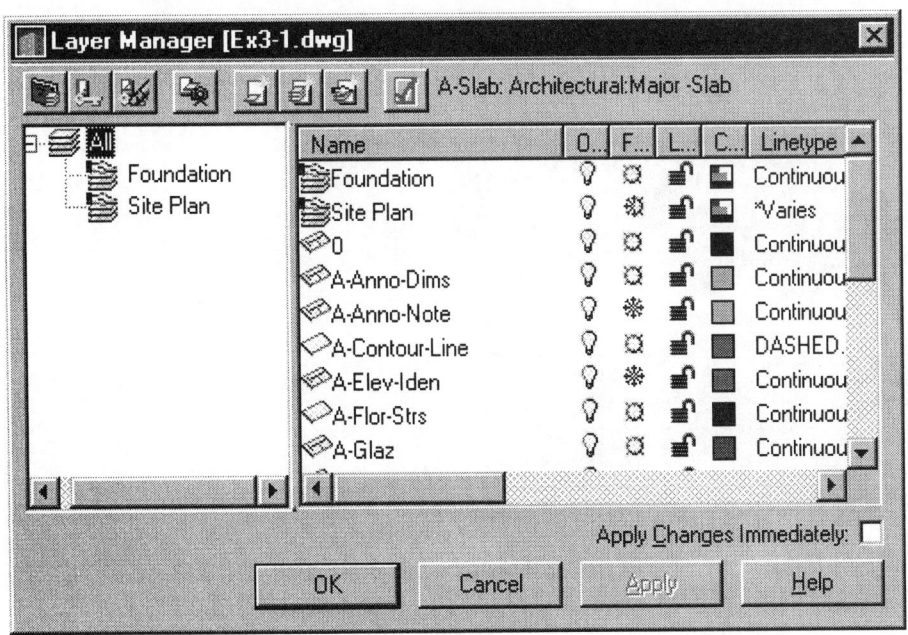

Note how our layers are organized.
It will now be easy to freeze or turn off a layer group.

Save as Lesson 3.dwg.

Notes:

Lesson 4
Floor Plans

The floor plan is central to any architectural drawing. We start by placing the exterior walls, then the interior walls, then doors, and finally windows.

Exercise 1:
Layer Filters

Drawing Name: Lesson 3.dwg
Estimated Time: 15 minutes

As you probably have noticed, our layer list is growing. It is not unusual for architectural drawings to have 100 or more different layers. Creating and using Layer Filter Groups is one way to help boost productivity.

A filter group contains layers that meet filter criteria that you specify for the group. Filter criteria can select layers according to layer states (on/off, frozen/thawed, locked/unlocked), properties, or names. For example, you can create a filter group that includes all of the red layers in the current drawing.

You can create two types of filter layer groups:

- **Dynamic filter groups:** Automatically update when you change the properties of layers that are part of the group. You cannot manually add layers or remove layers.

- **Static filter groups:** Do not automatically update, and include only the layers that met the filter criteria when the group was created. You can manually add layers and remove layers.

Open the Lesson 3.dwg file.

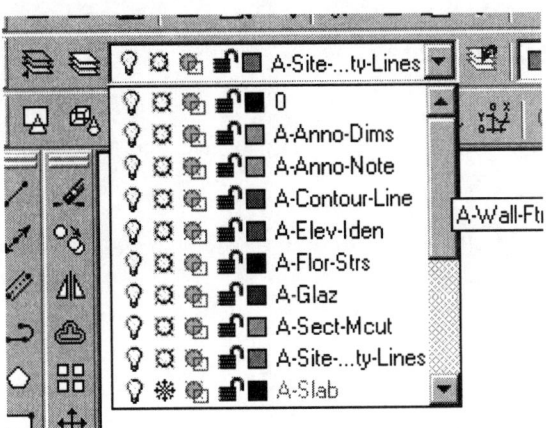

Check out our growing list of Layer Names. Most of these layers are automatically added as we add AEC objects into our drawing.

Activate the Layer Manager.

Highlight the word 'ALL' in the Layer Manager tree.

Select the 'New Filter' tool.

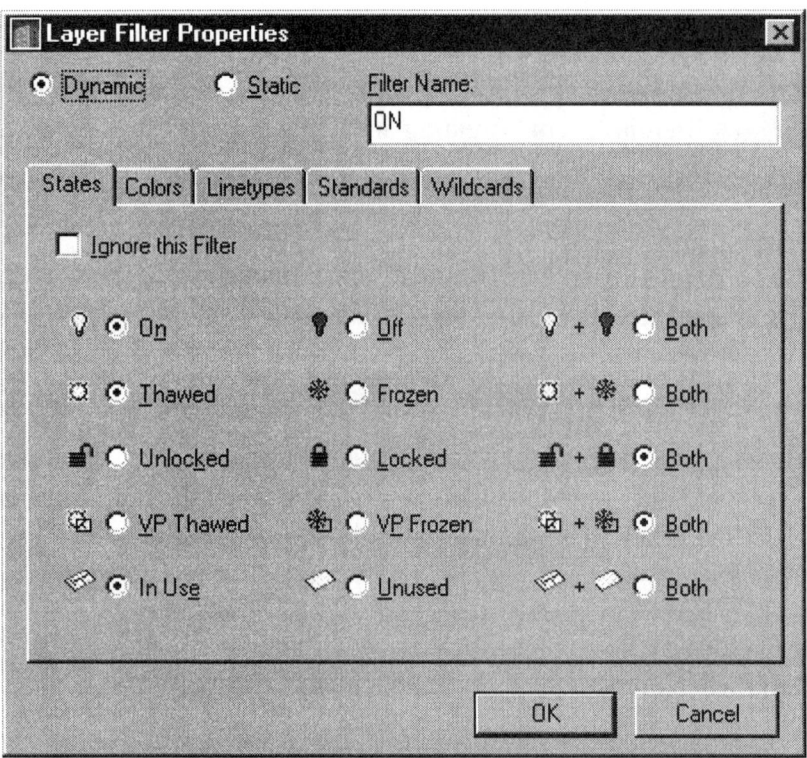

Enable 'Dynamic'.
Enter the 'Filter Name' 'ON' as shown.
Enable ON, THAWED, and IN USE.
This means we will only see layers that meet those criteria.

Lesson 4
Floor Plans

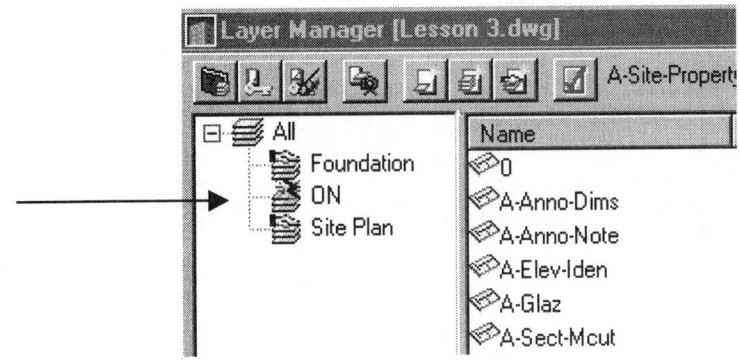

Our filter appears in our Layer Manager tree.

Exercise 2:
Snap Shots

Drawing Name: Lesson 3.dwg
Estimated Time: 15 minutes

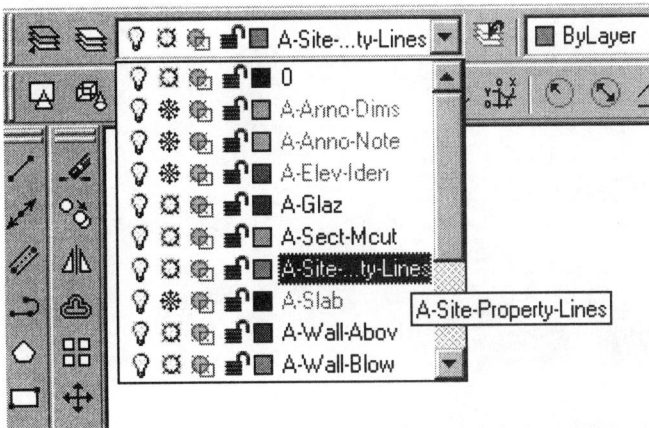

In the Layers Drop-down on the Object Properties toolbar, freeze A-Anno-Dims, A-Elev-Iden, and A-Slab as shown.

Lesson 4
Floor Plans

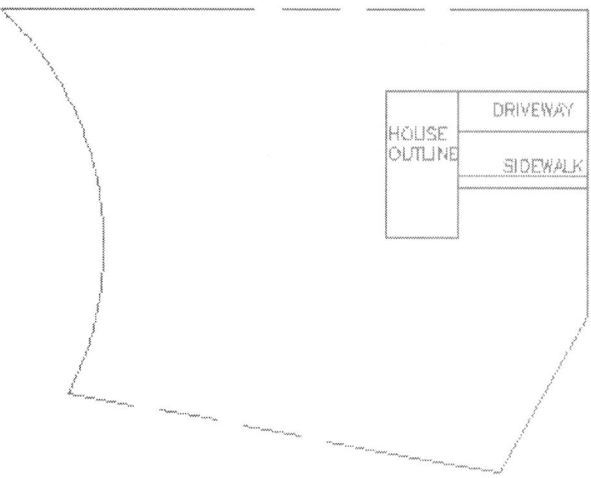

Now all that is visible on the screen is what you see above…the property lines and the house outline.

 Activate Layer Manager.

Select the 'Snapshots' tool.

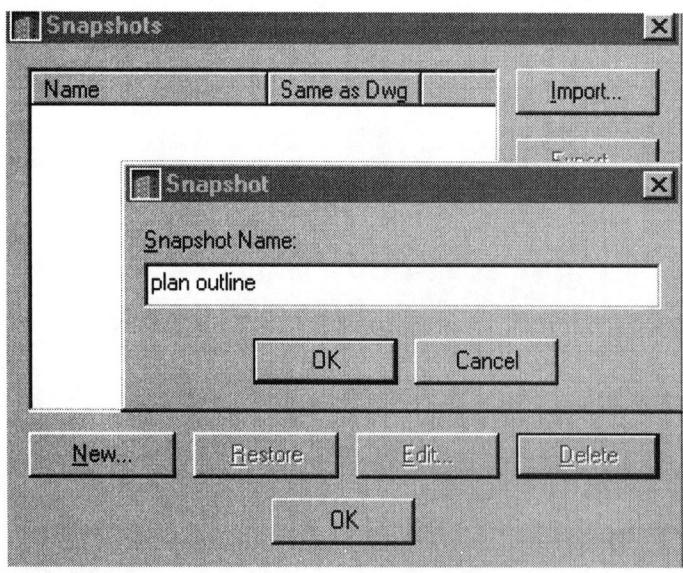

Press the 'New' button.
Type in 'plan outline' for our Snapshot Name.
Press 'OK'.

Close the Layer Manager dialog box.

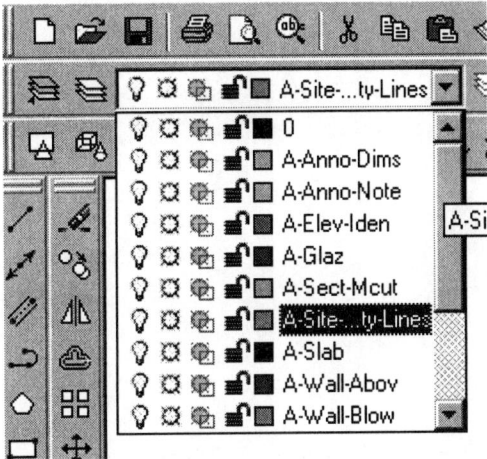

Thaw the layers.

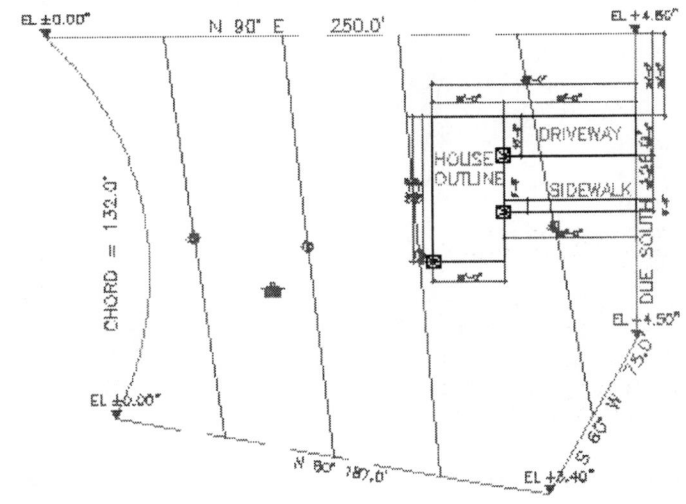

Our previous view is restored.

 Activate Layer Manager.

Select the 'Snapshots' tool.

Lesson 4
Floor Plans

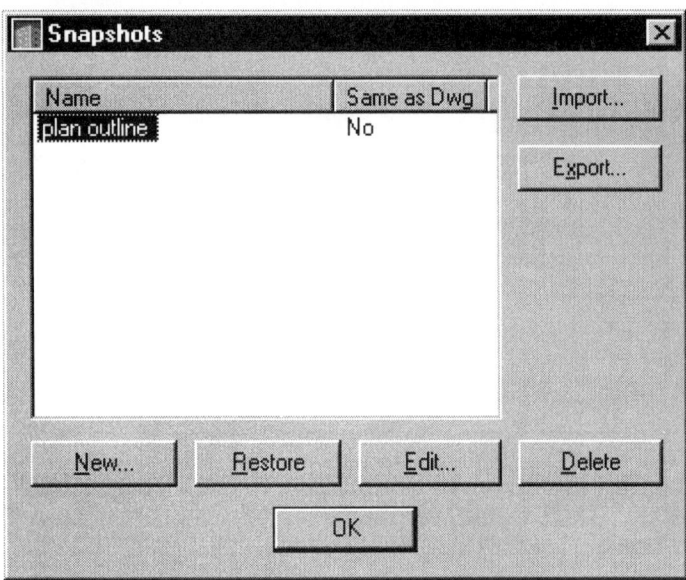

Note that in the column that says 'Same as Dwg' the word 'No' appears. This means the current drawing does not match our stored snapshot.

Press the 'Restore' button.

Press 'OK' twice to close the dialogs.

Our snapshot view is restored.

Lesson 4
Floor Plans

Exercise 3:
Creating Exterior Walls

Drawing Name: Lesson 3.dwg
Estimated Time: 15 minutes

This exercise reinforces the following skills:

   Create Walls

> **TIP:** Walls are automatically created on the correct layer.

Menu	Deisgn->Walls->Add Wall
Wall Toolbar	
Command Line	WallAdd

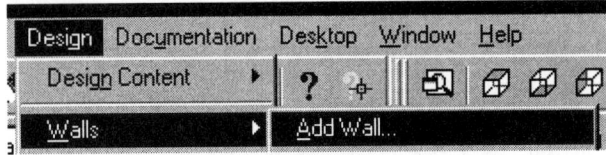

>
> **TIP:** If you draw a wall and the materials composing the wall are on the wrong side, you can reverse the direction of the wall. Simply select the wall, right click and select the Reverse option from the menu.

Lesson 4
Floor Plans

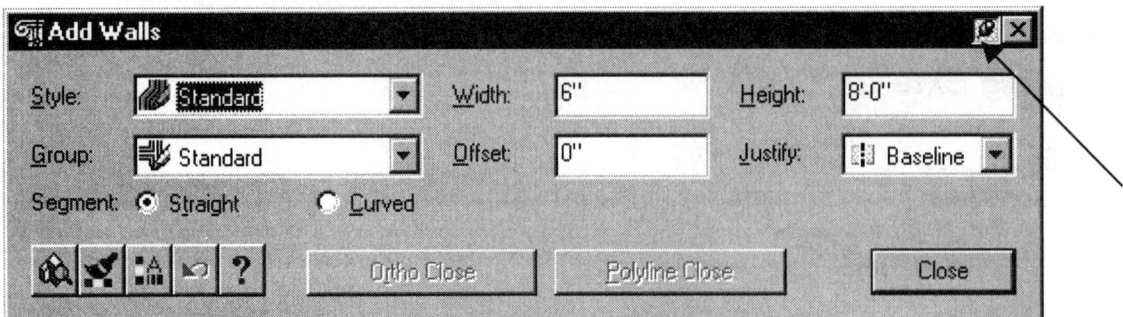

Start the Add Wall command.
Note the pushpin symbol in the upper right corner of the dialog. If this is toggled off, then the AEC Add Element dialogue boxes will minimize (but not down to the taskbar) when the cursor moves off the dialogue. This can save screen space on small monitors, but the dialogues can be moved off the screen and lost, or covered up by mistake. Enable the pushpin to keep the dialog in place while we are working.

We will use the Standard Wall Style as shown.

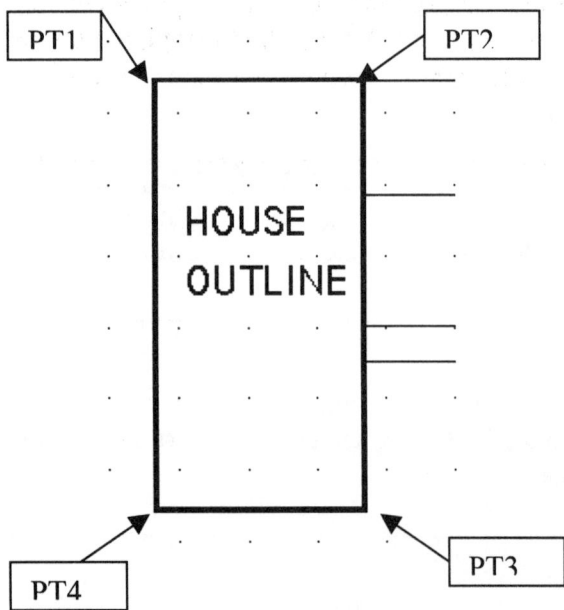

Select the four points indicated, then select the first point again. Right click and select 'Close'.
You notice a dark outline around the house outline.

Select your isometric view tool. If you're on the Work-3D tab the left viewport is already set to the SW isometric view.

You see that the walls you placed are really 3-dimensional.
Save your drawing as Ex4-3.dwg.

TIP: You can convert lines, arcs, circles, or polylines to walls. If you have created a floor plan in AutoCAD and want to convert it to 3D, open the floor plan drawing inside of ADT. Use the Convert to Walls tool to transform your floor plan into walls.

Lesson 4
Floor Plans

Exercise 4:
Convert to Walls

Drawing Name: Lesson 4.dwg
Estimated Time: 60 minutes

This exercise reinforces the following skills:

- Convert to Walls
- Creating Interior Walls

Walls are placed on layer A-Wall. Color is Yellow. Linetype is Continuous.

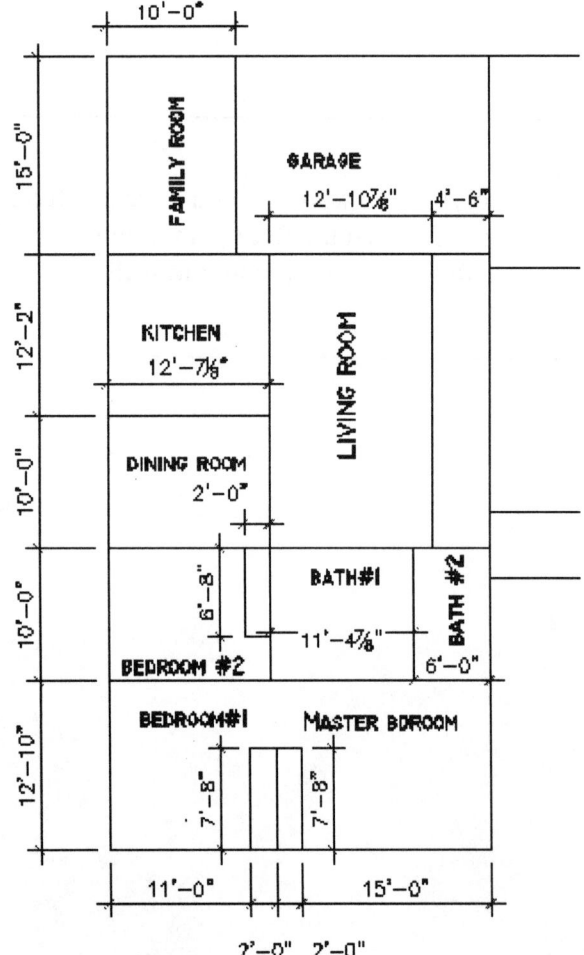

Create the layout shown using lines or polylines. (It may be helpful to turn off the A-Walls layer while you are working and create your lines on layer 0).
Do not add the dimensions or the text to your drawing. They are there to help you place the lines only.

4-10

Don't draw using rectangles if you are going to use the CONVERT method or you will get duplication of lines over lines, which will affect the wall creation.

If you do not want to spend time creating the floor plan, you can download the drawing 'floor plan.dwg' from www.schroff.com.

Menu	Design->Walls->Convert to Walls
Walls Toolbar	Convert to Walls
Command Line	WallConvert

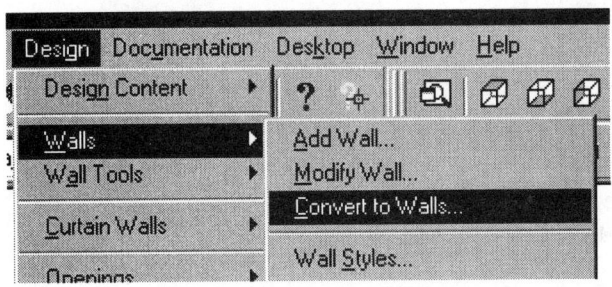

Initiate the Convert to Walls command.

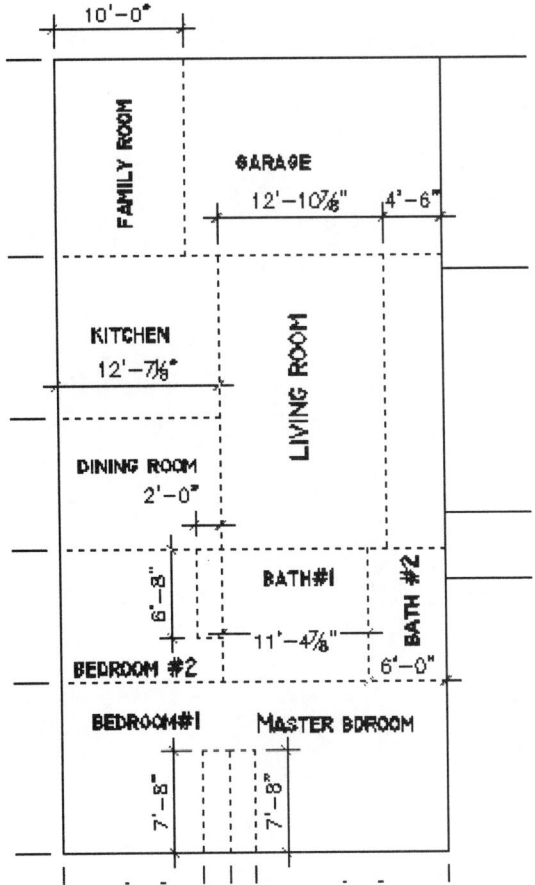

Lesson 4
Floor Plans

Select all the interior polylines you just created.

```
Erase layout geometry? [Yes/No] <N>: Y

15 new wall(s) created.
Command: Regenerating model.
```

You are prompted if you want to erase the layout geometry. Type Y for Yes.

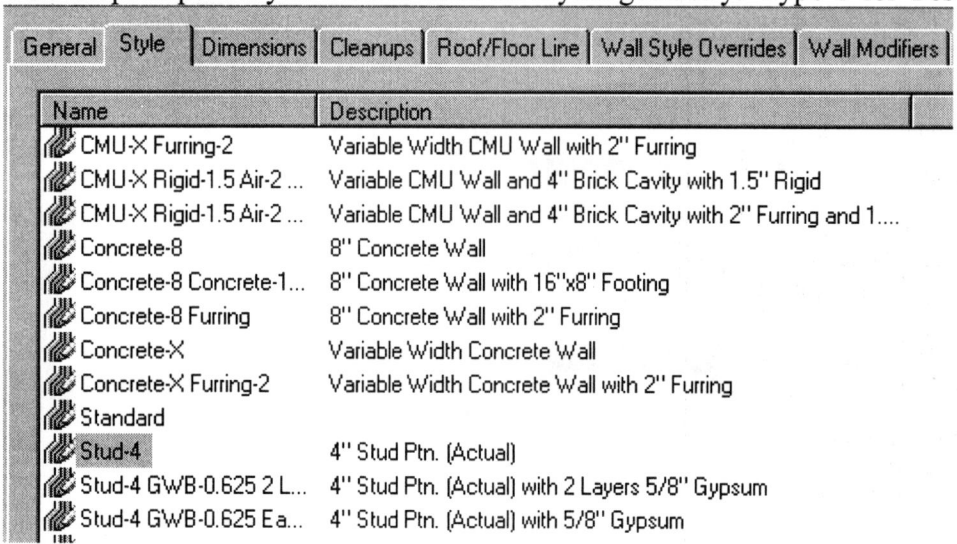

Select the Stud-4 for the wall style.

Press 'OK'.

Lesson 4
Floor Plans

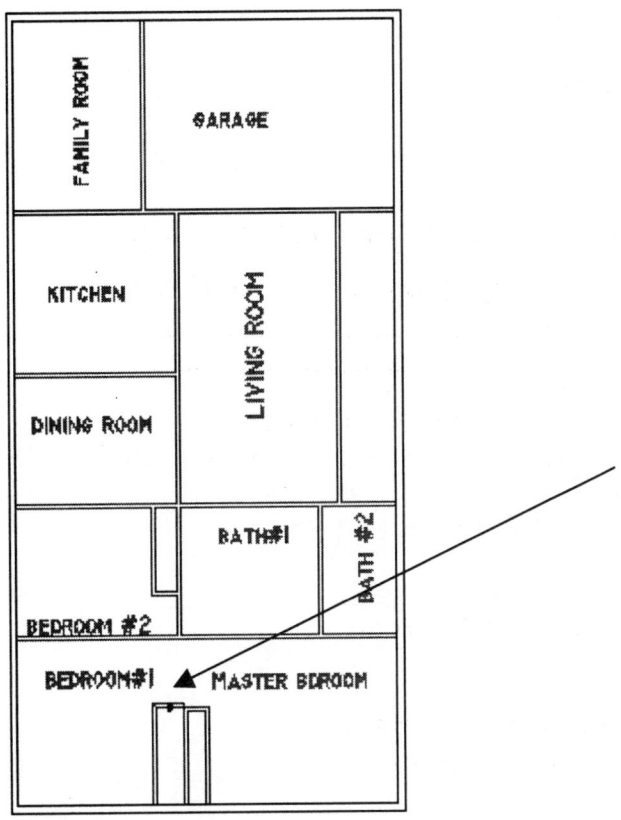

Your walls appear. Some walls may not be aligned with others, and wall cleanup error symbols may appear as well.

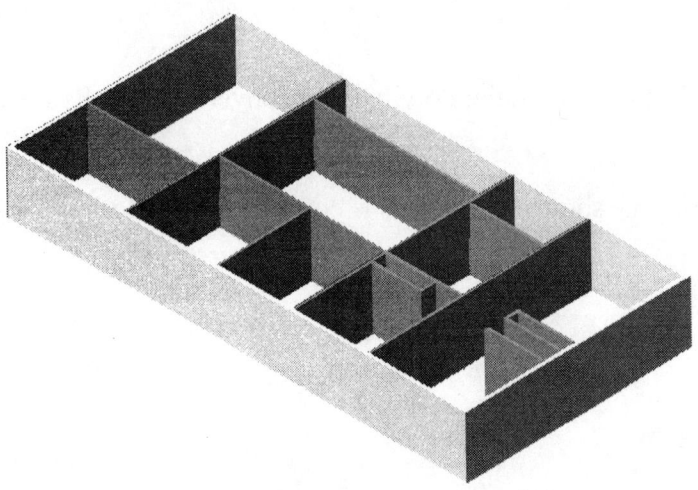

An isometric view shows you how your house looks in 3D.
Save the file as Ex4-4.dwg.

Lesson 4
Floor Plans

Exercise 5:
Wall Cleanup

Drawing Name: EX4-4.dwg
Estimated Time: 15 minutes

This exercise reinforces the following skills:

- Modifying Walls
- Wall Tools

If your walls are all clean, you can skip this exercise.

Our closet walls are not even. This is because the Polylines we converted to walls were drawn in different directions.

When you see a red circle like the one shown above, this is considered a "defect condition".

Lesson 4
Floor Plans

Right click on the wall with the red circle, select Plan Tools/Reverse and the wall will change direction. (You might have to regen the viewport.)

Menu	Design->Wall Tools->Reverse Wall
Walls Tools Toolbar	
Select & Right click	Plan tools->Reverse

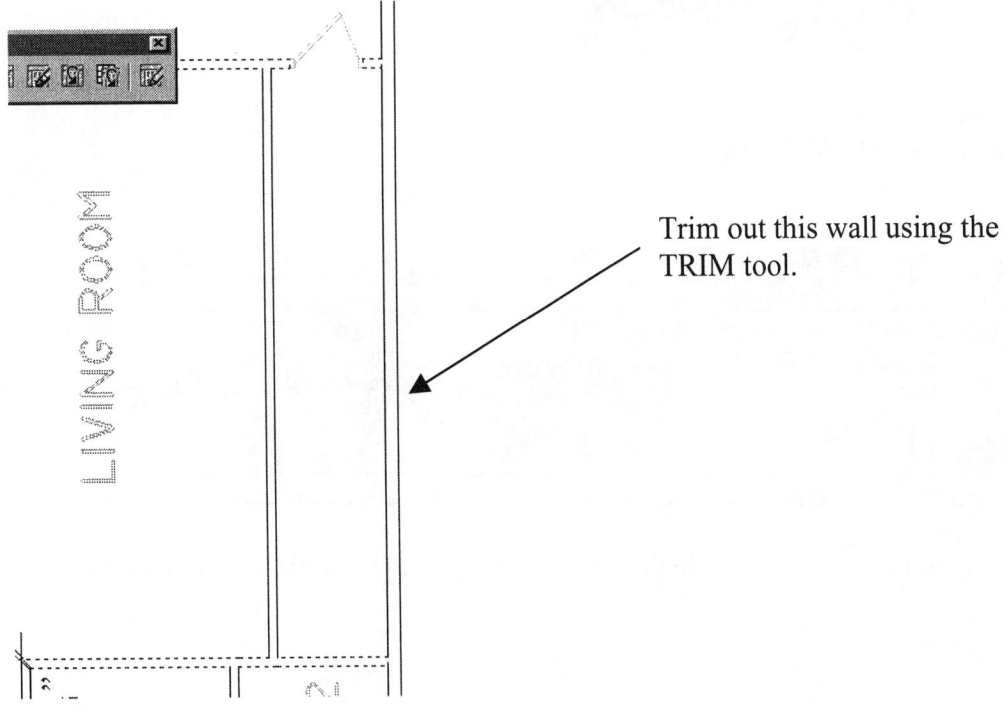

Trim out this wall using the TRIM tool.

4-15

Lesson 4
Floor Plans

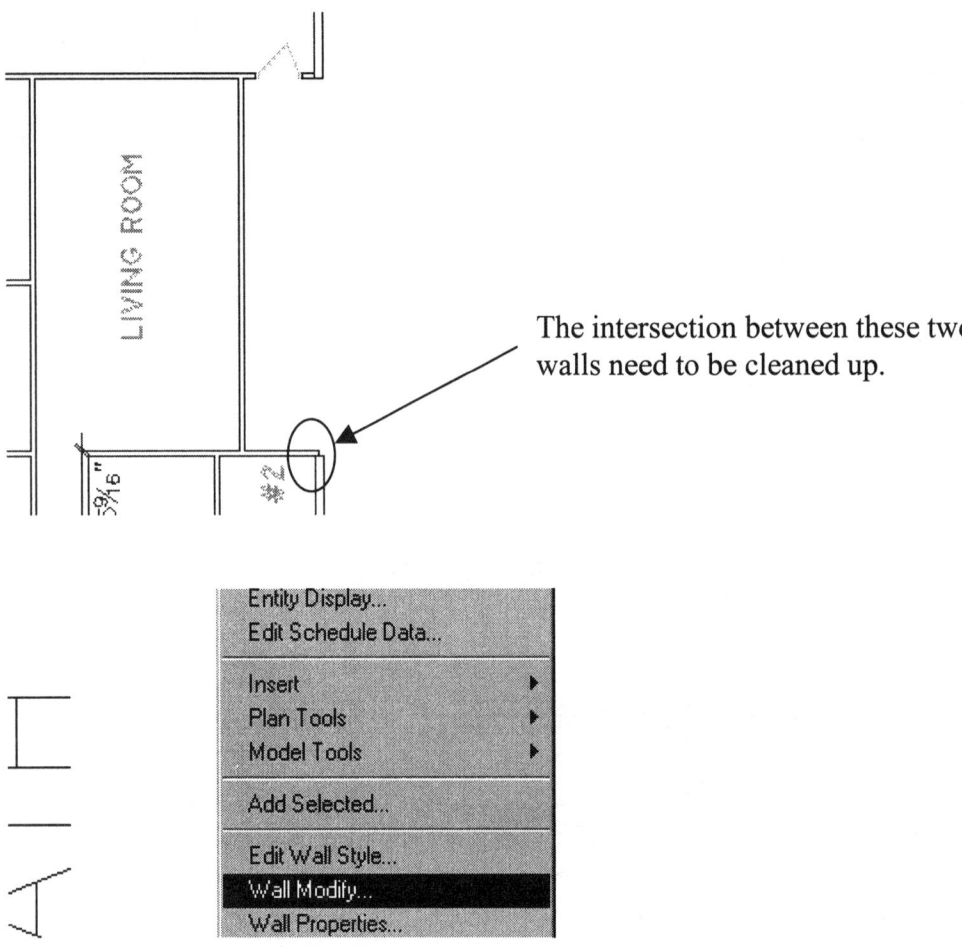

The intersection between these two walls need to be cleaned up.

Select the horizontal wall.
Right click and select 'Wall Modify'.

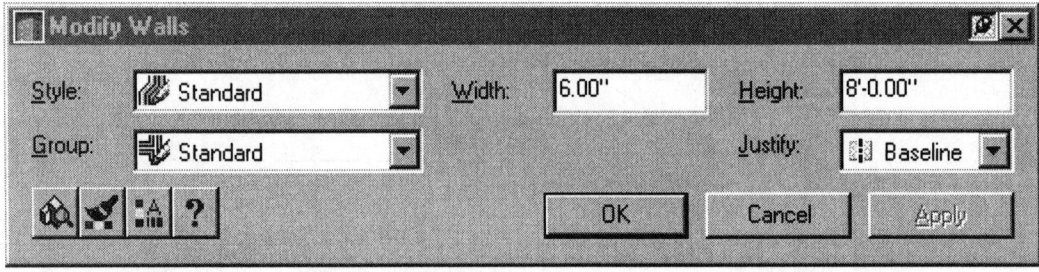

You can modify the wall by changing the Style setting or the Justification to clean up the intersection.

4-16

Lesson 4
Floor Plans

Exercise 6:
Adding Closet Doors

Drawing Name: Ex4-4.dwg
Estimated Time: 30 minutes

This exercise reinforces the following skills:

❑ Adding Doors

Doors are placed on Layer A-Door. Color is Cyan. Linetype is Continuous.

Menu	Design->Doors->Add Door
Doors Toolbar	
Command Line	DoorAdd

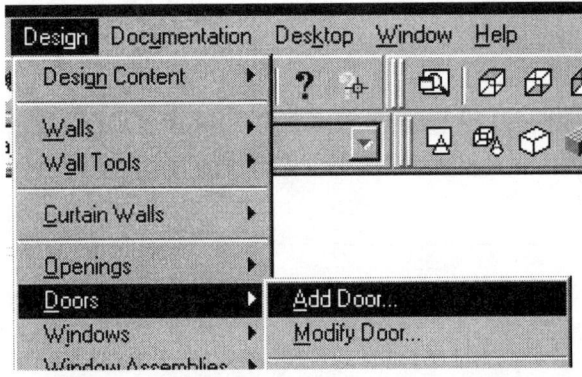

TIP: To create a freestanding door, press the ENTER key when prompted to pick a wall. You can then use the grips on the door entity to move and place the door wherever you like.

To move a door along a wall, use Door->Reposition->Along Wall. Use the OSNAP From option to locate a door a specific distance from an adjoining wall.

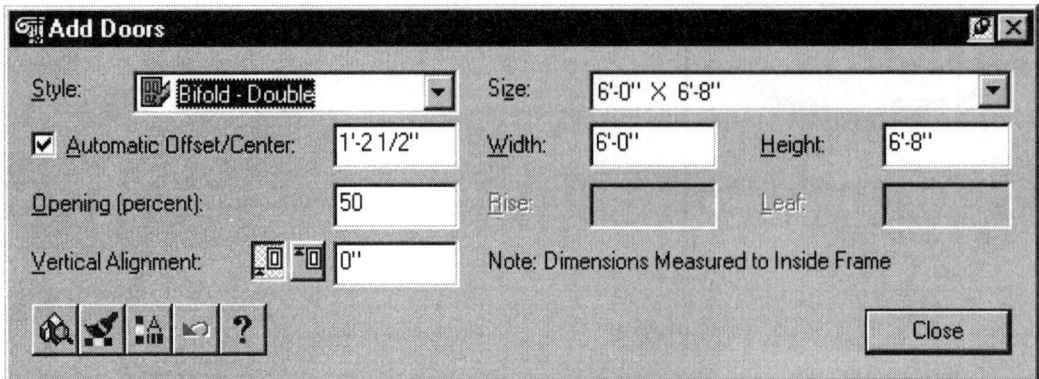

Locate the Bifold-Double door in the Style dropdown.
Enable Automatic Offset/Center and set to 1'-2.5". (This will center the closet doors along the wall.
Set the width to 6'.
The Opening is set to 50%. The value of the opening determines the angle of the arc swing. A 50% value indicates the door will appear half-open at a 90-degree angle. A 25% opening will show a door swing at a 45-degree angle.

TIP: Note the vertical alignment button. It defaults to a threshold height of 0" for doors and a head height of 6'8" for windows. You will need to adjust these defaults in multi-story buildings.

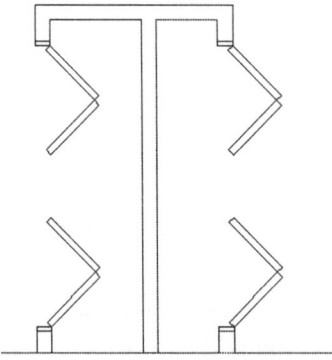

Select the two closet walls between the Master bedroom and Bedroom #1.

One of the closet doors is facing the wrong way. If you made this mistake as well, you can fix it easily.

Lesson 4
Floor Plans

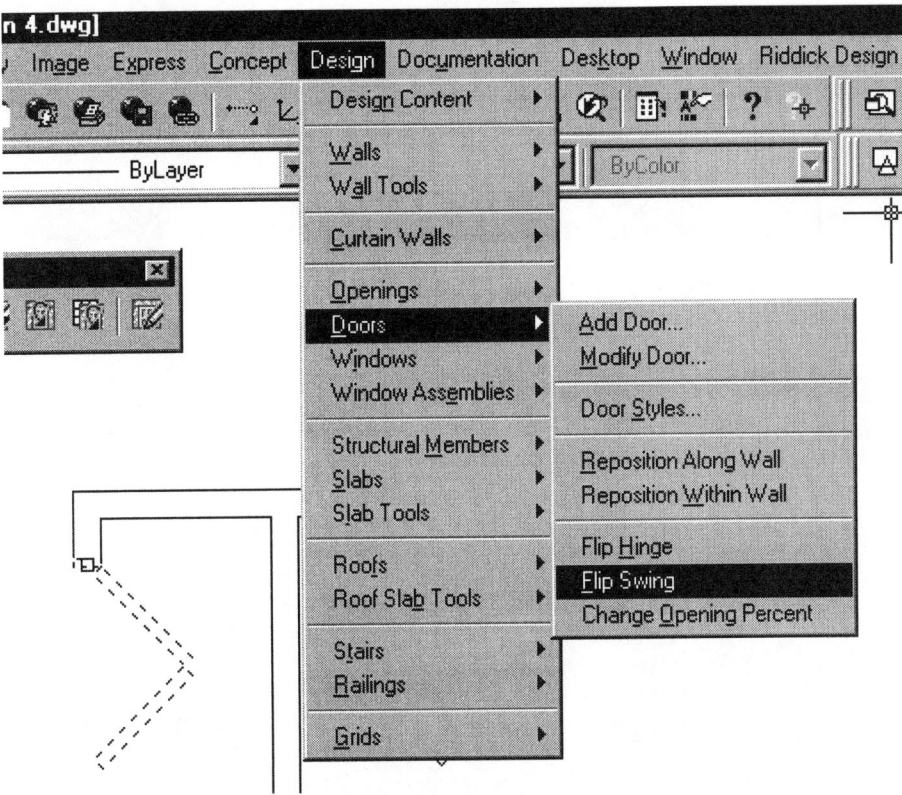

Go to Design->Doors->Flip Hinge.
Select the door that needs to be flipped.

NOTE: You can select the door to flip before or after you select Flip Swing from the menu.

You can also grip-edit the swing and hinge on any door or window.

To use Grips:

Pick the door you want to flip.

Lesson 4
Floor Plans

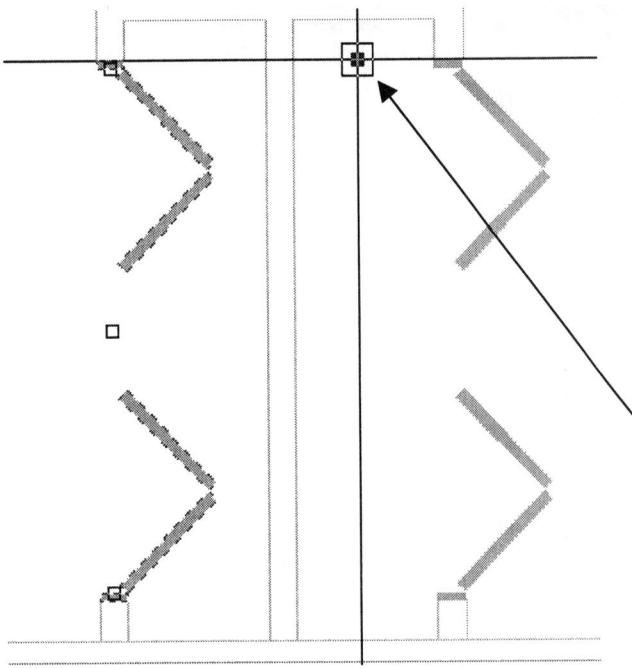

Activate the grip as shown. (When a grip is activated, it is HOT. This is indicated by the color RED.)

Then drag the mouse in the direction of the desired swing and pick.

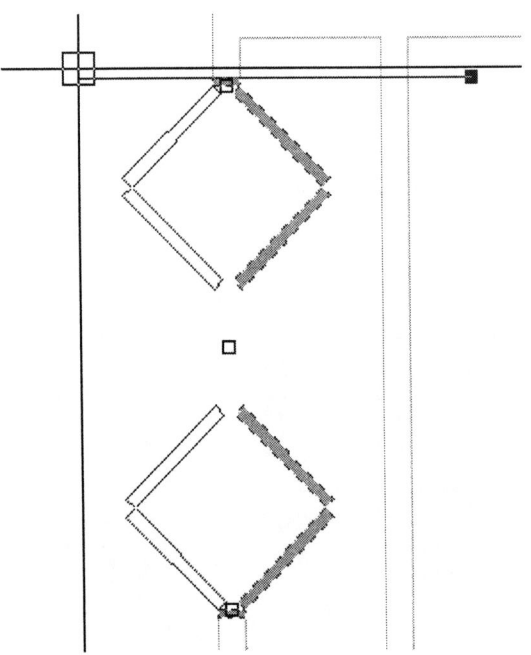

To release the selection, press ESCAPE or right click and select 'Deselect All' from the popup menu.

4-20

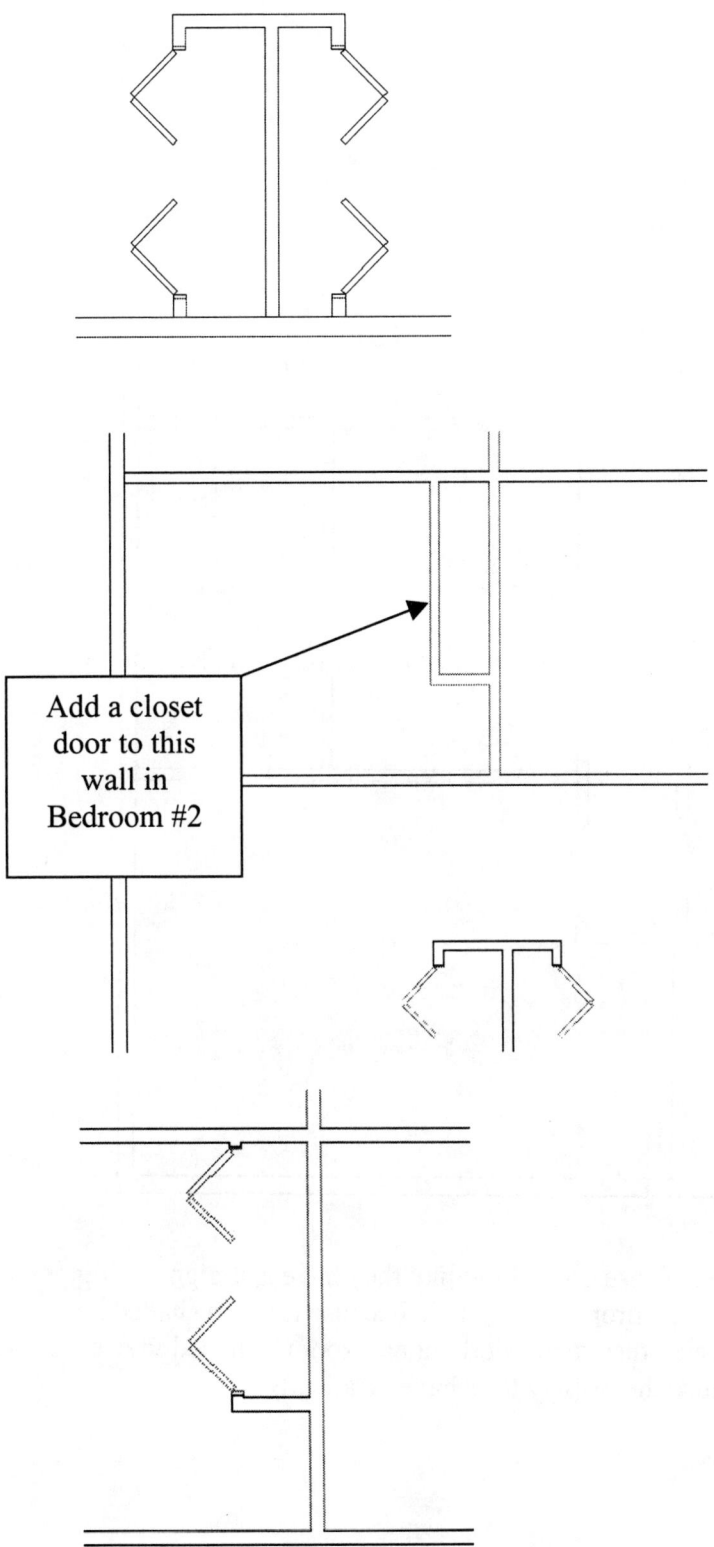

Add bi-fold closet doors using an offset of 4".

Lesson 4
Floor Plans

Exercise 7:
Adding Interior Doors

Drawing Name: Lesson 4.dwg
Estimated Time: 30 minutes

This exercise reinforces the following skills:

- Adding Doors

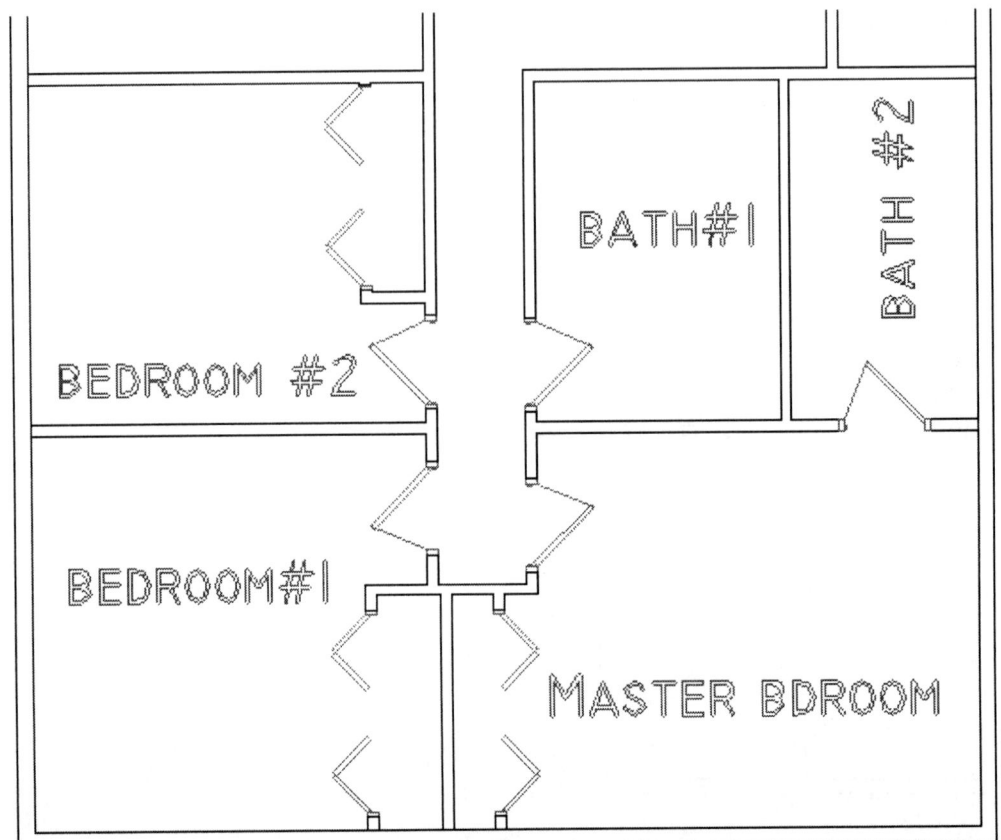

Place the single hinge doors as shown. These doors look like they have a straight swing, which is an override to the default display properties. That's because this is a shaded view—in wire-frame view the swing arcs display properly at any zoom. Shaded views are good for visualization, but can cause the display to behave erratically.

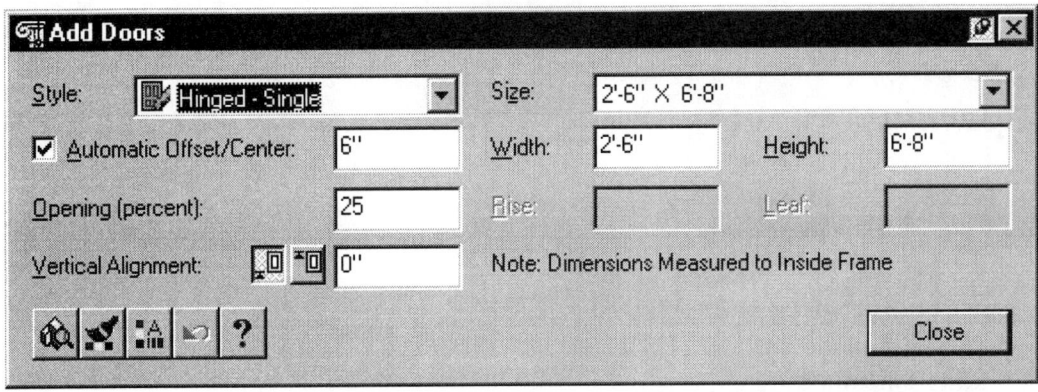

You will need to do some wall clean up to get the rooms to look proper.

Use AddWall, Extend, and Trim.

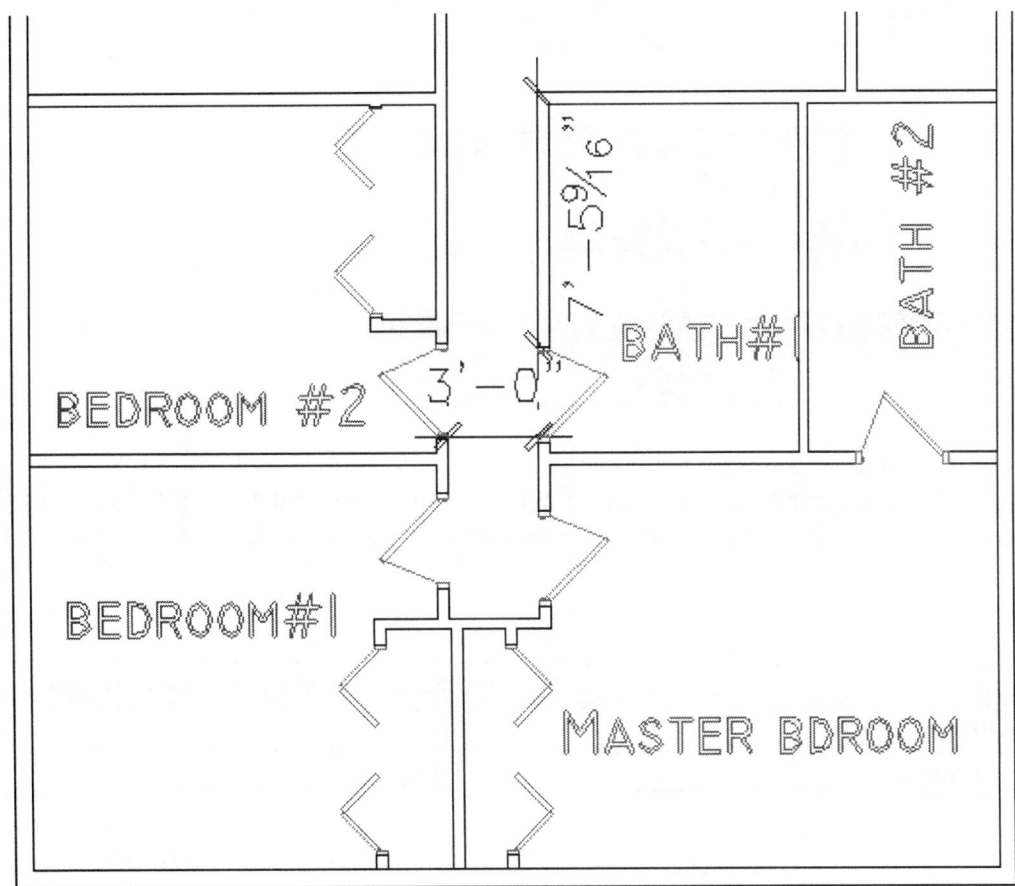

Try to keep the walls so they line up to keep the floor plan looking clean.

Lesson 4
Floor Plans

Exercise 8:
Add Opening

Drawing Name: Lesson 4.dwg
Estimated Time: 30 minutes

This exercise reinforces the following skills:

❑ Adding Openings

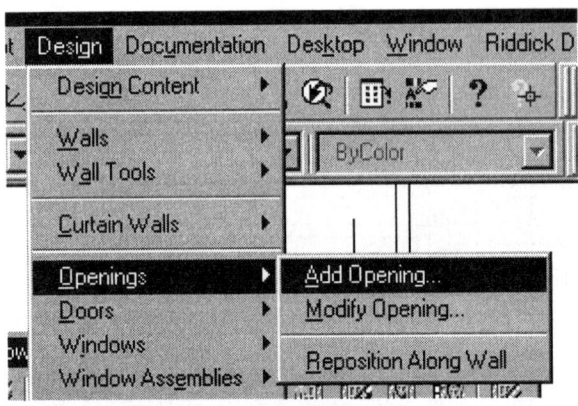

Openings can be any size and elevation. They can be applied to a wall or be freestanding. Openings are placed on Layer A-Wall-Open and are Dark Blue. The Add Opening dialog allows the user to either select a Pre-defined shape for the opening or use a custom shape.

Menu	Design->Openings->Add Opening
Openings Toolbar	
Command Line	OpeningAdd

4-24

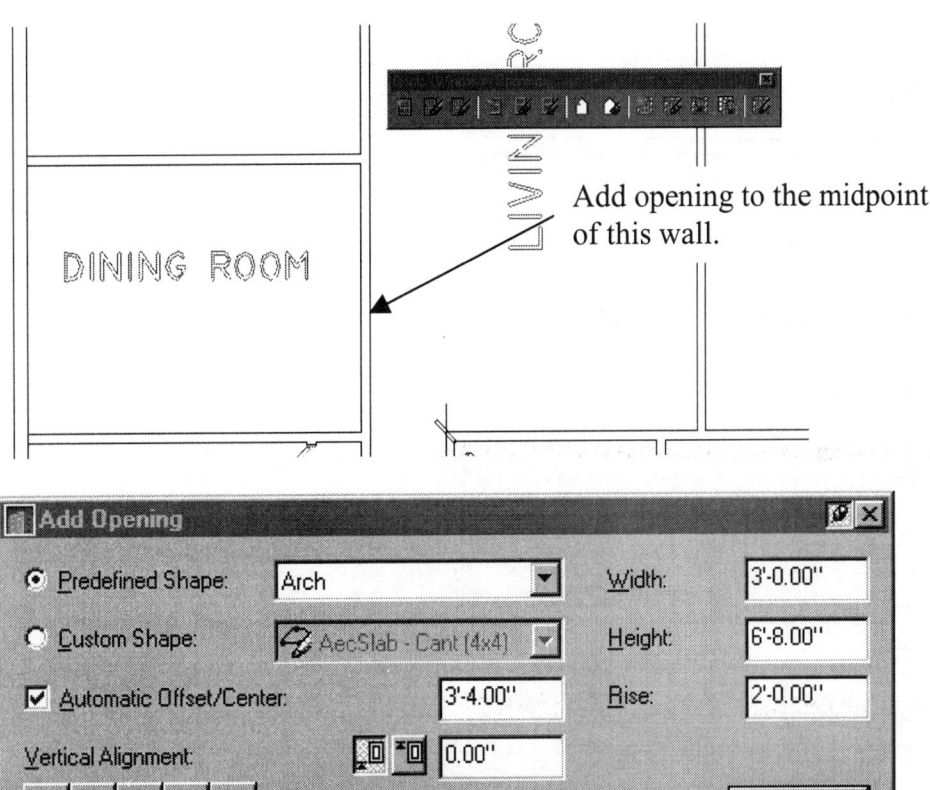

Initiate the Opening Add command.
Set the Predefined Shape to Arch.
Set the width to 3'.
Enable the Automatic Offset/Center and set it to 3'-4".
Set the Rise to 2'-0". This is the radius of the Arch.

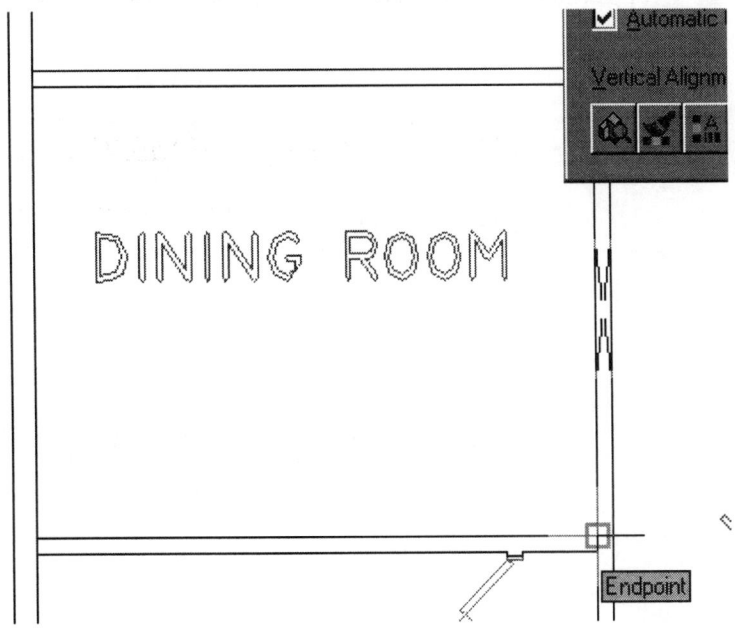

Select the wall shown and then pick the endpoint shown to establish the offset.

Use DIST to verify that the opening is centered properly in the wall.

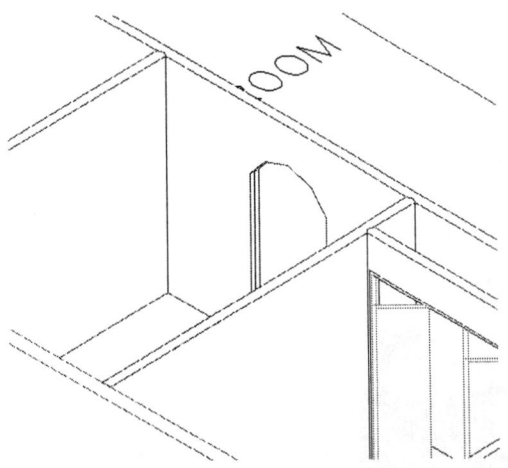

Our arched doorway is shown here in isometric view.
The View Mode is Hidden.

Lesson 4
Floor Plans

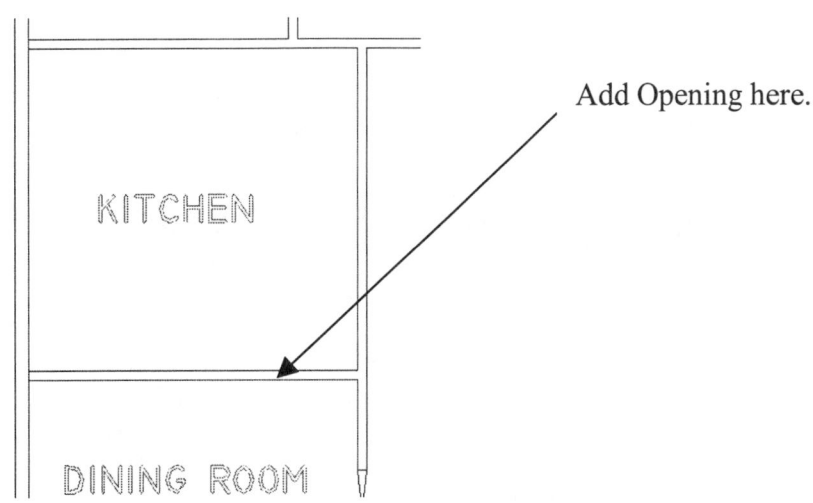

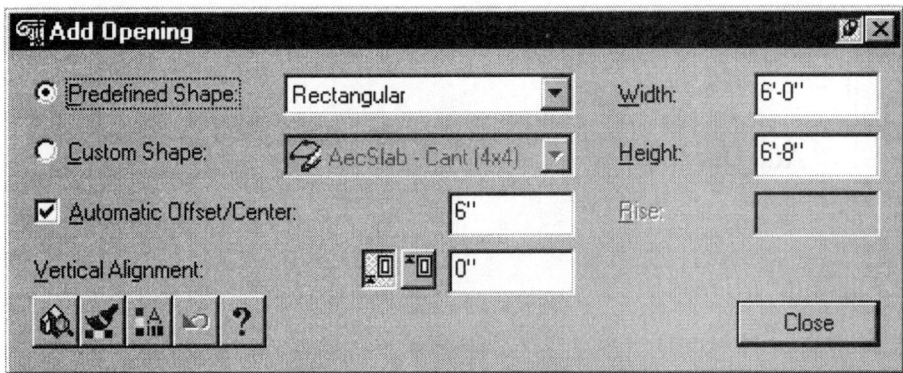

Set the opening to Rectangular. Set the Offset to 6". Set the Width to 6'.

Select the wall. Then select the endpoint shown to set the offset correctly.

Lesson 4
Floor Plans

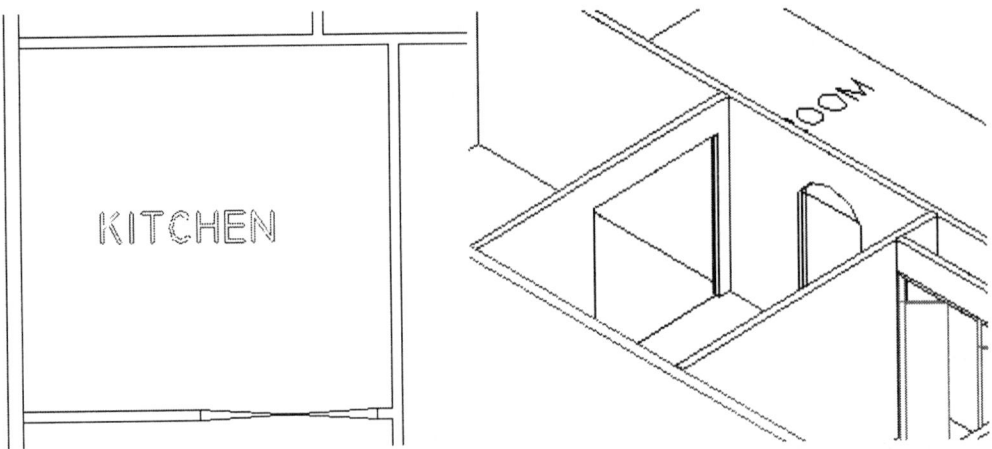

The door is shown in the TOP View and the Isometric View.

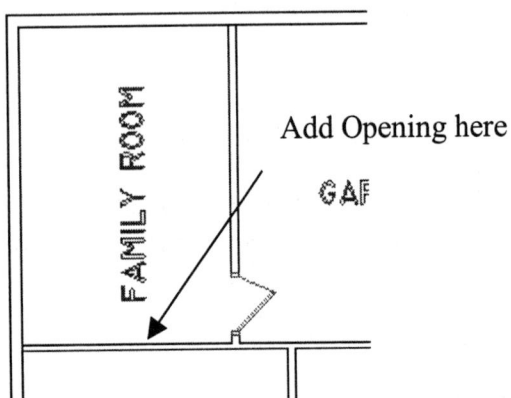

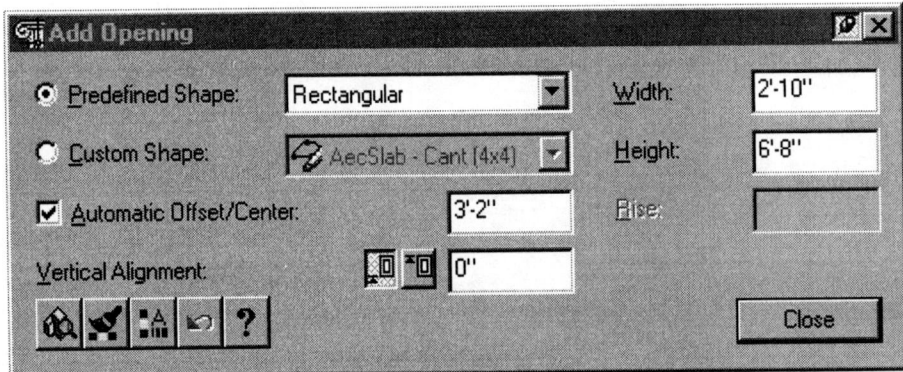

Set the Predefined Shape to Rectangular. Set the width to 2'-10". Set the Offset to 3'-2".

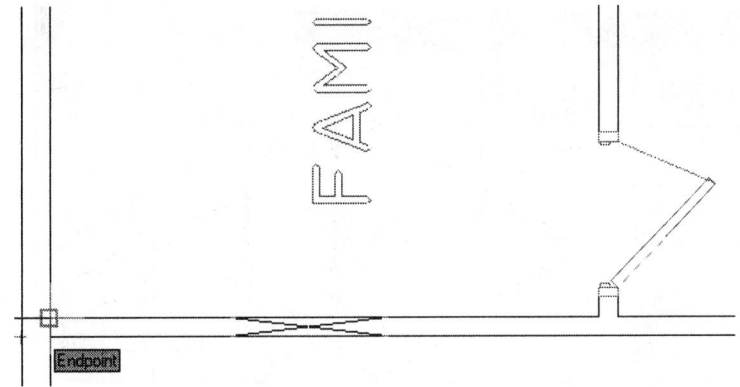

Select the wall and endpoint shown to place the opening.

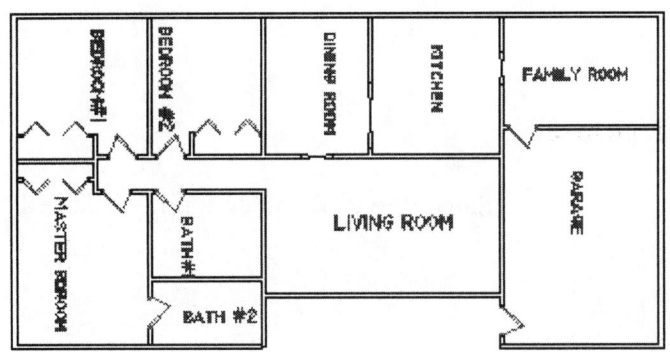

Our floor plan so far.

(This view is rotated 90 degrees to make it easier to see.)

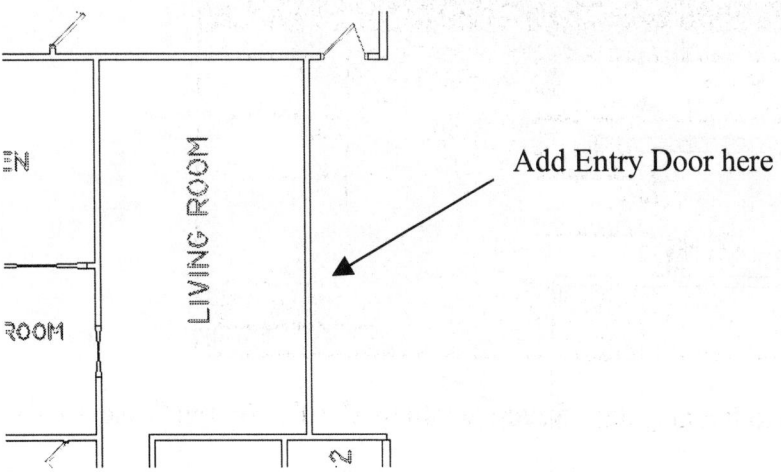

Add Entry Door here

Lesson 4
Floor Plans

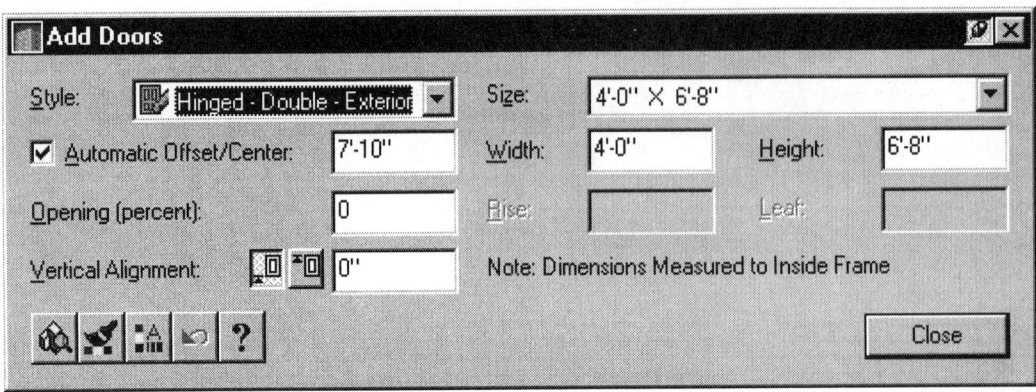

Set the entry door to be Hinged-Double-Exterior. Set the Width to 4' and height to 6'-8". Set the Offset to 7'-10". Set the Opening to 0. This means that the door will show as closed.

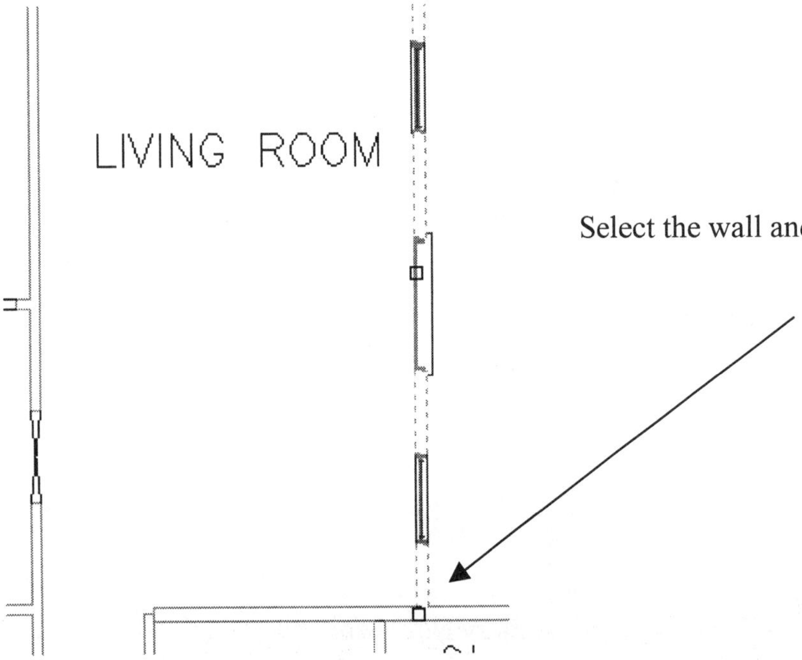

Select the wall and endpoint indicated.

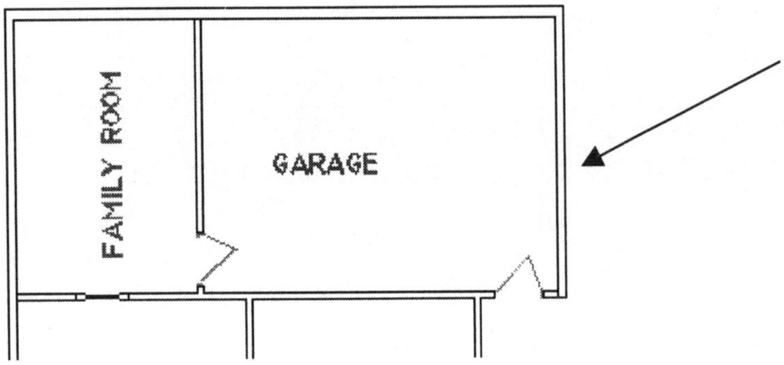

Add Garage Door here.

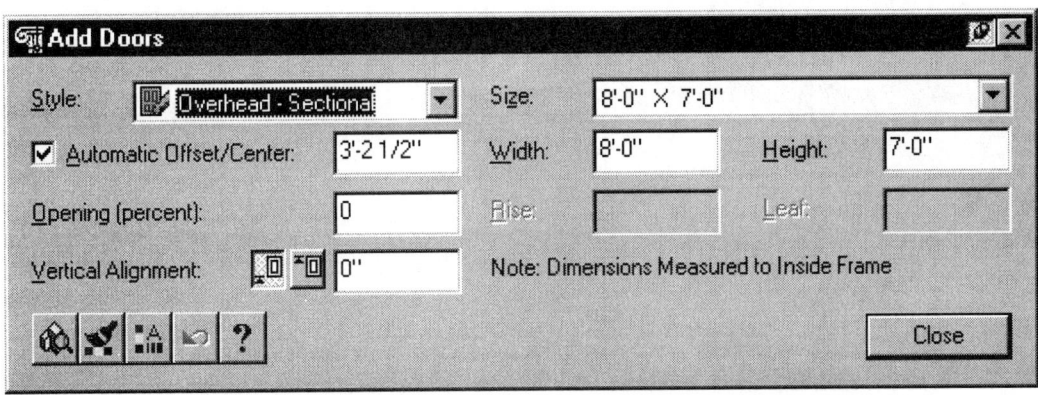

Set the garage door using Overhead-Sectional. Set the Offset to 3'-2 1/2".
Set the Width to 8'. Set the Height to 7'.

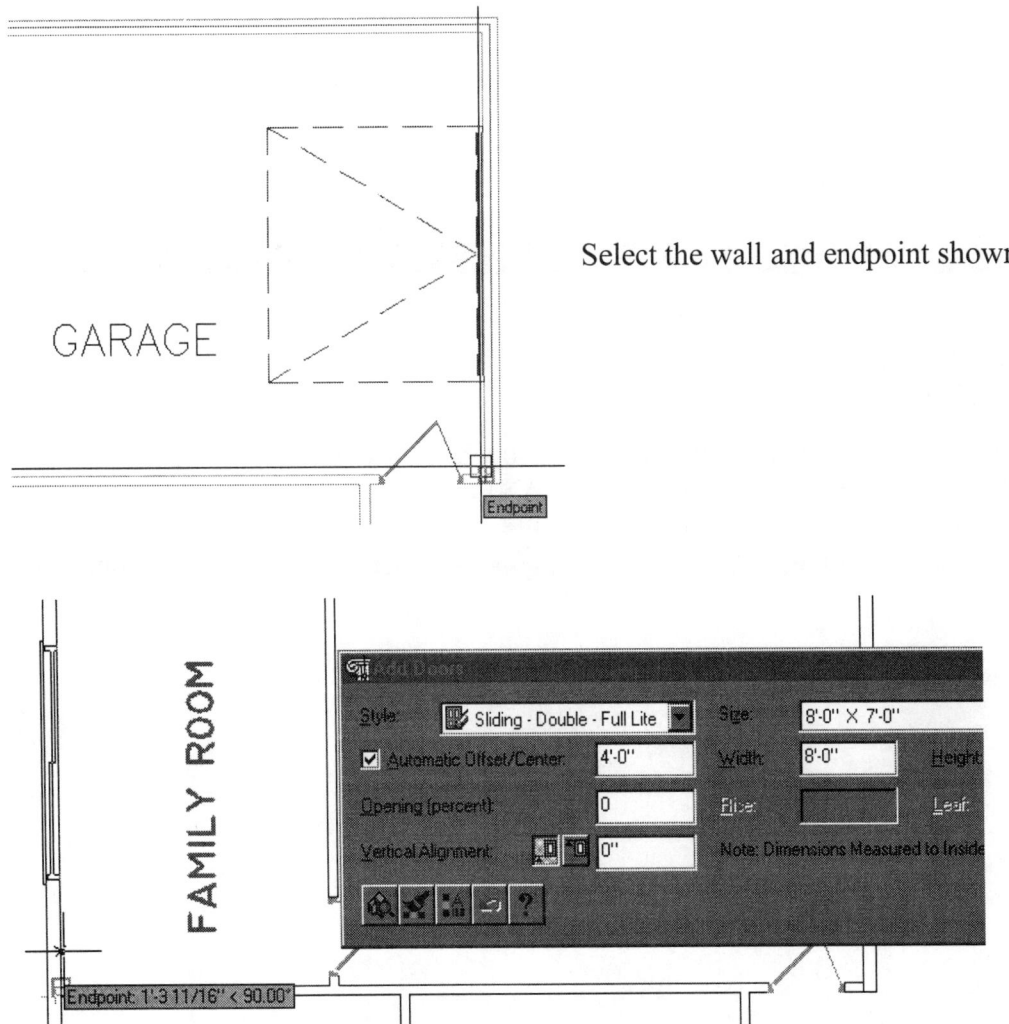

Select the wall and endpoint shown.

Add a Sliding Door –Double Full Lite to the family room.
Set the Size to 8' x 7'. Set the Offset to 4'.
Select the wall and endpoint shown.

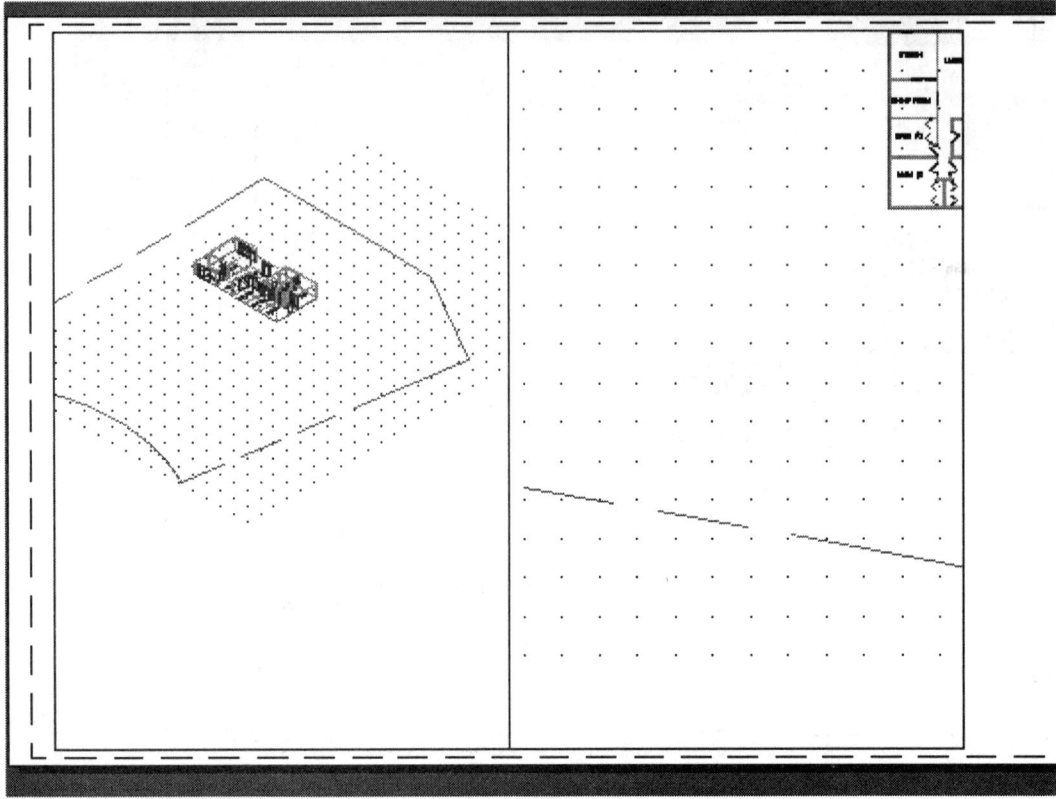

Select the Work-3D tab. This is set up with two viewports. The left viewport is 3D isometric. The right viewport is our top (PLAN) view.

To activate a viewport, pick inside the rectangular boundary. You can then zoom and pan.

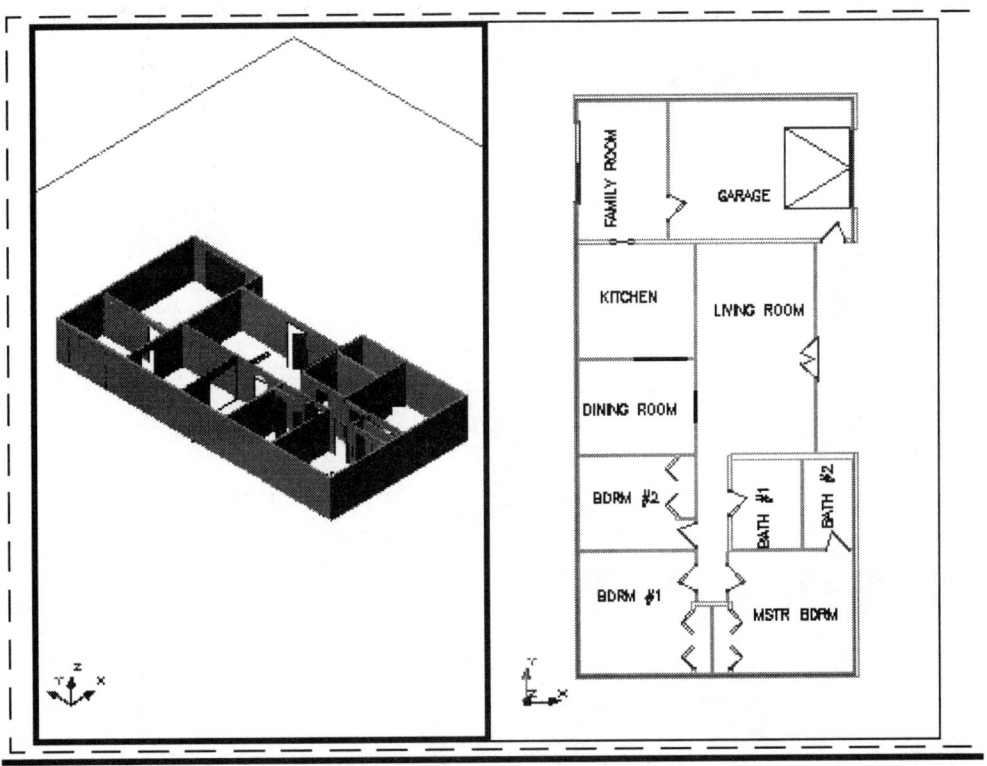

Our floor plan shown so far.
Save as Ex4-8.dwg.

Lesson 4
Floor Plans

Exercise 9:
Add Window Assemblies

Drawing Name: Lesson 4.dwg
Estimated Time: 30 minutes

This exercise reinforces the following skills:

- Add Windows

Menu	Design->Windows->Add Window
Windows Toolbar	
Command Line	WindowAdd

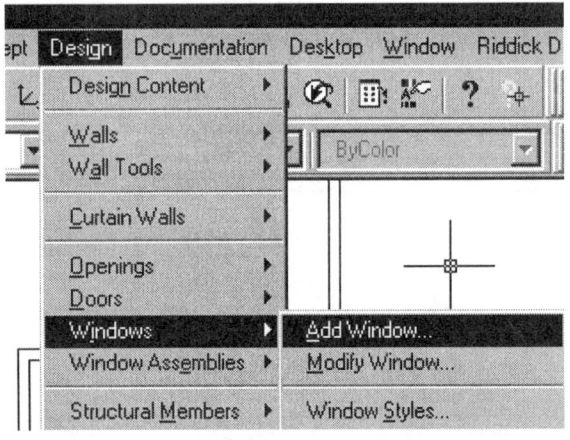

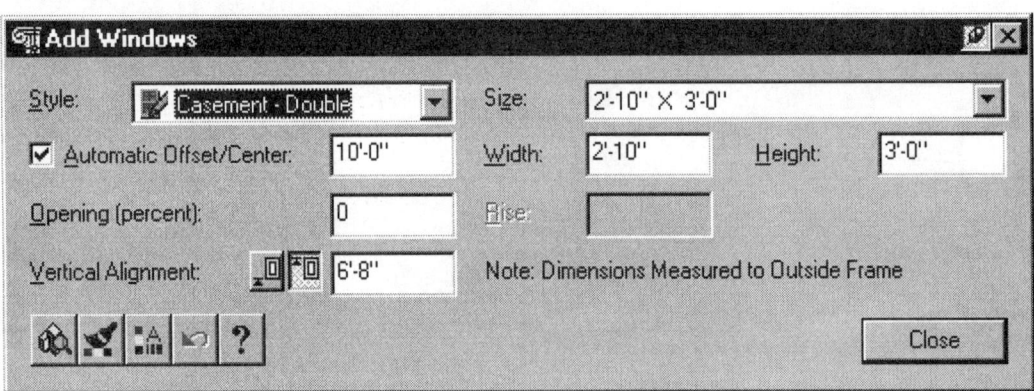

Set the Window Style to Casement-Double. Set the size to 2'-10" x 3'.
Set the Offset to 10'.
Note the vertical alignment controls shifts to head height, at 6'8".

4-34

Lesson 4
Floor Plans

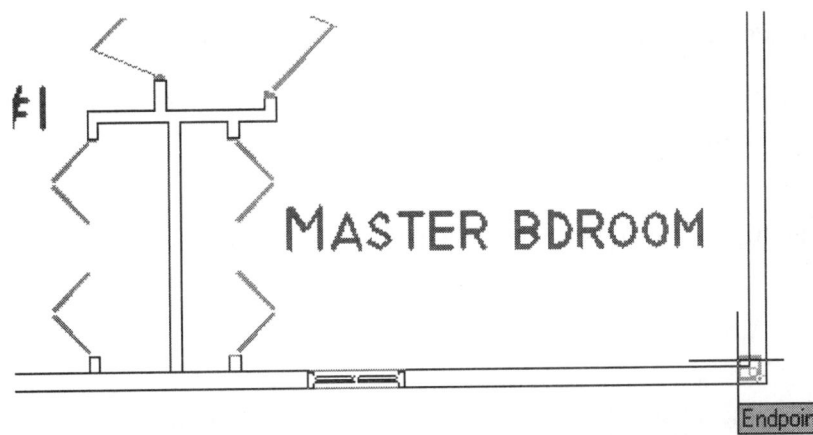

Select the wall shown and the endpoint indicated.

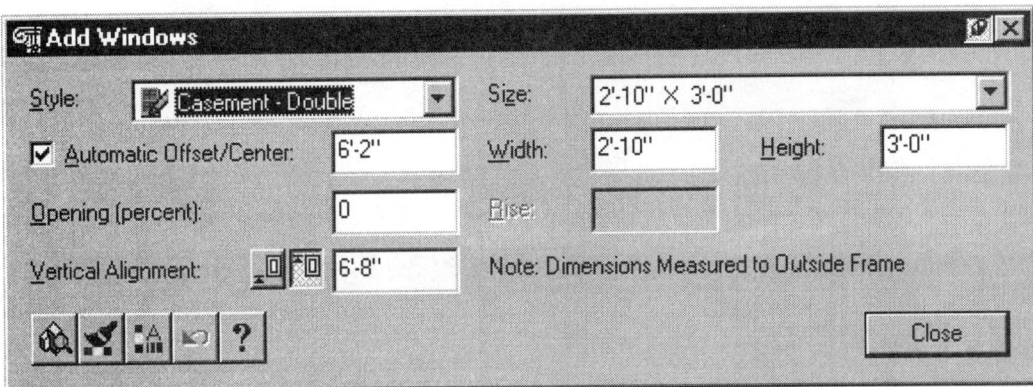

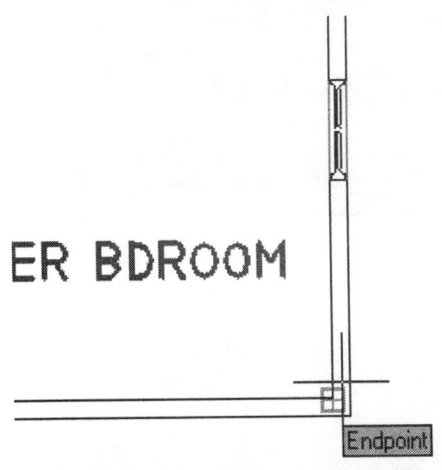

Set the Offset to 6'-2".
Select the wall and endpoint shown.

Remember – if you don't like the position of any of the Windows, you can reposition them.
Just select the window, right click, and select 'Reposition Along Wall.'

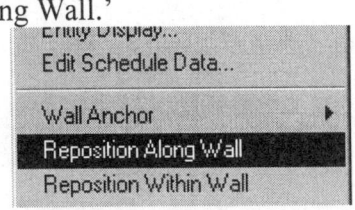

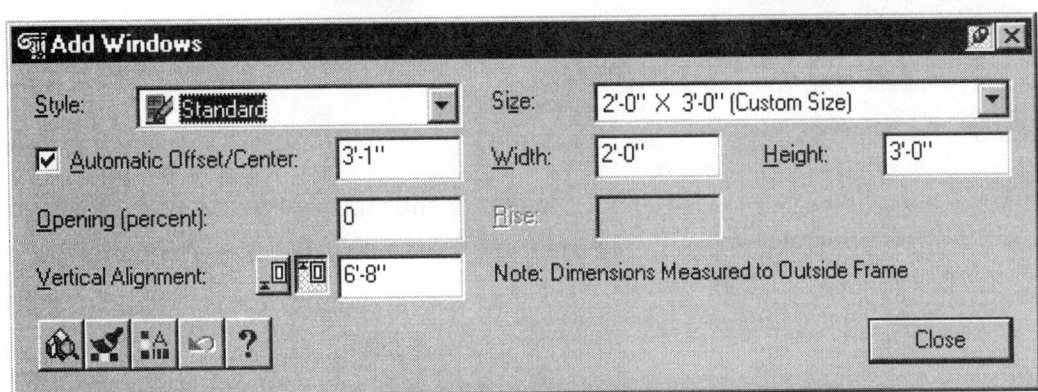

4-35

Lesson 4
Floor Plans

Set the Window Style to Standard. Set the size to 2' x 3'. Set the Offset to 3'-1".

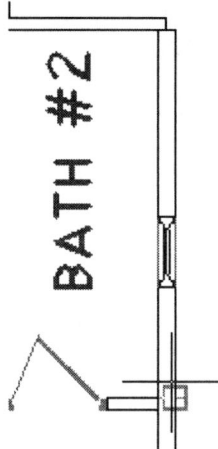

Select this wall and endpoint indicated.

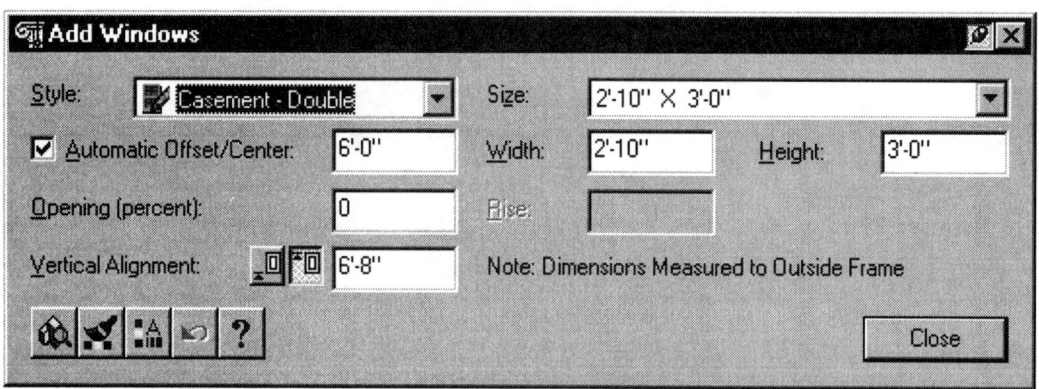

Set the Style to Casement-Double. Set the Size to 2'10"x 3'. Set the Offset to 6'.

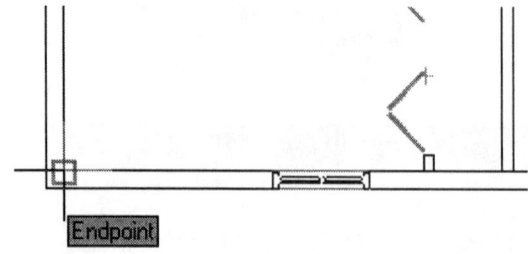

Select the back wall for Bedroom #1 and the endpoint shown.

4-36

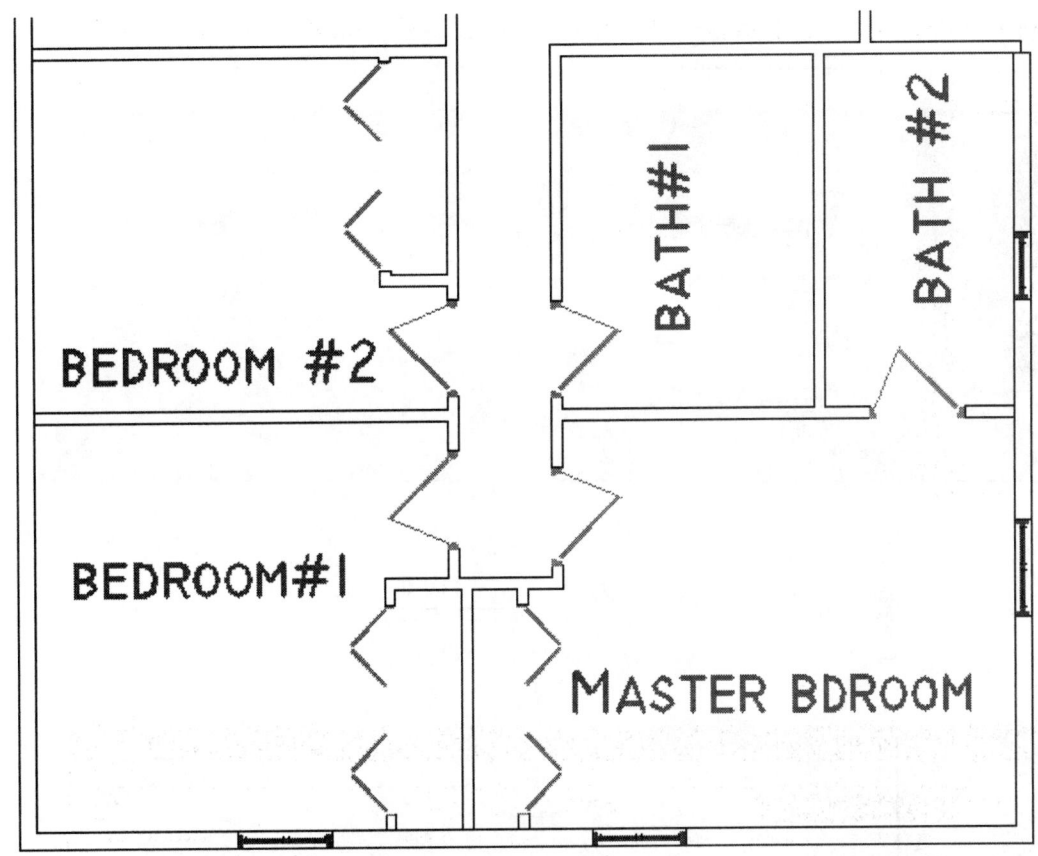

Our floor plan so far.

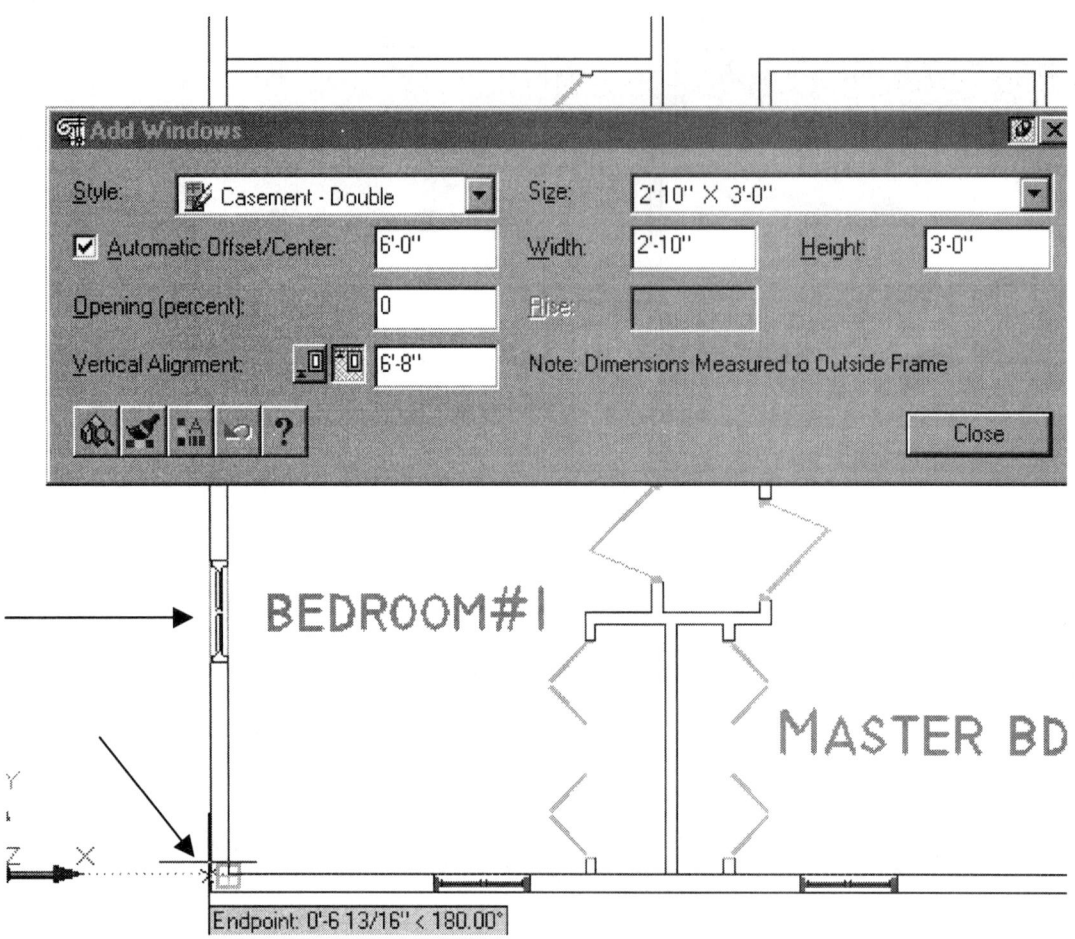

Add a second Double Casement window to Bedroom #1 as shown.

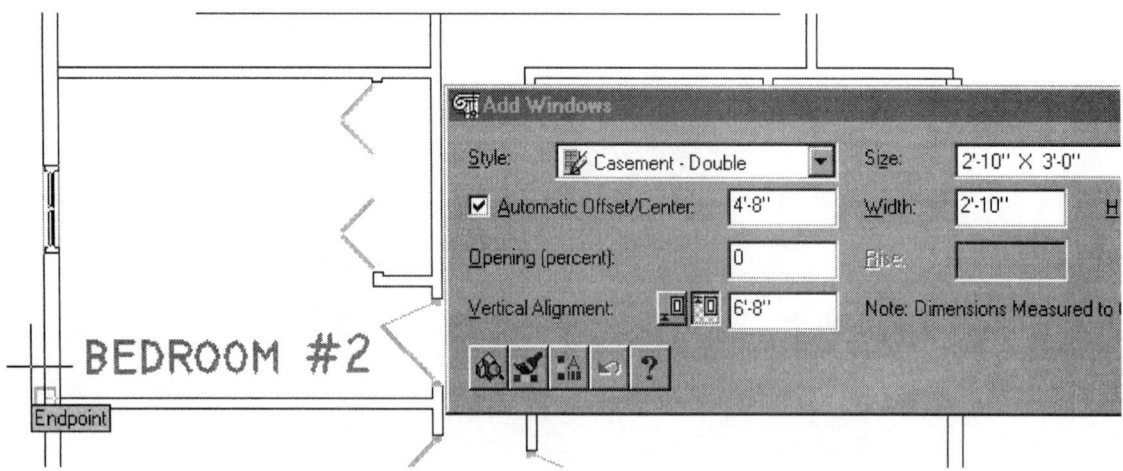

Add a Casement-Double to Bedroom #2 using an Offset of 4'-8" as shown.

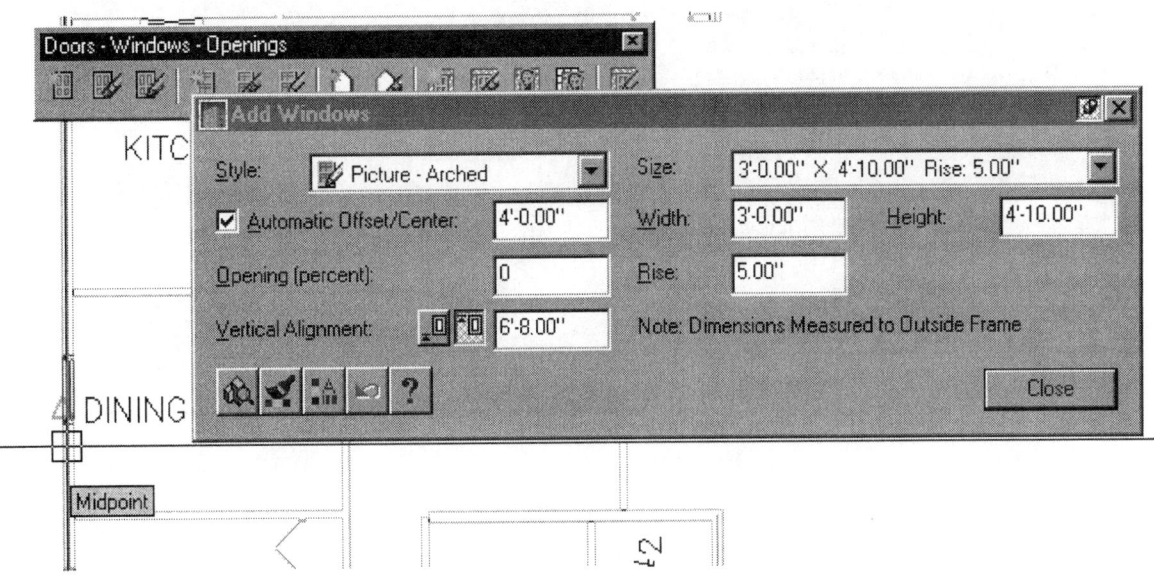

Add a 3' x 4'-10" Picture- Arched Window to the Dining Room using Offset of 4'-0" as shown..

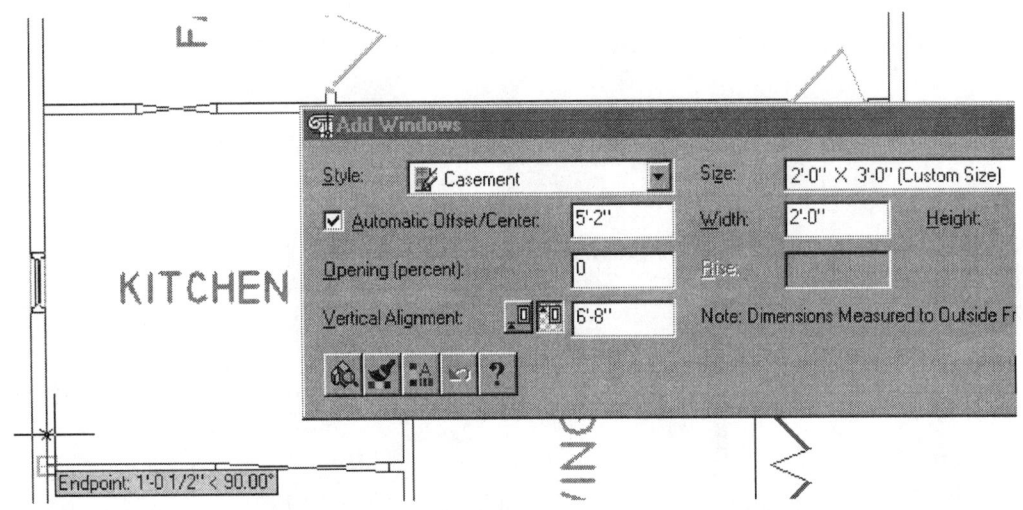

Add a 2' x 3' Casement Window to the Kitchen. Set the Offset to 5'-2".

Lesson 4
Floor Plans

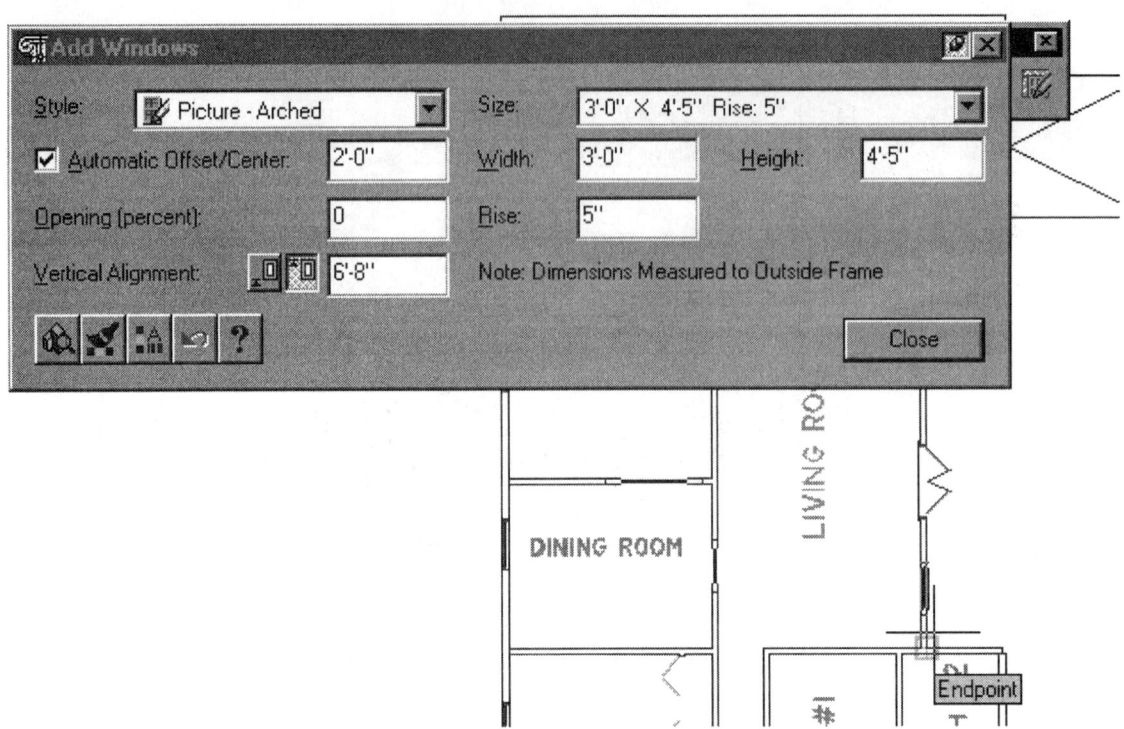

Add a 3' x 4'-5" Rise: 5" Picture-Arched to the Living Room. Set the Offset to 2'.

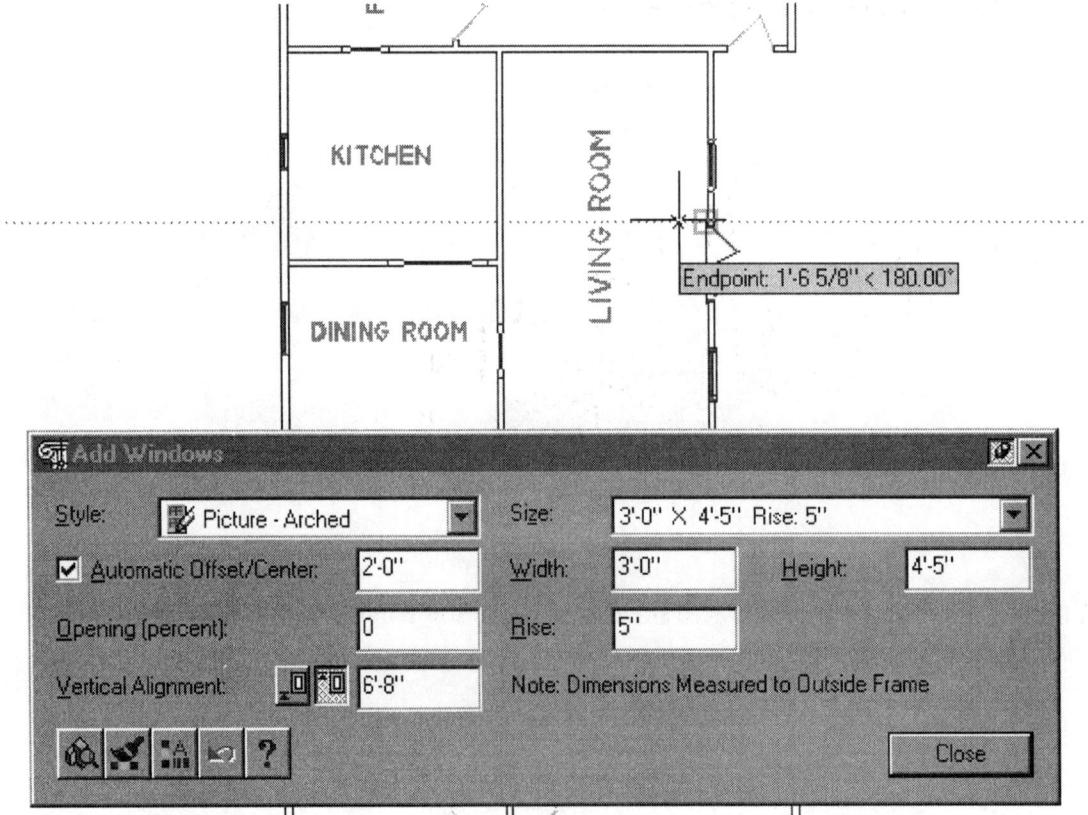

Add a 3' x 4'-5" Rise: 5" Picture-Arched to the Living Room. Set the Offset to 2'.

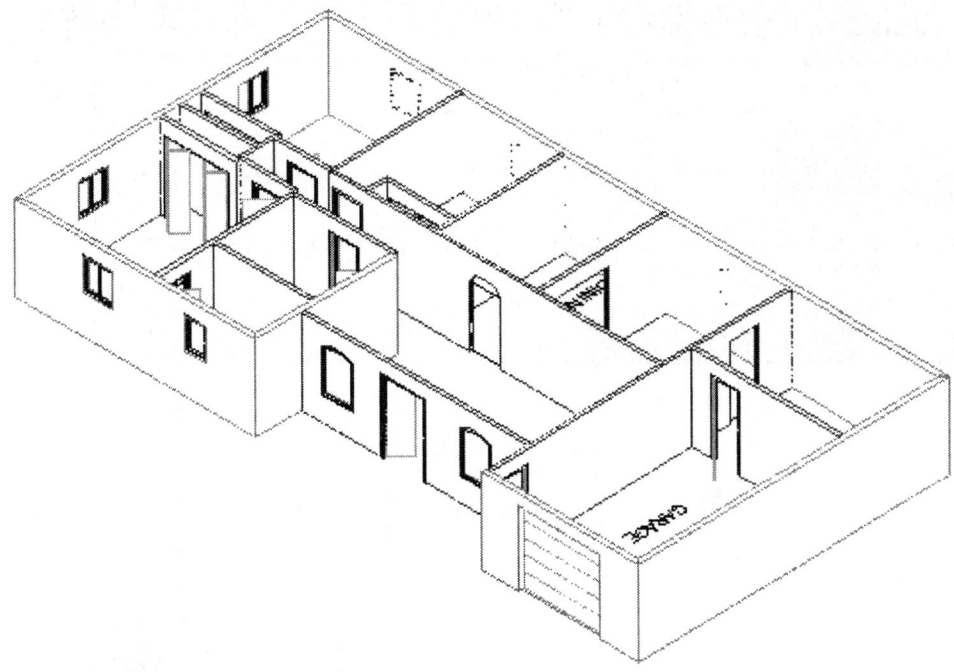

Our floor plan so far.

(This is shown using NE Isometric View.)

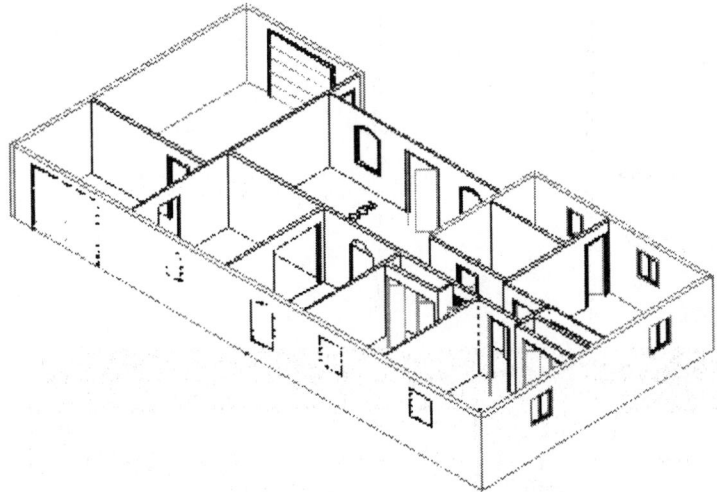

When we look at the SE view, we see that some of the windows are not positioned properly within the walls.

Lesson 4
Floor Plans

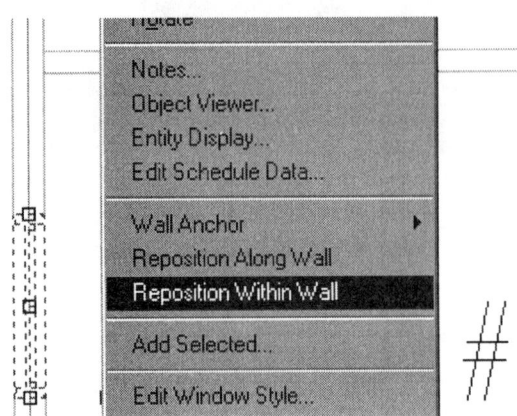

Select any windows or doors that do not appear properly.
Right click and select the 'Reposition Within Wall' option.
When prompted for the side to offset, select the outside of the house.

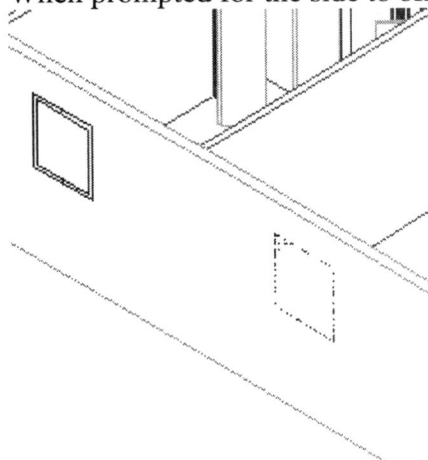

The window on the left is positioned correctly within the wall.
The window on the right needs to be repositioned.

Exercise 10:
Adding a Fireplace

Drawing Name: Lesson 4.dwg
Estimated Time: 30 minutes

This exercise reinforces the following skills:

- Adding Complex Geometry
- Adding Openings
- Design Center

ADT currently doesn't come with any fireplaces. However, you can use Mass Elements to create a Fireplace Shape and then use that to add to your model.

You can download a fireplace drawing from the publisher's website to complete this exercise at www.schroff1.com, create your own fireplace (See Lesson 1, Exercise 6), or you can opt to skip this exercise.

Activate the Design Center.

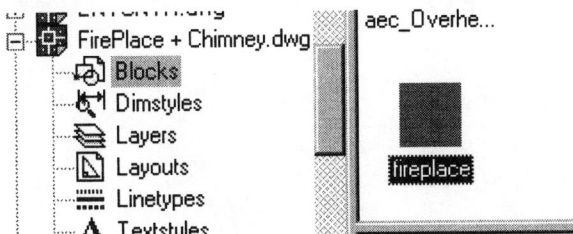

Locate the Fireplace + Chimney drawing. Under blocks, you will see a block called Fireplace.

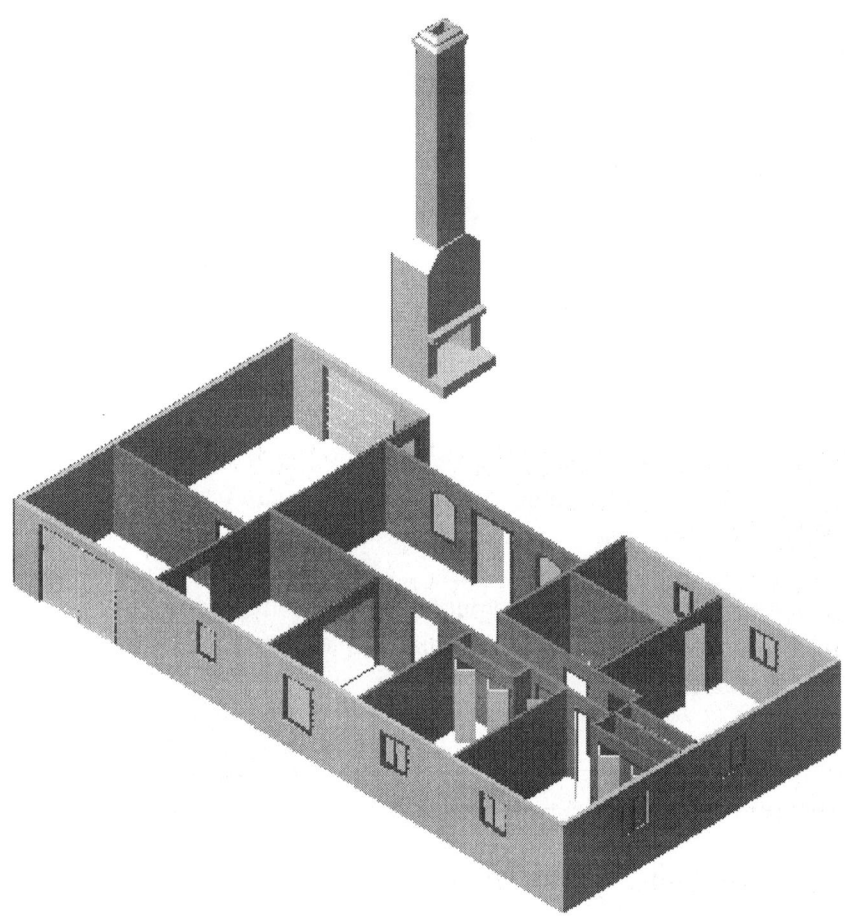

Drag and drop the fireplace block into your drawing.

Scale the block to ½ size.

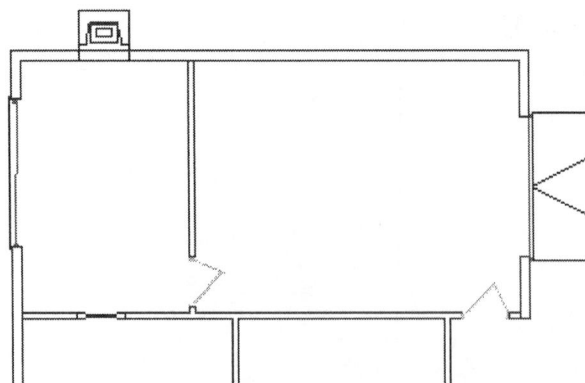

Locate the fireplace in the family room as shown.

If you look at it from an isometric view, you see that we need to create an opening for the fireplace.

Lesson 4
Floor Plans

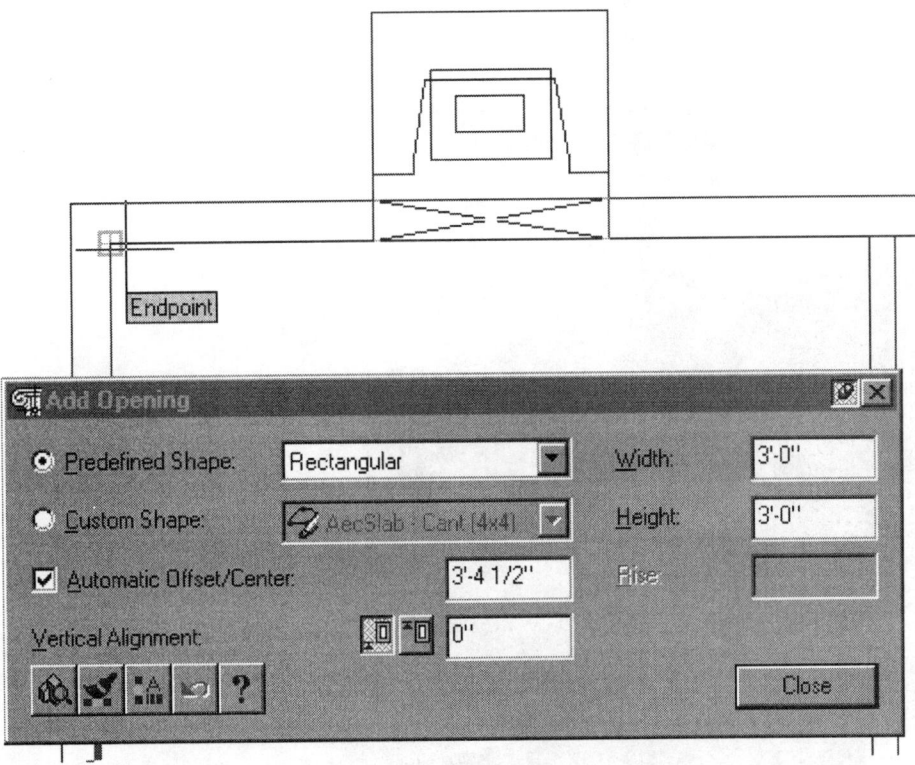

Set the opening to 3'x 3' rectangular with an offset of 3'-4.5".

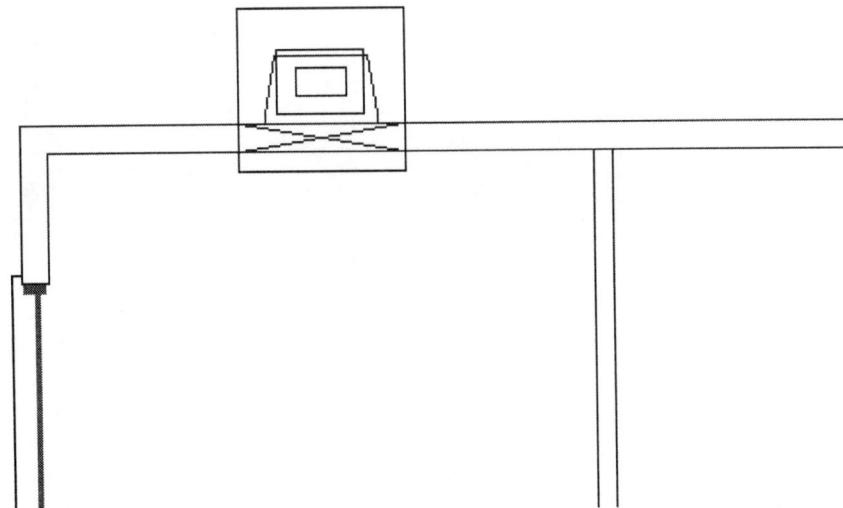

Move the fireplace in slightly to eliminate any gaps between the fireplace and the wall.

You will need to use the 3D orbit tool to inspect your fireplace placement to ensure that it is OK.

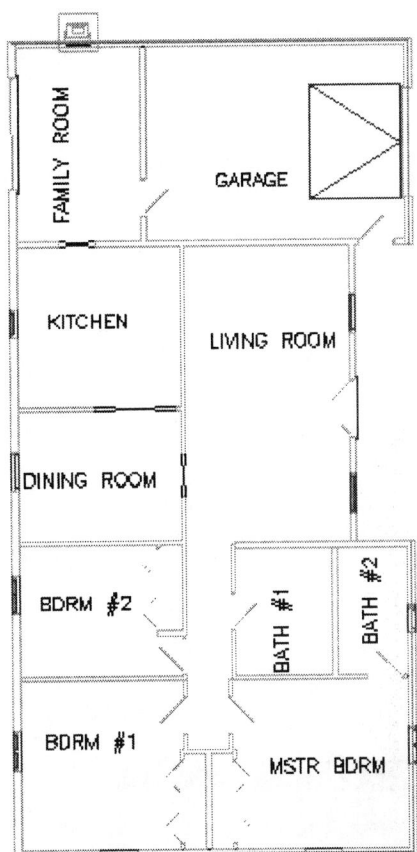

Our completed floor plan.

 Activate the Layer Manager.

Highlight the word 'ALL' in the Layer Manager tree.

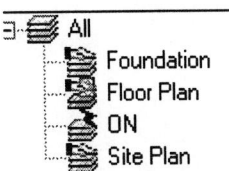

Select the New User Group tool.

```
⊟ All
    Foundation
    Floor Plan
    ON
    Site Plan
```

Let's call this group 'Floor Plan'.

Lesson 4
Floor Plans

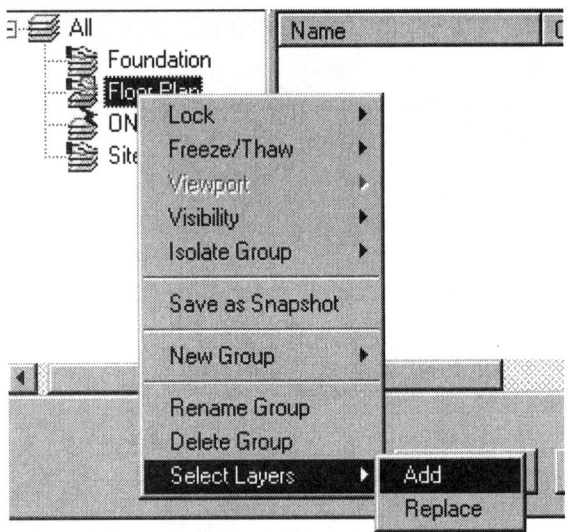

Highlight 'Floor Plan'.
Right click and select 'Select Layers->Add'.

Window around your floor plan.

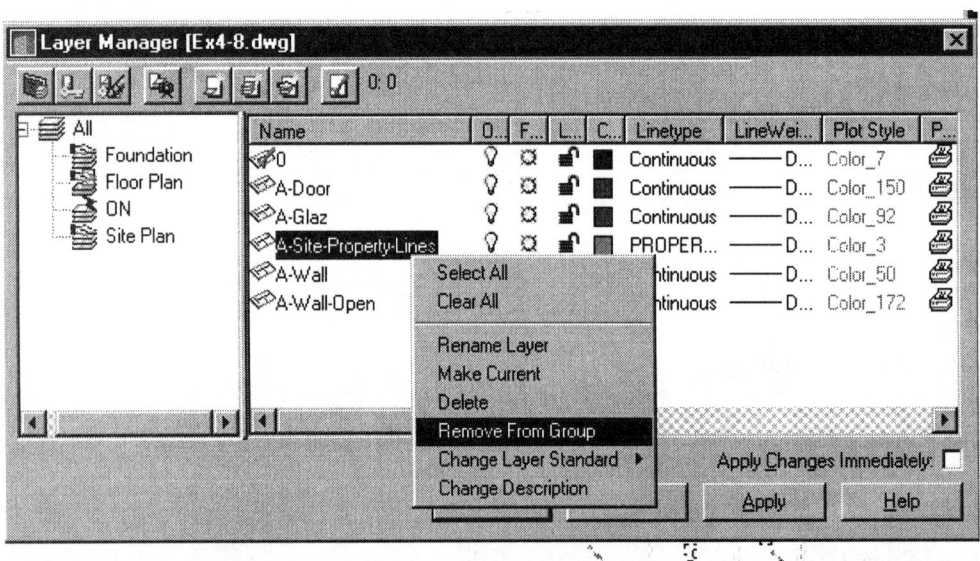

We don't want the A-Site-Property-Lines layer in this group.
Highlight that layer. Right click and select 'Remove From Group'.

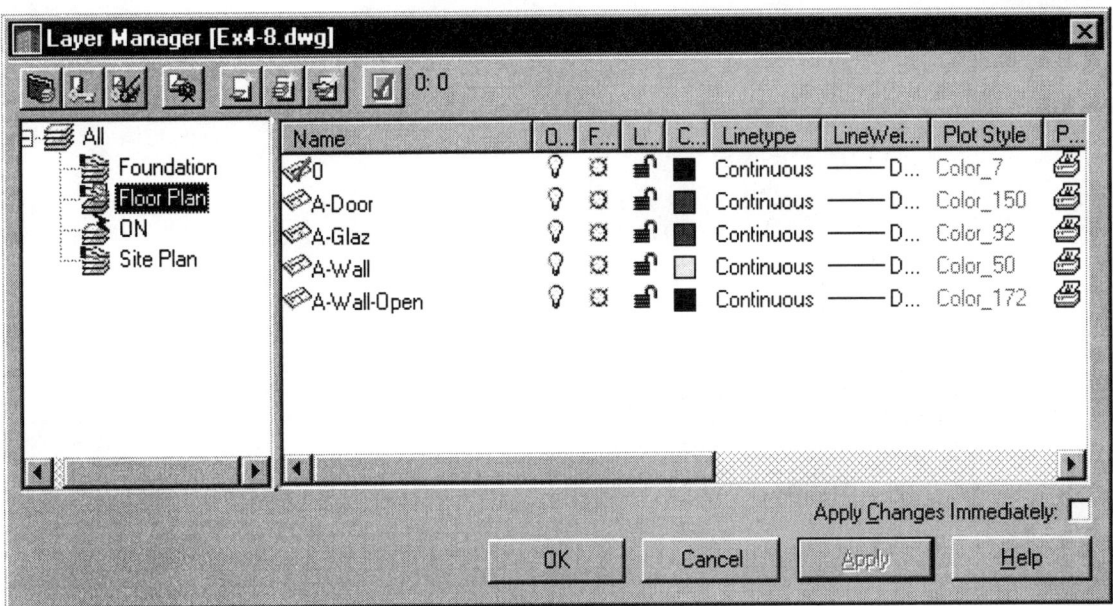

The layers for our floor plan group are shown above.

Save the file as Lesson 4.dwg.

Notes:

Quiz 2

True or False

1. The two types of Layer Filter Groups are Dynamic and Stationary.
2. You can create Layer Filter Groups using the Layer Style Manager Dialog.
3. A Snapshot saves the layer settings.
4. Once a door or window is placed, it can not be moved or modified.
5. Openings can be any size and any elevation.
6. The Offset value when placing a door/window/opening determines how far the door/window/opening is placed from a selected point.
7. Door, window and opening Dimensions can be applied using the Design Center.

Multiple Choice

8. The icon shown is:

   A. ADD DOOR
   B. ADD WINDOW
   C. ADD OPENING
   D. ADD WALL

9. The icon shown is:

   A. ADD DOOR
   B. ADD WINDOW
   C. ADD OPENING
   D. ADD WALL

10. The icon shown is:

    A. ADD DOOR
    B. ADD WINDOW ASSEMBLY
    C. ADD OPENING
    D. ADD WALL

11. Select the entity type that can NOT be converted to a wall:

    A. Line
    B. Polyline
    C. Circle
    D. Spline

12. Doors are automatically placed on this layer:

    A. A-OPENING
    B. A-DOOR
    C. A-WINDOW
    D. A-WALL-OPENING

13. A HOT grip is indicated by this color:

    A. GREEN
    B. BLUE
    C. RED
    D. YELLOW

14. Openings are placed on this layer:

    A. A-OPENING
    B. A-DOOR
    C. A-WINDOW
    D. A-WALL-OPENING

ANSWERS:

1) F; 2) F 3) T; 4) F; 5) T; 6) T; 7) T; 8) A; 9) C; 10) B; 11) D; 12) B; 13) C; 14) D

Lesson 5
Space Planning

A residential structure is divided into three basic areas:

- Bedrooms: Used for Sleeping and Privacy
- Common Areas: Used for gathering and entertainment, such as family rooms and living rooms, and dining area
- Service Areas: Used to perform functions, such as the kitchen, laundry room, garage, and storage areas

When drawing your floor plan, you need to verify that enough space is provided to allow placement of items, such as beds, tables, entertainment equipment, cars, stoves, bathtubs, lavatories, etc.

ADT comes with Design Content to allow designers to place furniture to test their space. Additional architectural content is provided at no charge from the Schroff Development Corporation's website at www.schroff1.com for readers of this textbook.

Exercise 1:
Creating AEC Content

Drawing Name: Lesson 4.dwg
Estimated Time: 15 minutes

This lesson reinforces the following skills:

- Design Center
- AEC Content
- Customization

Download the following files from the Schroff website at www.schroff.com:

Bedking.dwg
Bedtwin.dwg

Lesson 5
Space Planning

Copy these drawings into the Imperial/Design/Furniture/Bed subfolder:

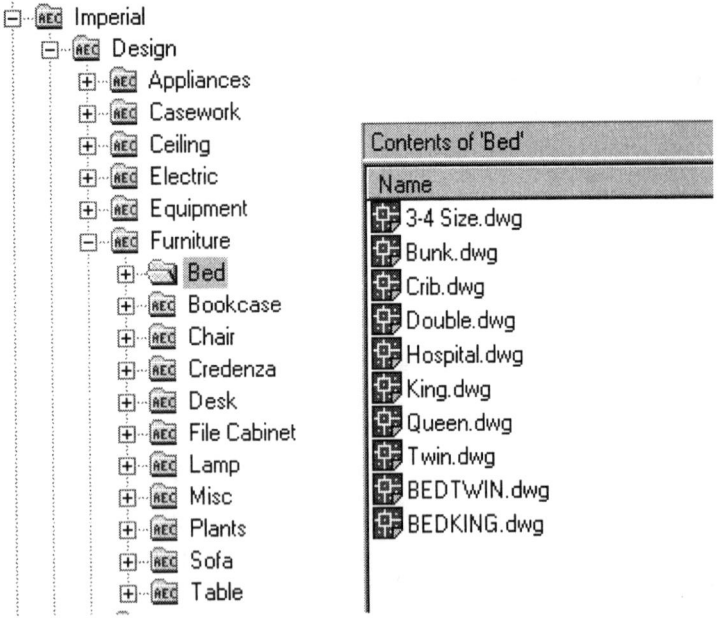

TIP: Place custom content in the correct standard content subfolder. Then when you drag and drop, the content will automatically be placed on the correct layer. To ensure the correct layers are utilized, the Layer Standard must be set.

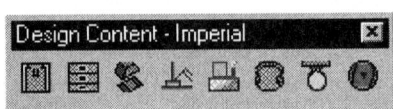

Activate the Design Content- Imperial Toolbar.

Select the Furniture tool.

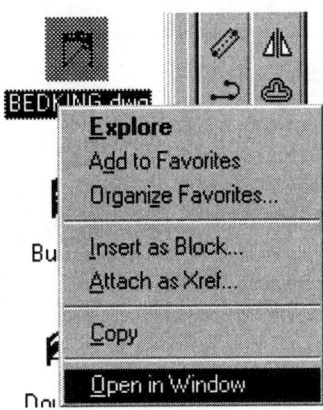

Locate the Bed subfolder.
Highlight the BEDKING.dwg.
Right click and select 'Open in Window'.

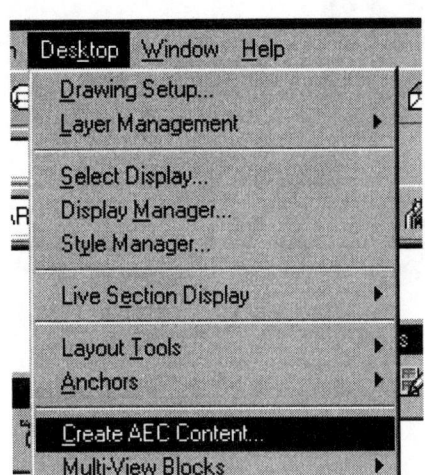

Go to Desktop->Create AEC Content.

Lesson 5
Space Planning

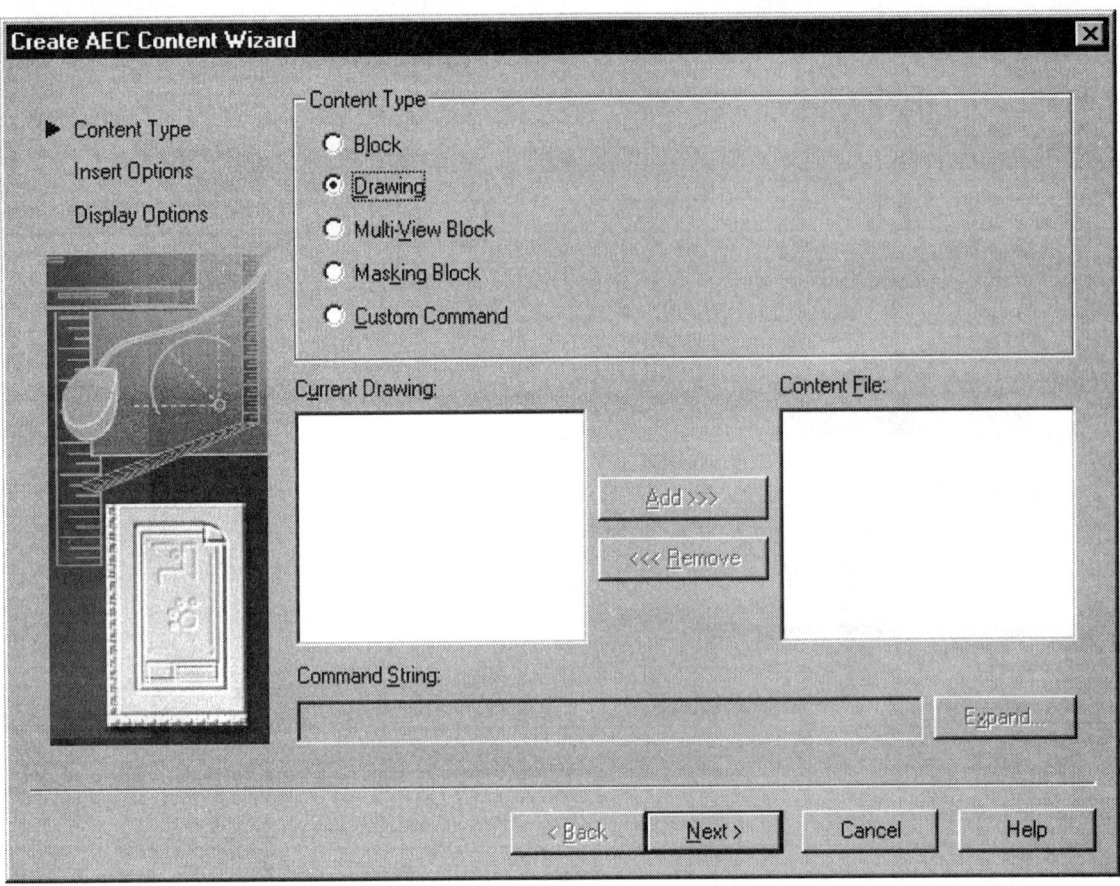

The Create AEC Content Wizard dialog appears.

Enable 'Drawing.'
Press 'Next'.

Lesson 5
Space Planning

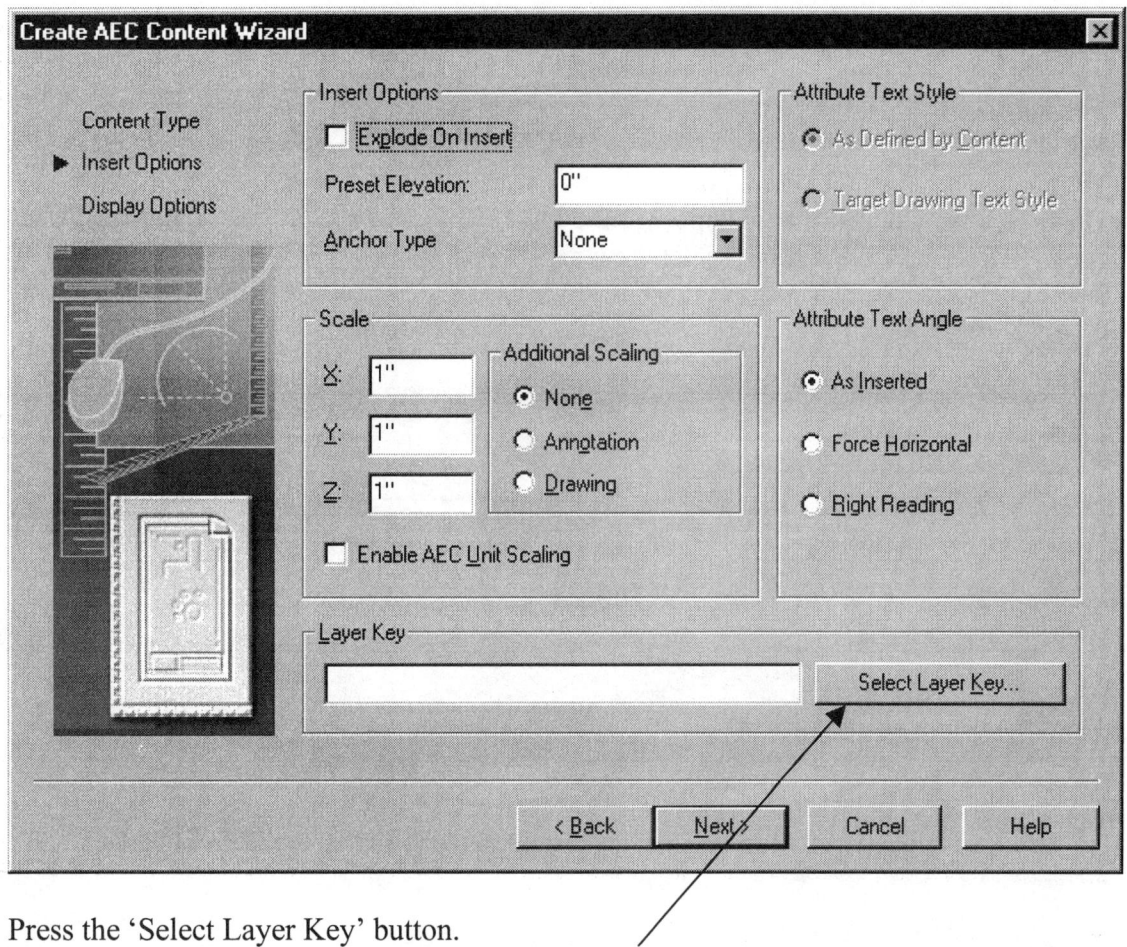

Press the 'Select Layer Key' button.

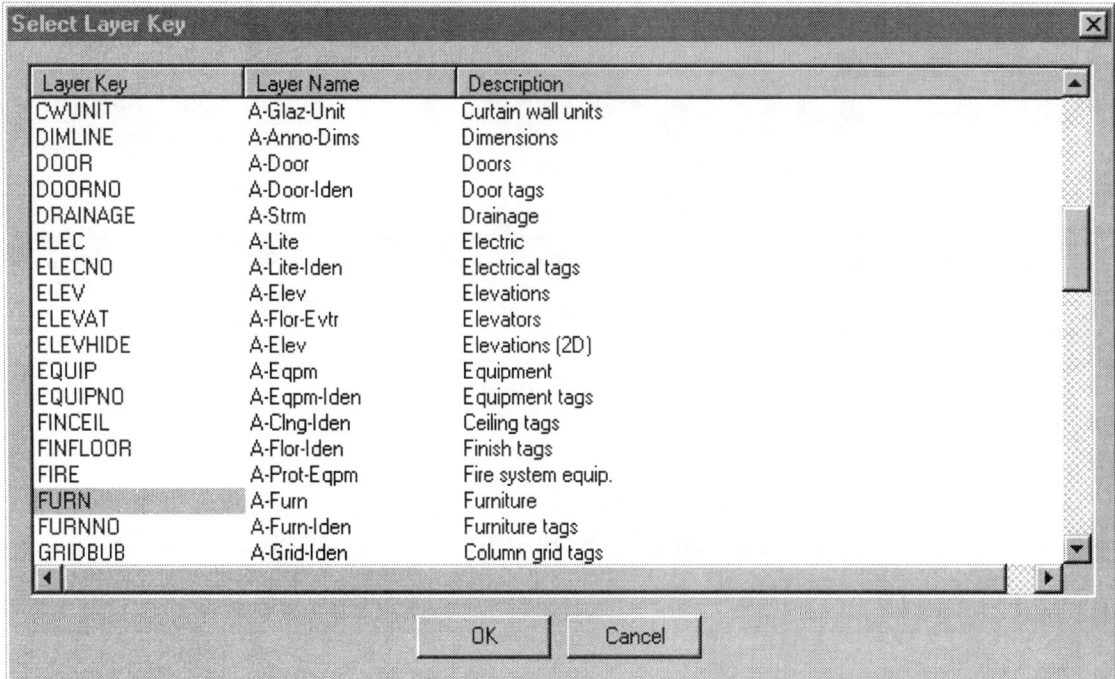

Locate the Layer called 'FURN'. Press 'OK'.

Lesson 5
Space Planning

The layer you selected should appear in the Layer Key edit box. If you do not see the word FURN, then you did not select the correct layer. You can edit the word and just type FURN.
Press 'Next'.

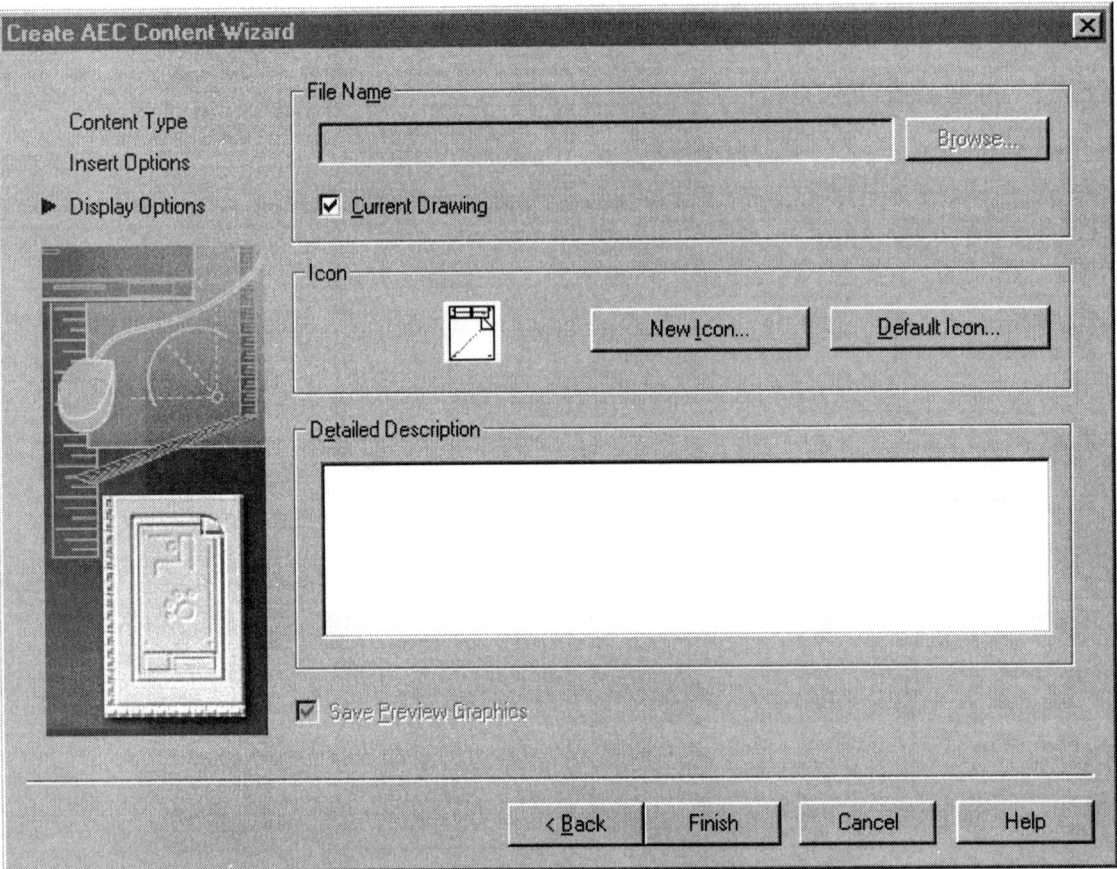

Enable 'Current Drawing'.
Press 'Finish'.
Save and close the drawing.

What you did:
You enabled the layer key setting so that when this drawing is dragged and dropped from the Design Center, it will automatically be placed on the correct layer A-FURN.

Lesson 5
Space Planning

Exercise 2:
Furnishing the Bedrooms

Drawing Name: Lesson 4.dwg
Estimated Time: 30 minutes

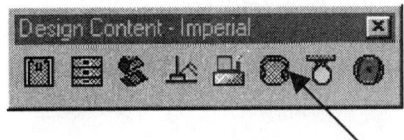

Furniture is one of the categories for Design Content.

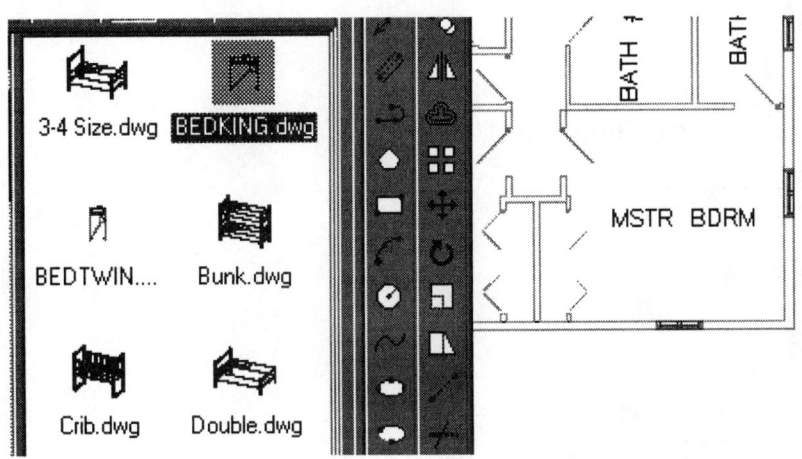

Locate the file called Bedking.dwg.

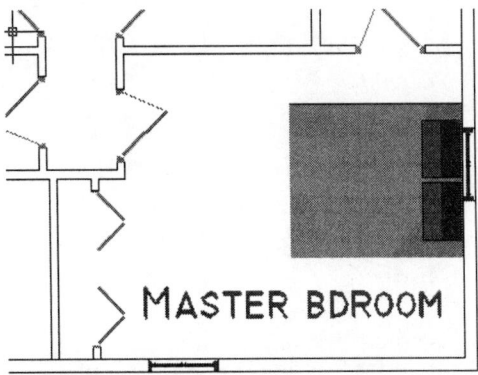

Place the bed as shown using Drag and Drop.

Lesson 5
Space Planning

Switching to a 3D view, you see that the bed you just placed is a 3-dimensional bed.

Select the bed. Right click and select 'Properties'.

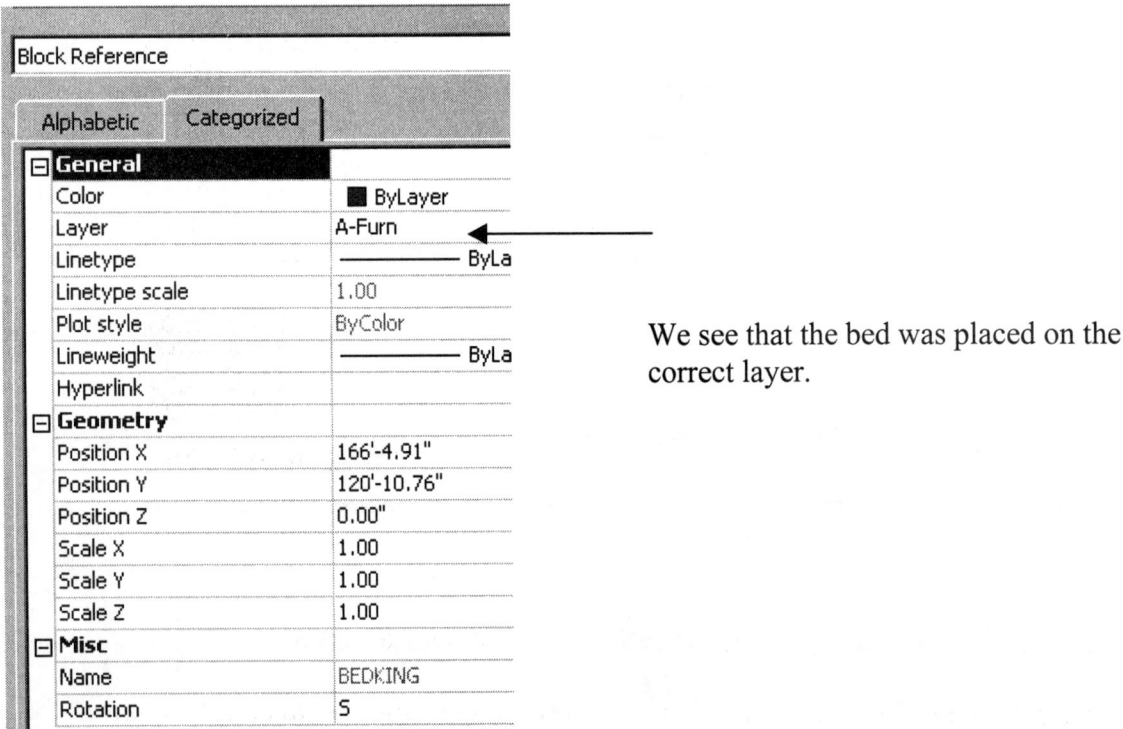

We see that the bed was placed on the correct layer.

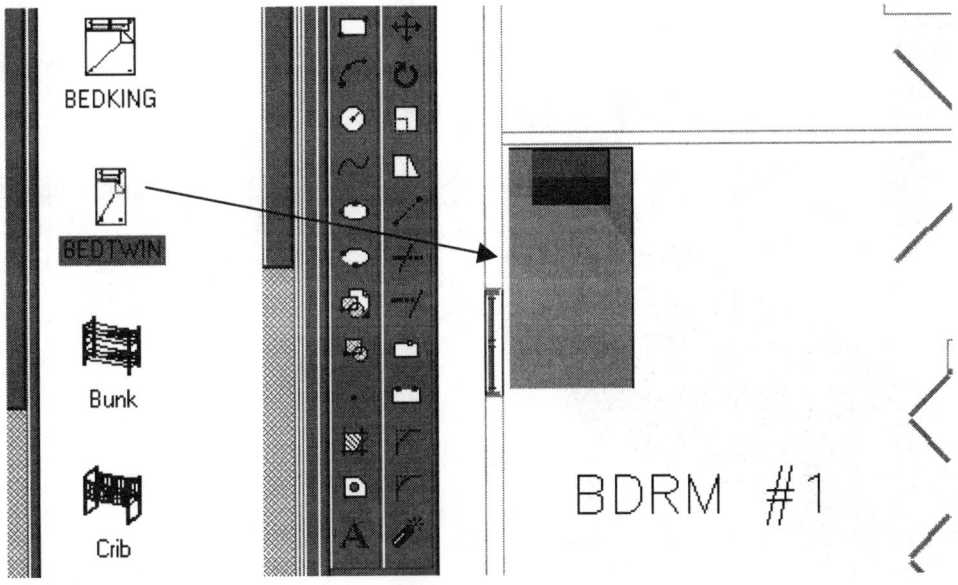

Locate the bedtwin.dwg and place it in Bedroom#1.

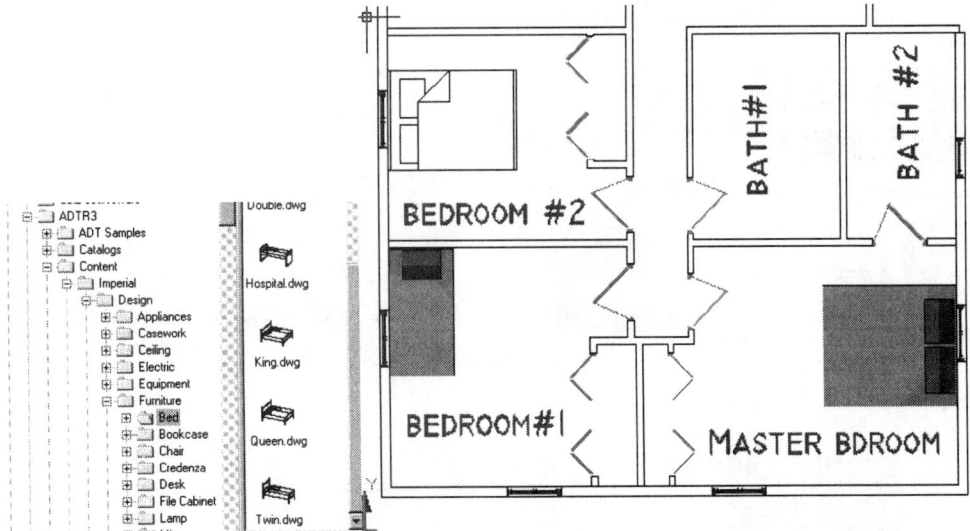

Locate the Queen.dwg. Drag and drop it into Bedroom #2 as shown.

Switching to a 3D View and using your 3D orbit tool, you should be able to see all the furniture you have placed in 3D.

Lesson 5
Space Planning

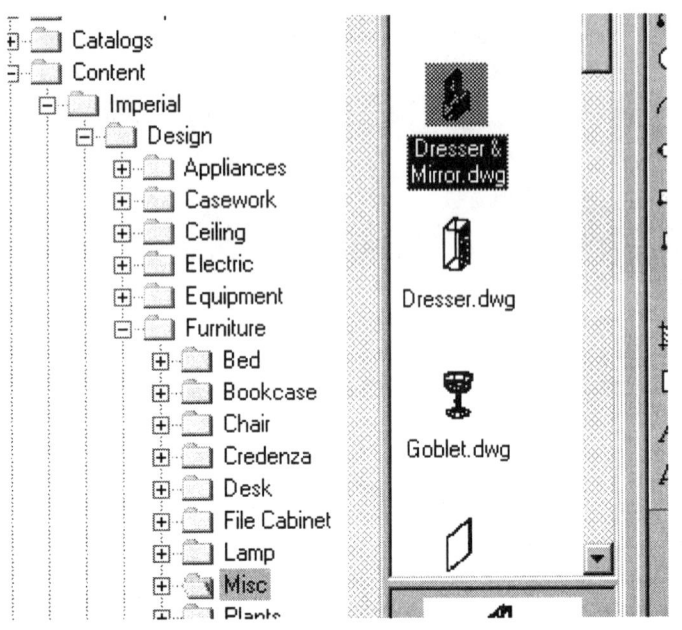

Under /Design/Furniture/Misc, we find a Dresser & Mirror.dwg which will go perfectly in the Master Bedroom. Drag and drop it in as shown.

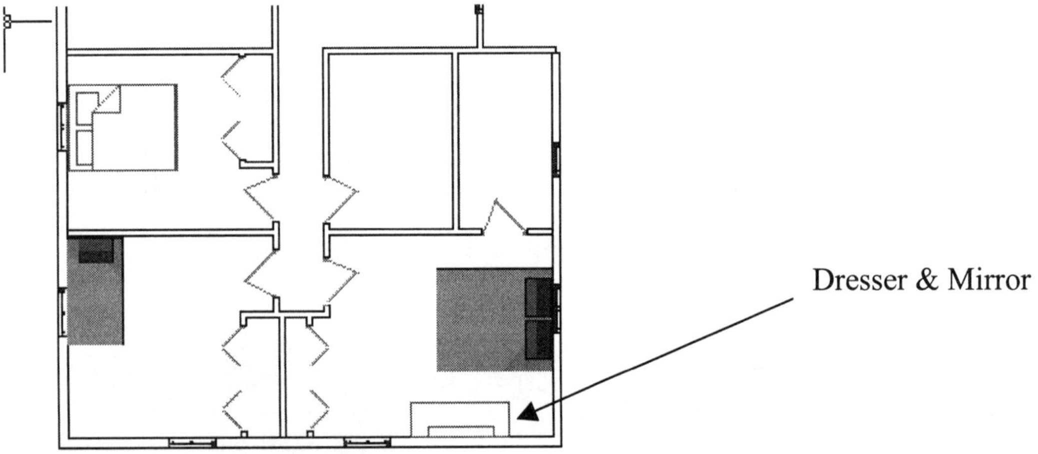

Dresser & Mirror

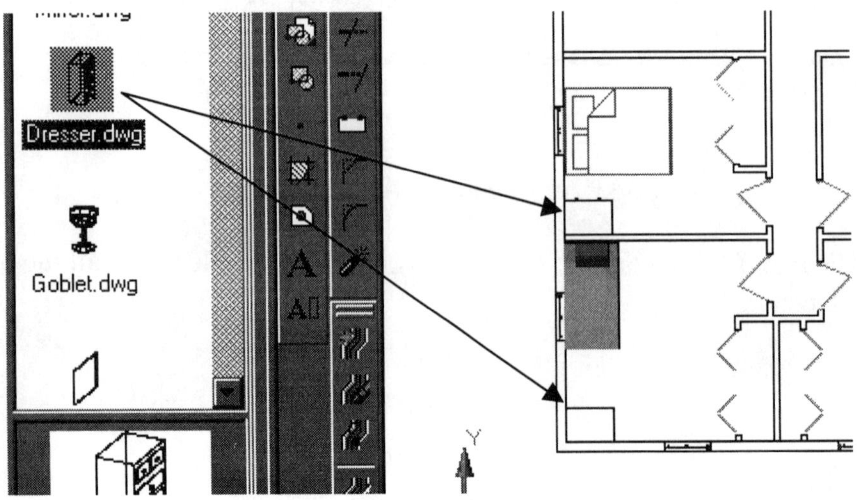

Add a dresser to each of the other bedrooms.

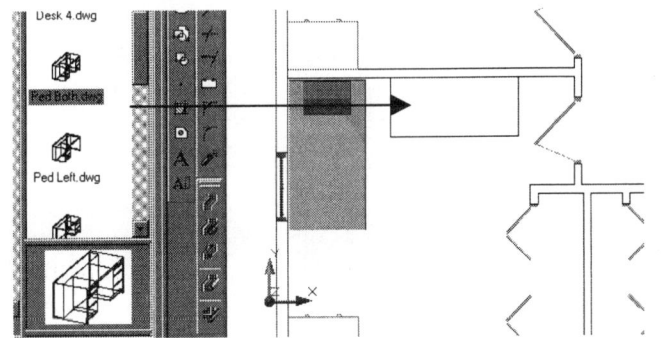

Add the Ped Both.dwg to Bedroom#1. This is a desk with drawers on both sides. This is located in the Desk subfolder. Verify that you have placed the desk with the drawers facing out.

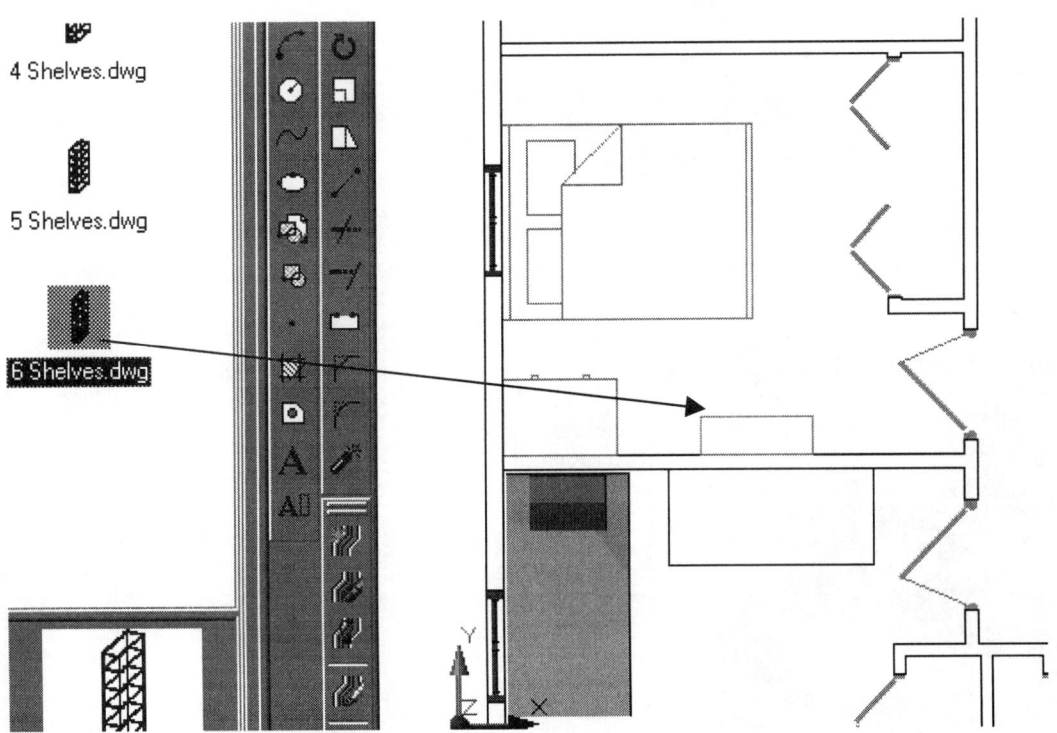

Place a 6 shelves bookcase in Bedroom #2.
This is located in Design/Furniture/Bookcase.

Lesson 5
Space Planning

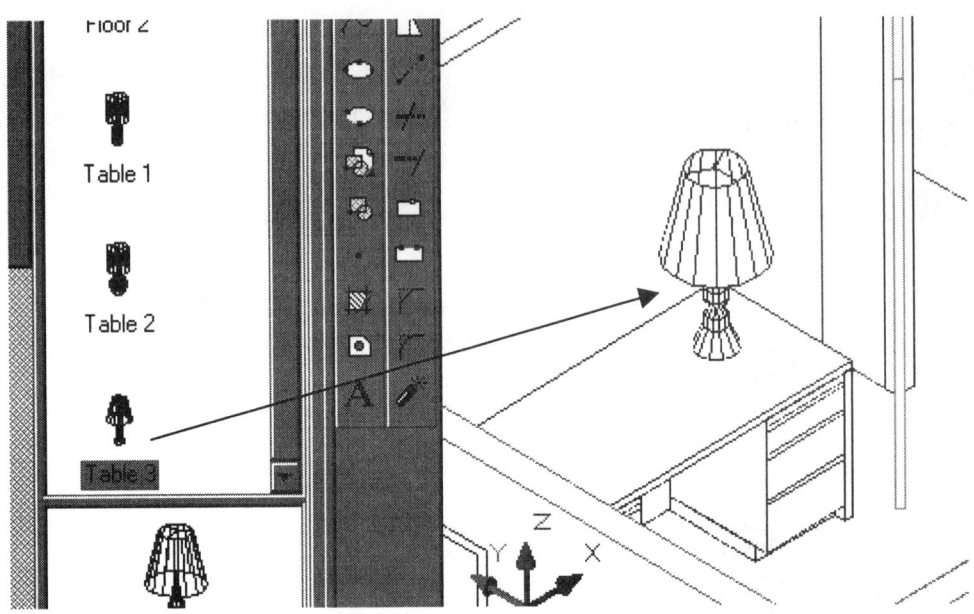

A table lamp is placed on top of the desk to provide light for the diligent student.
To place the lamp on top of the desk, switch to a 3D view.

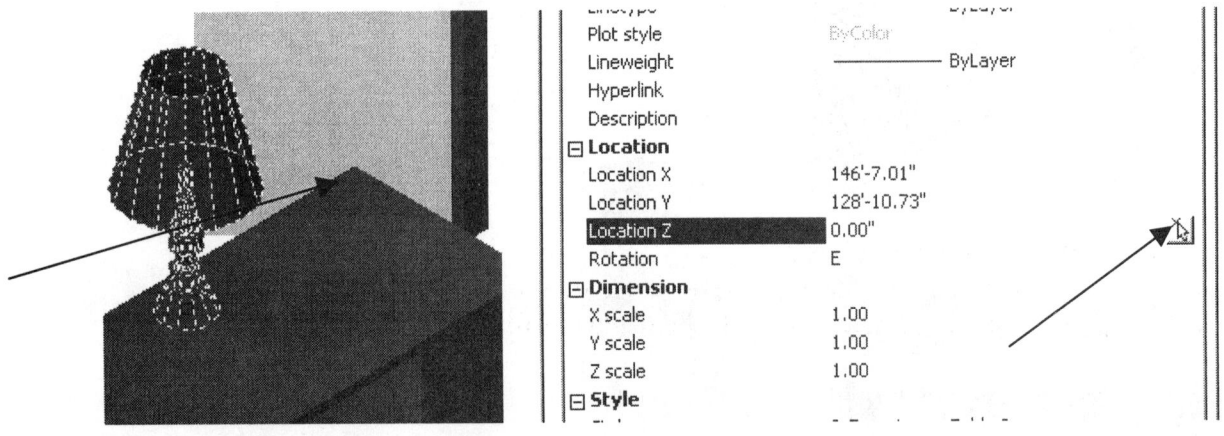

Select the lamp, right click and select 'Properties'.
In the field for Location Z, select the 'Pick' button.

```
Command:
Pick a point in the drawing.
_from Base point: _endp of <Offset>: @-1',-1'
```

Select the FROM OSNAP. Select the upper right corner of the desk.
Set the Offset as @-1', -1'.

This will place the lamp in a good position.

Save as Ex5-2.dwg

Exercise 3:
Equipping the Bathrooms

Drawing Name: Ex5-2.dwg
Estimated Time: 30 minutes

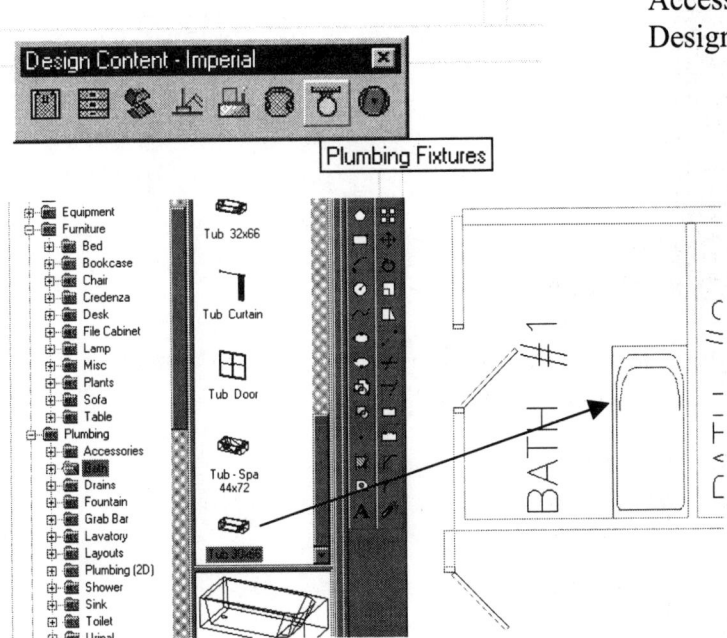

Access Plumbing Fixtures from the Design Center.

Go to the Plumbing/Bath Subdirectory.
Locate the Tub 30x66 and place it as shown. Be sure to orient the tub so the drain is closest to the wall. Otherwise, you'll have to add a wall to carry the faucet plumbing.

Note that the tub is automatically placed on the A-FLOR-PFIX layer.

TIP: The Space Planning process is not just to ensure that the rooms can hold the necessary equipment, but also requires the drafter to think about plumbing, wiring, and HVAC requirements based on where and how items are placed.

Lesson 5
Space Planning

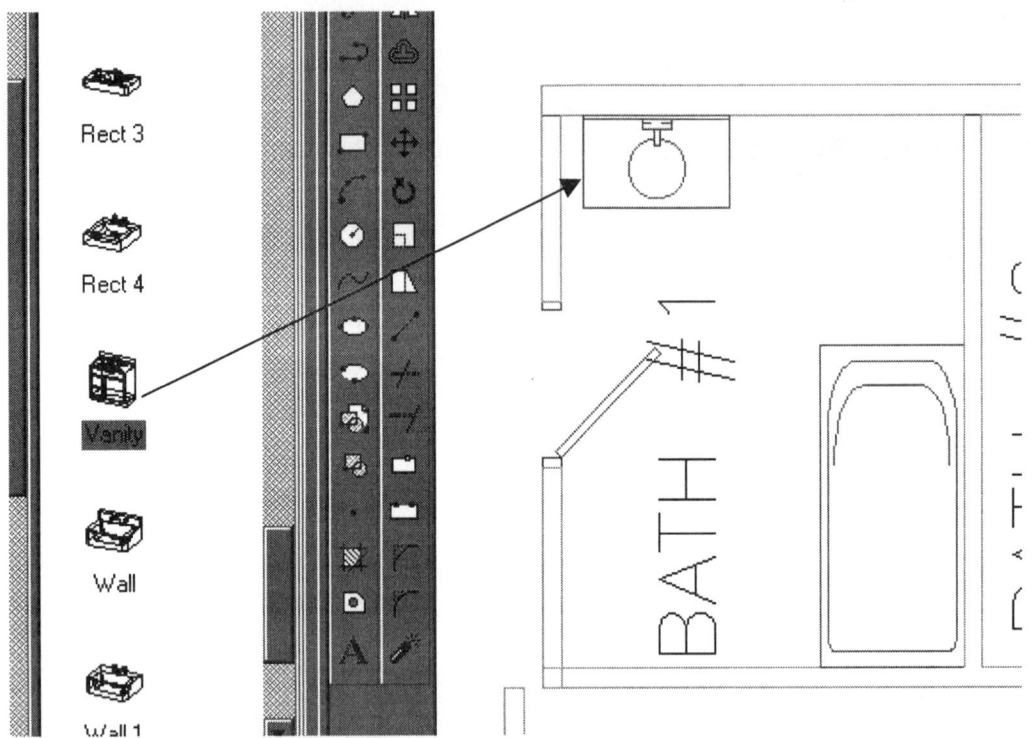

Go to the Plumbing/Lavatory subdirectory.
Locate the Vanity and place as shown.

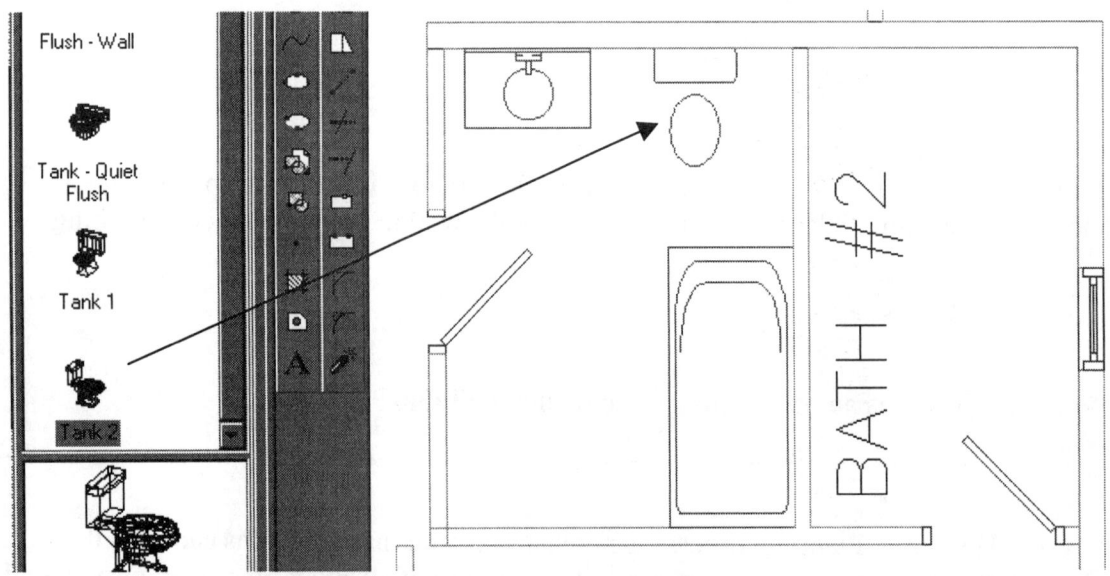

Go to the Plumbing/Toilet subdirectory.
Locate the Tank 2 and place as shown.

Lesson 5
Space Planning

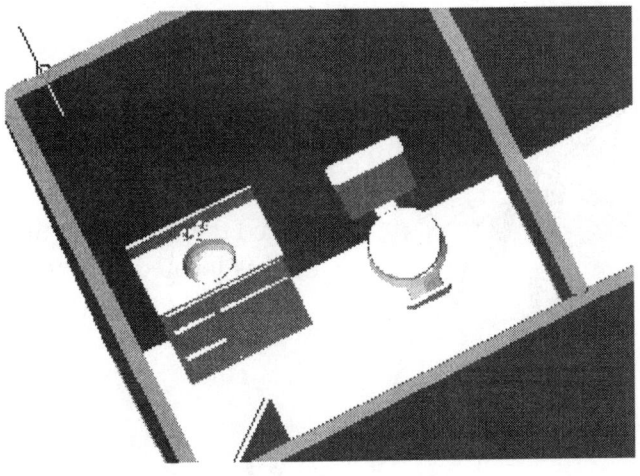

You can use 3D Orbit to verify placement of the items.

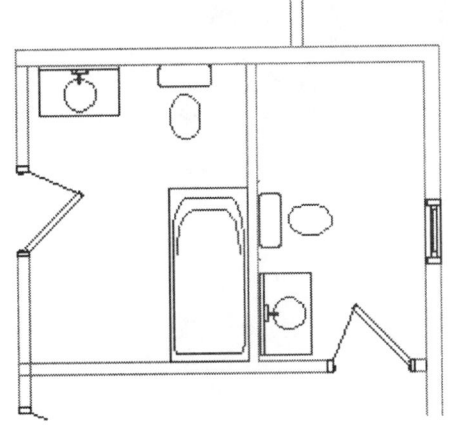

Use COPY, MOVE, ROTATE to copy and place the sink and the toilet to the master bathroom as shown.

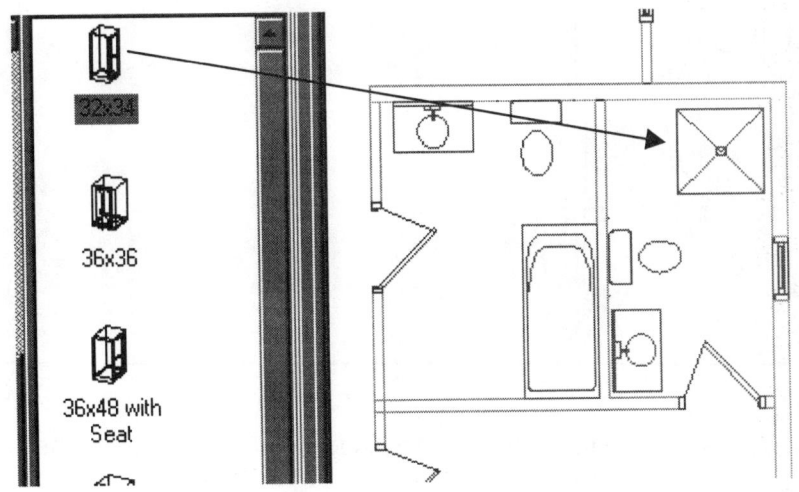

Go to the Plumbing/Shower subdirectory.
Locate the 32 x 34 Shower and place as shown.

Lesson 5
Space Planning

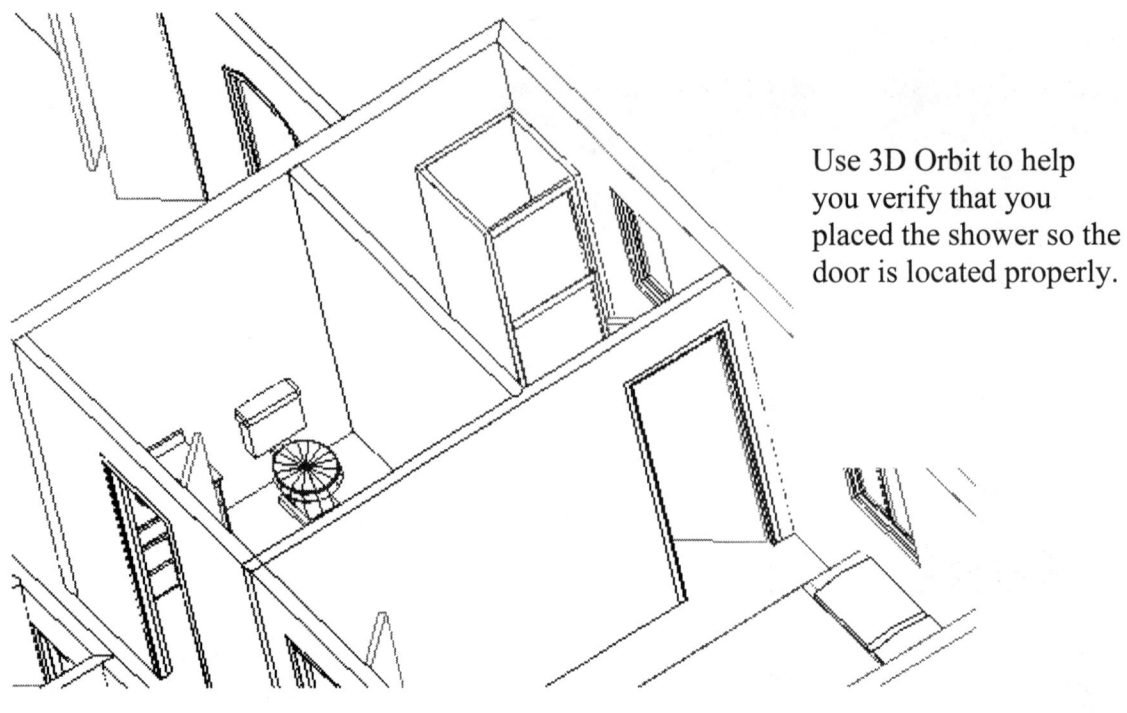

Use 3D Orbit to help you verify that you placed the shower so the door is located properly.

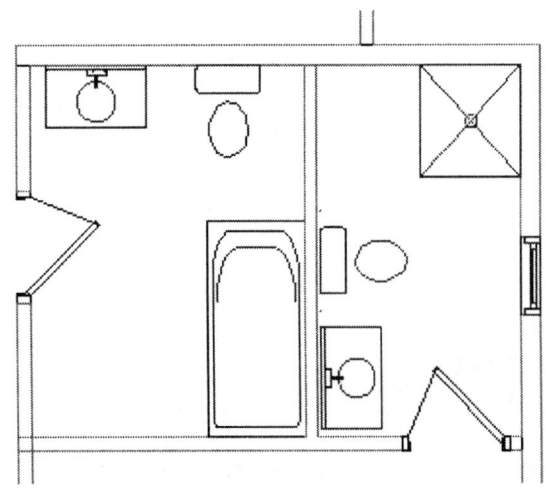

Our completed bathrooms.

Save as Ex5-3.dwg

TIP: As an additional exercise, place towel bars, soap dishes and other items in the bathrooms.

Lesson 5
Space Planning

Exercise 4:
Furnishing the Common Areas

Drawing Name: Ex5-4.dwg
Estimated Time: 30 minutes

Common areas are the Living Room, Dining Room, and Family Room.

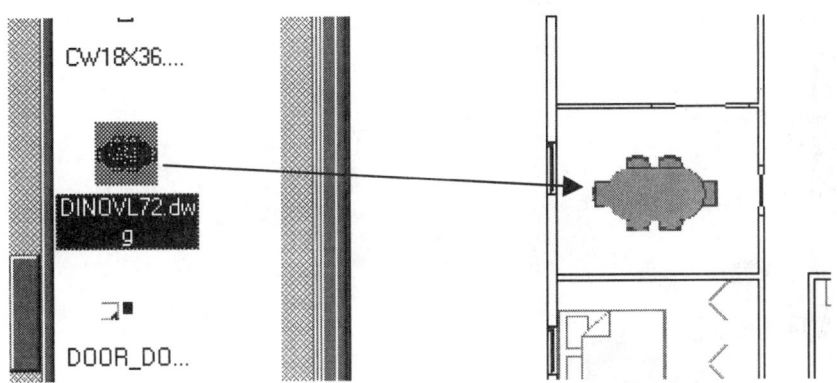

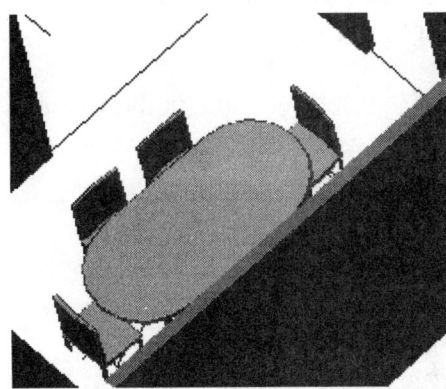

You can download a dining set from the www.schroff1.com website to be placed in the Dining Room. The file is called dinovl72.dwg. We placed this drawing in the Furniture/Misc. subfolder.

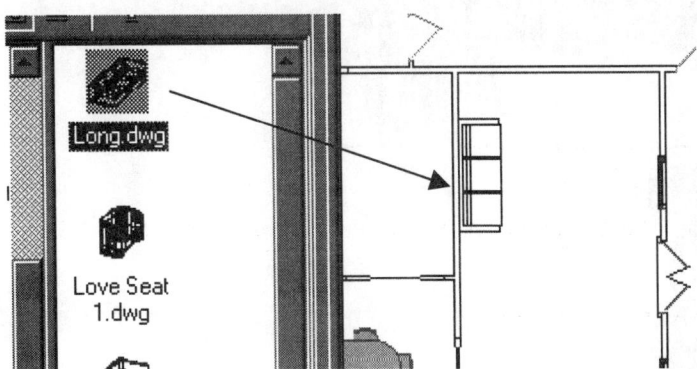

Sofas are located in the Design/Furniture/Sofa subdirectory.

Place a sofa in the living room as shown.

Lesson 5
Space Planning

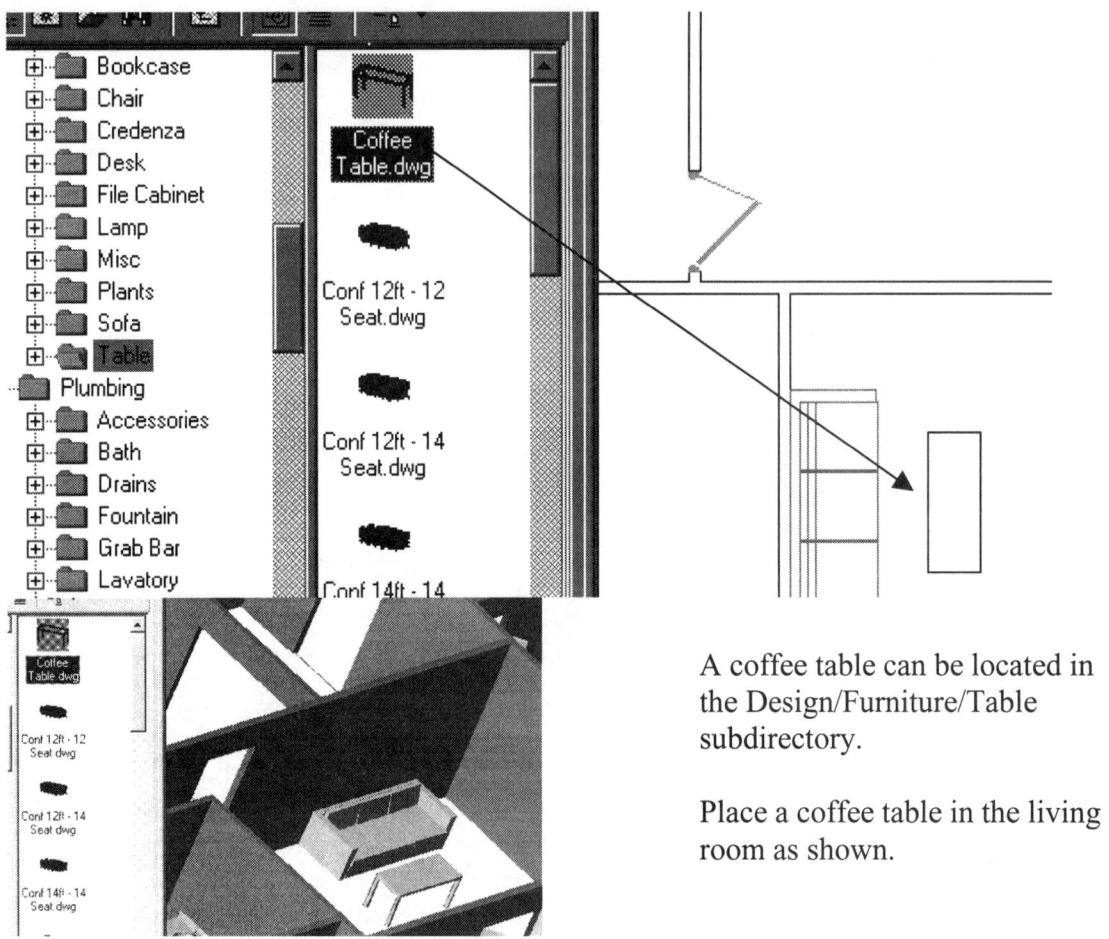

A coffee table can be located in the Design/Furniture/Table subdirectory.

Place a coffee table in the living room as shown.

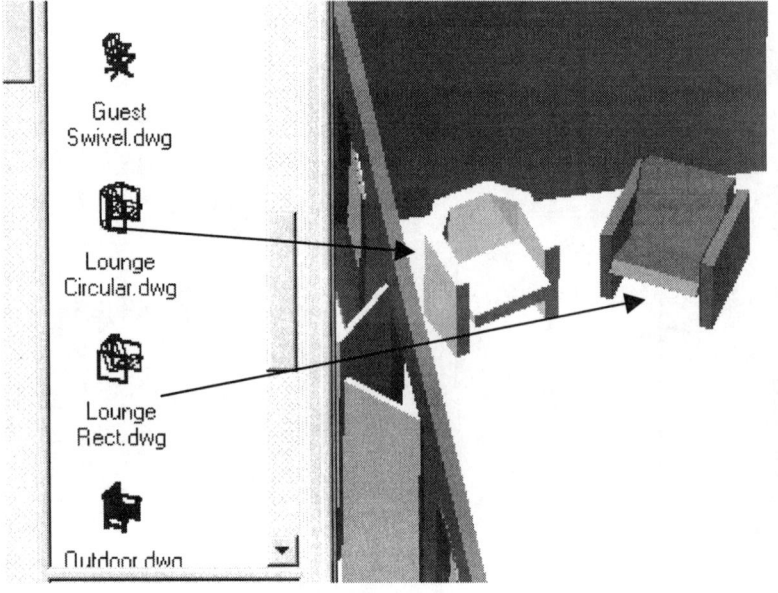

Add some lounge chairs to the Living Room area.

Lesson 5
Space Planning

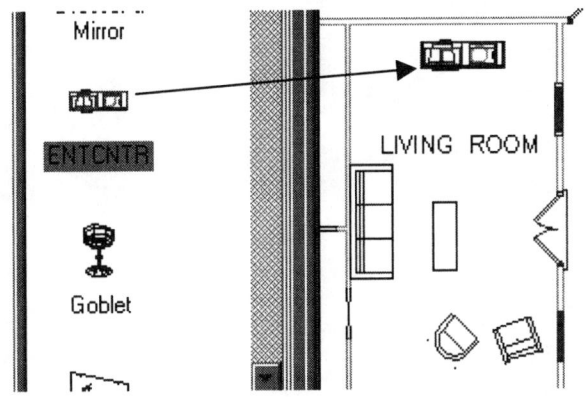

You can download a fully equipped entertainment center, complete with television, VCR, and stereo from the www.schroff1.com site. Place in the living room as shown.

We placed the ENTCNTR.dwg in the Furniture/Misc. subfolder.

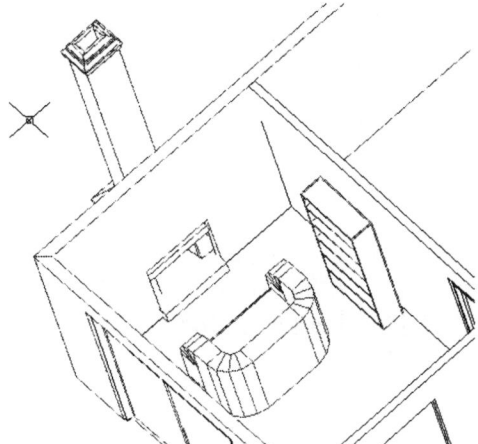

Place a loveseat and bookcase in the family room.

The loveseat is located in the Sofa subdirectory and the bookcase is located in the Bookcase subdirectory under Furniture.

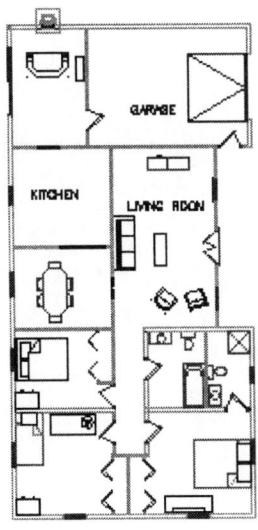

Our furnished house so far.
Save the file as Ex5-4.dwg.

Lesson 5
Space Planning

Exercise 5:
Adding to the Service Areas

Drawing Name: Ex5-4.dwg
Estimated Time: 30 minutes

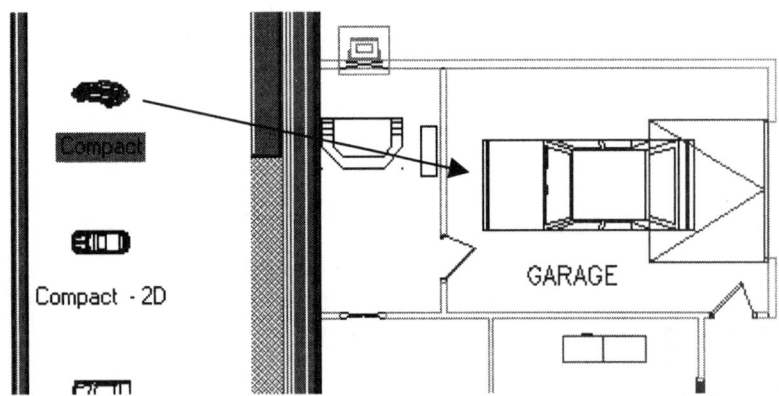

To place a vehicle in the garage, you can select a vehicle from the Design/Site/Vehicle subdirectory. There are additional vehicles available for download from the Schroff website.

Vehicles are automatically placed on the A-Pkng-Cars layer.

TIP: If the preview shown in the Design Center does not appear as a 3D object, then it is a 2D object. Don't select it for use in your model.

Lesson 5
Space Planning

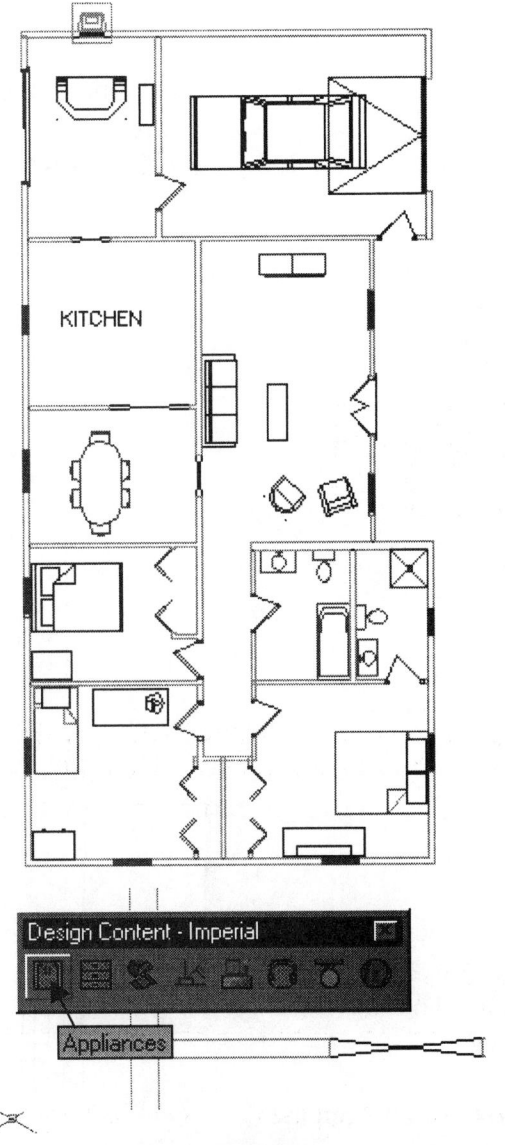

All that remains is the kitchen area.

Kitchen appliances, such as refrigerators, ovens and dishwashers can be located under Appliances on the Design Content toolbar.

You can download this kitchen sink from www.schroff1.com. Locate it centered underneath the kitchen window.

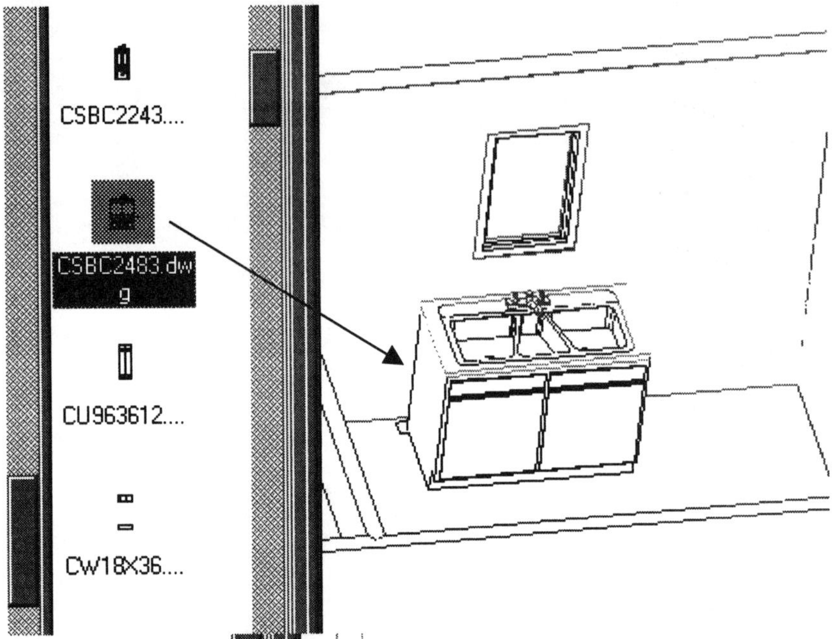

Lesson 5
Space Planning

The name of the file is csbc2483.dwg.

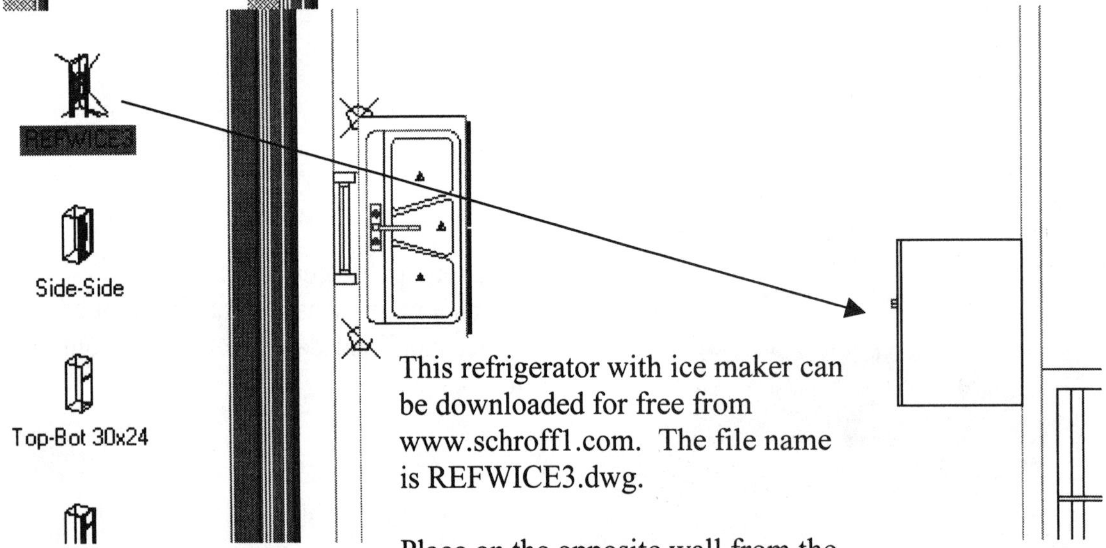

This refrigerator with ice maker can be downloaded for free from www.schroff1.com. The file name is REFWICE3.dwg.

Place on the opposite wall from the sink.

Remember that you need to place any drawings you download into the Content/Design subfolders in order for the blocks to place on the correct layers when you drag and drop into your drawing.

Appliances are placed on the A-Flor-Appl layer.

Lesson 5
Space Planning

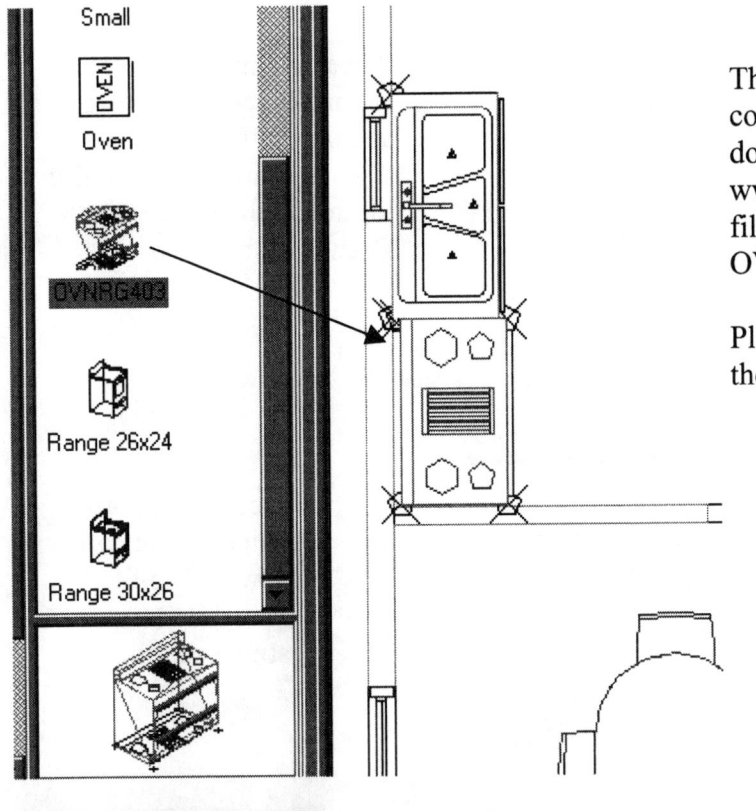

This oven/range combination can be downloaded for free from www.schroff1.com. The file name is OVNRG403.dwg.

Place on the same wall as the sink.

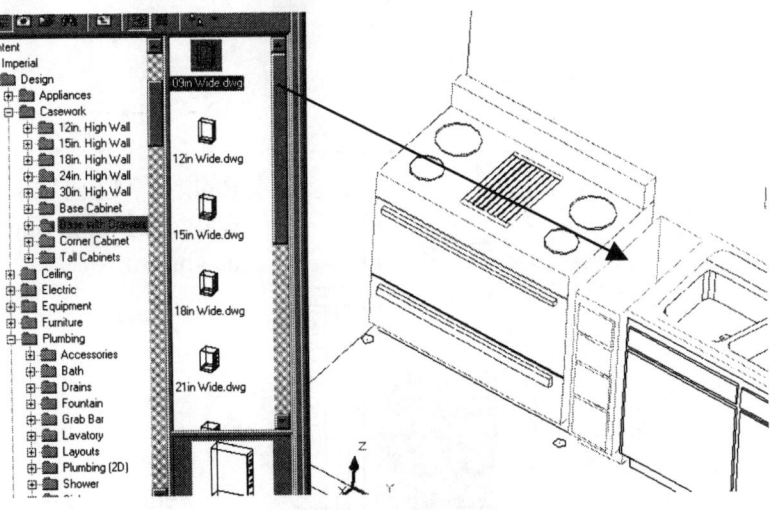

A base cabinet can be placed between the stove and the sink. You can use the file named 09in wide.dwg located in the casework\base with drawers subdirectory.

Lesson 5
Space Planning

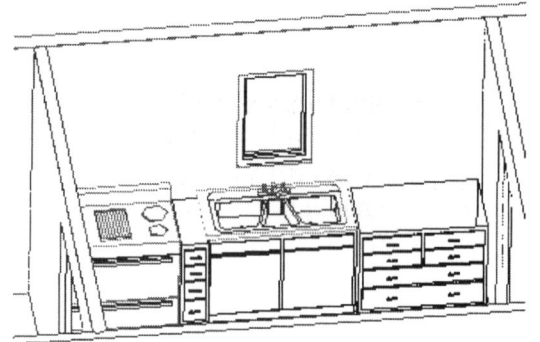

Place a 42 in wide base cabinet between the sink and the wall as shown. Use the 42in Wide.dwg file located in the Base with Drawers subdirectory.

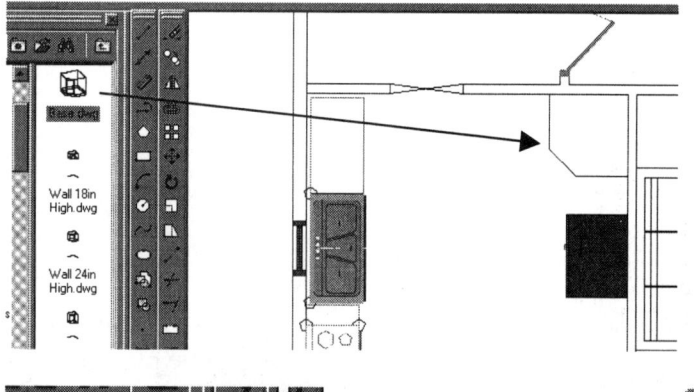

Place a base corner cabinet in the corner as shown. The file is located in the corner cabinet subdirectory.

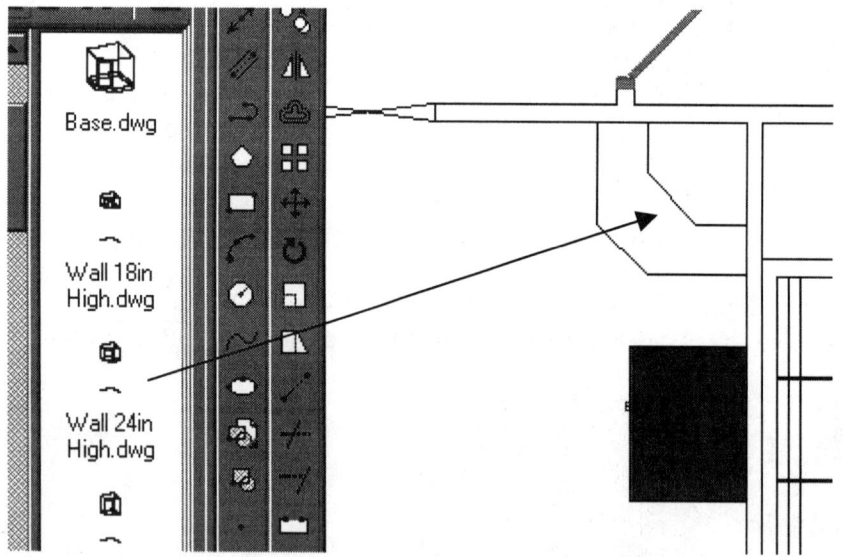

Place a 24 in High wall cabinet is placed as shown.

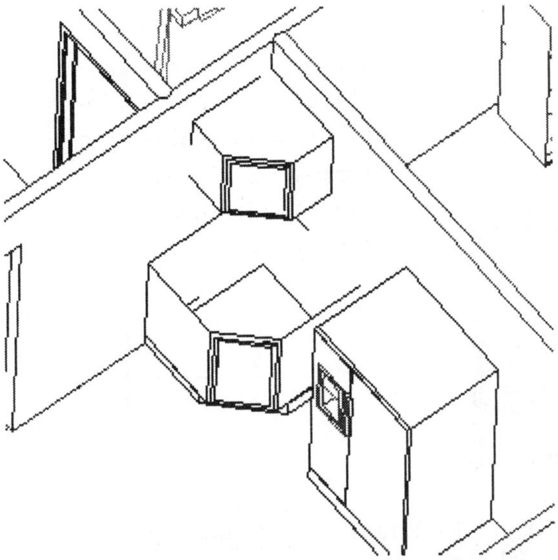

When you place the wall cabinet, it is automatically located at the correct height, so you do not need to worry about modifying the elevation.

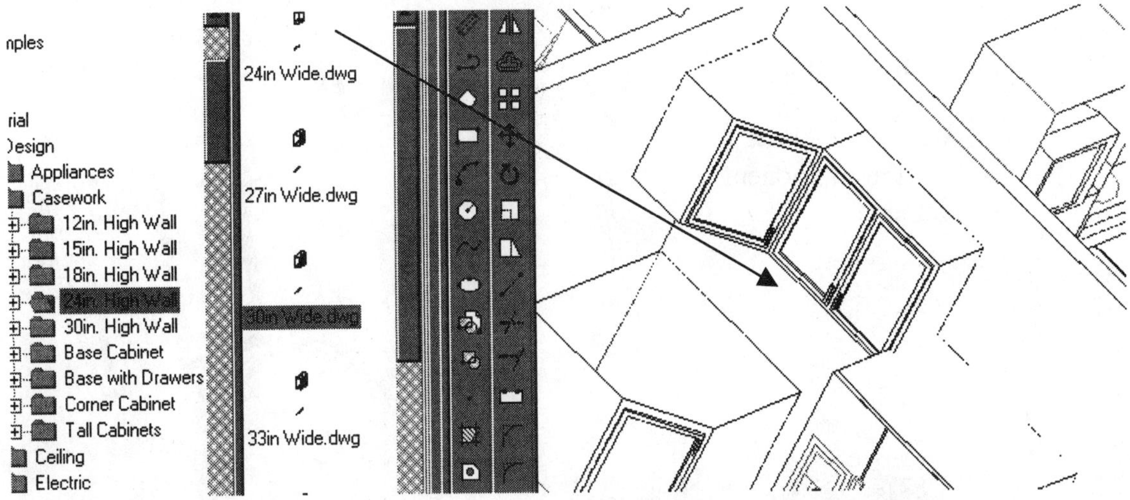

To place an additional wall cabinet, select a 30 in wide cabinet under the 24 in high subdirectory. You may have to shift the refrigerator over to eliminate any interference.

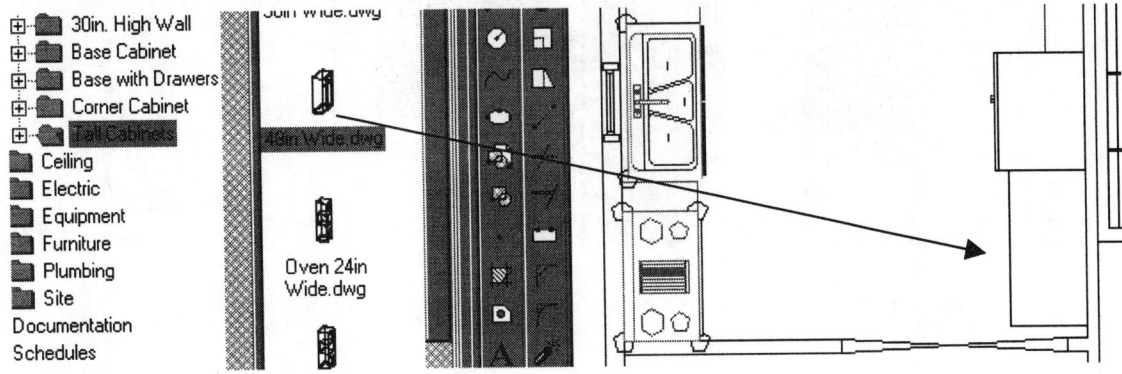

Place a 48-inch Wide Tall Cabinet next to the refrigerator as shown.

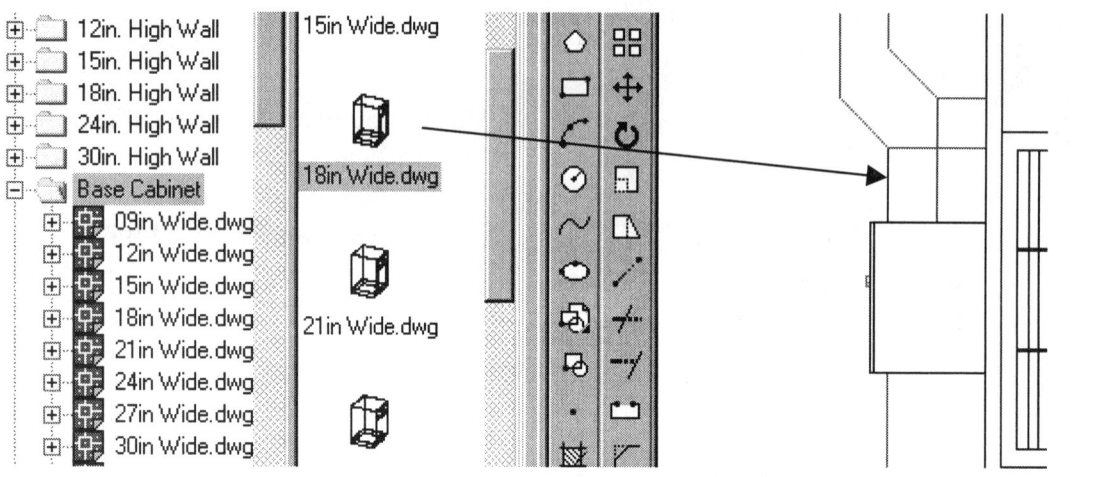

Place an 18in wide base cabinet between the refrigerator and the corner base cabinet.

Our kitchen so far.

Use 3D orbit to check to see how the components fit together.

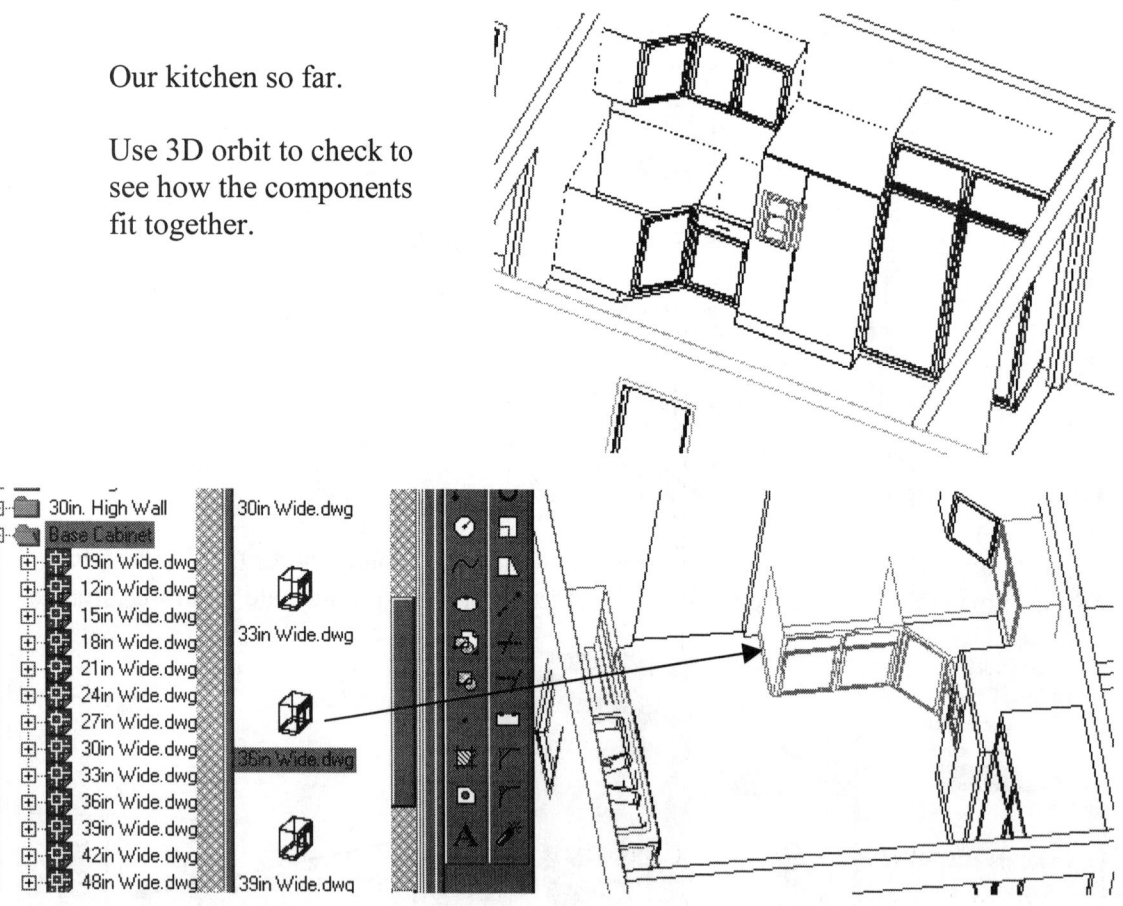

Place a 36inch wide base cabinet as shown.

Lesson 5
Space Planning

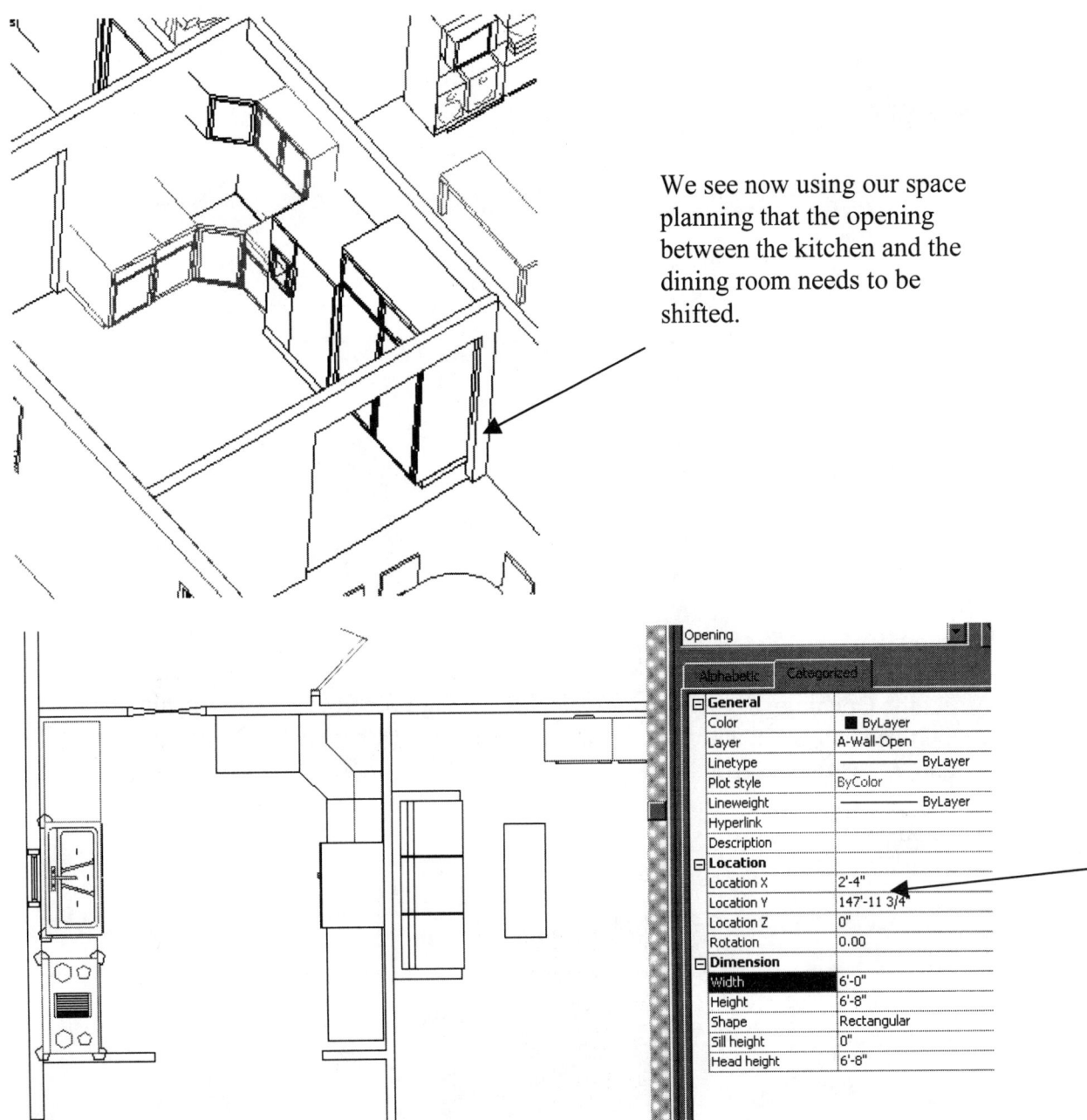

We see now using our space planning that the opening between the kitchen and the dining room needs to be shifted.

Method 1:

Highlight the opening. Right click and select Properties.
Change the value of "Location X" to 2'-4". The opening will dynamically shift so you can see if it looks OK.

Lesson 5
Space Planning

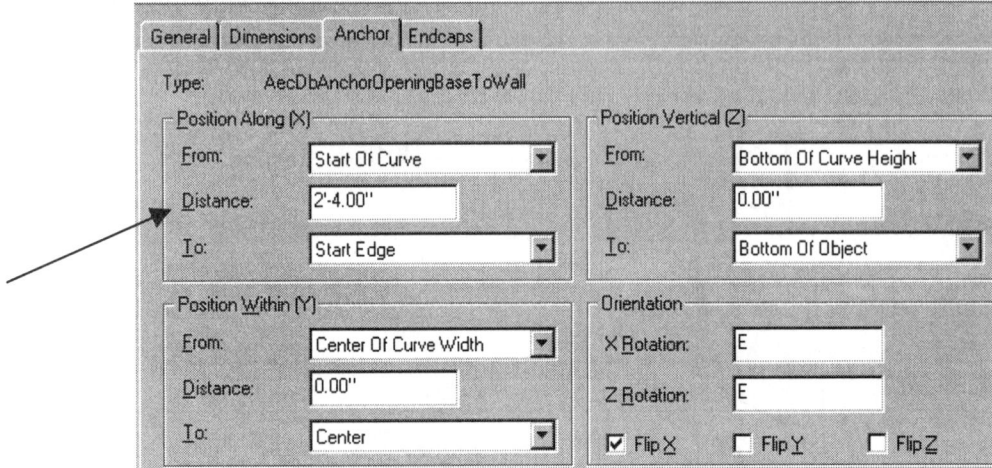

Method 2:

Select the Opening. Right click and select 'Opening Properties'. The Opening Properties Dialog will activate. Select the Anchor tab. Change the Distance Value to 2'-4" under the Position Along (X).

Method 3:

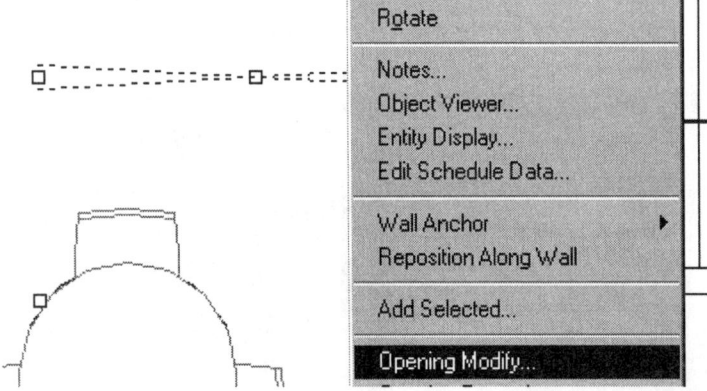

Select the opening, right click and select 'Opening Modify'.

Lesson 5
Space Planning

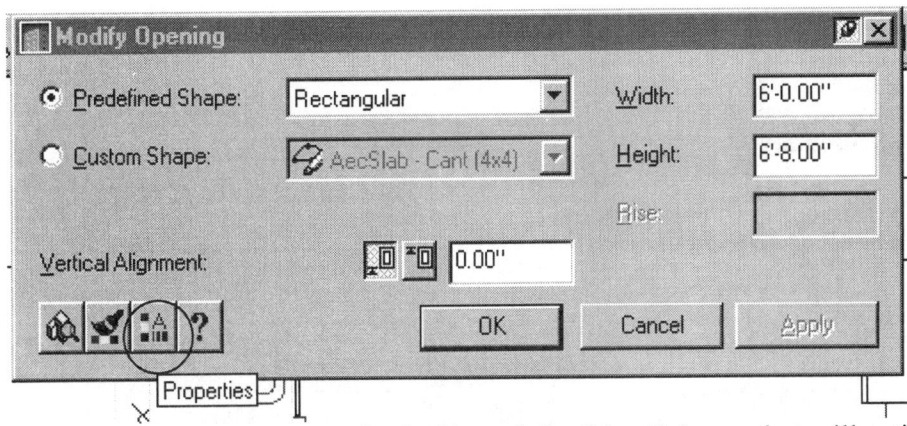

There is a Properties Button in the lower left of the dialogue that will activate the Opening Properties dialogue.

Save the file as Ex5-5.dwg.

Exercise 6:
Adding Decorator Touches

Drawing Name: Ex5-5.dwg
Estimated Time: 30 minutes

Add plants, lamps, pictures, rugs, etc. to add visual interest to your space plan. You can find existing content in the Content/Imperial/Design/Furniture subdirectory.

Once you have done with your decorating, create a Layer User group for all the appliances, furniture, casework, etc. that we added during this lesson.

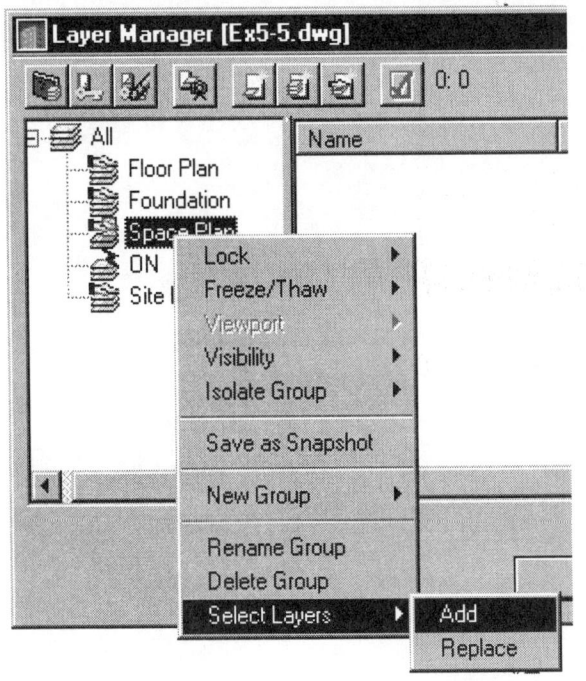

Call our Layer User Group 'Space Plan'.
Highlight, Right click, select 'Select Layers->Add'.

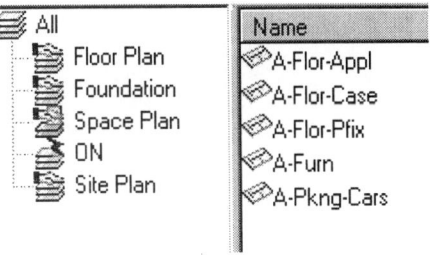

Lesson 5
Space Planning

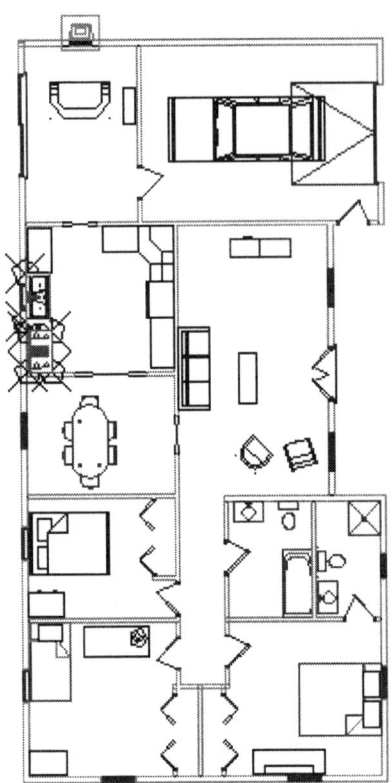

You may also want to create a SNAPSHOT for future use.

Save our drawing as Lesson 5.

Lesson 6
Roofs

Roofs can be created with single or double slopes, with or without gable ends, and with or without overhangs. Once you input all your roof settings, you simply pick the points around the perimeter of the building to define your roof outline. If you make an error, you can easily modify or redefine your roof.

You need to pick three points before the roof will begin to preview in your graphics window. There is no limit to the number of points to select to define the perimeter.

To create a gable roof, uncheck the gable box in the Roof dialog. Pick the two end points for the sloped portion of the roof. Turn on the Gable box. Pick the end point for the gable side. Turn off the Gable box. Pick the end point for the sloped side. Turn the Gable box on. Pick the viewport and then press ENTER. A Gable cannot be defined with more than three consecutive edges.

Roofs can be created using two methods: ROOFADD places a roof based on points selected or ROOFCONVERT which converts a closed polyline or closed walls to develop a roof.

Menu	Design->Roofs->Add Roof
Roofs Toolbar	
Command Line	ROOFADD

TIP: If you opt to use ROOFCONVERT and use existing closed walls, be sure that the walls are intersecting properly. If your walls are not properly cleaned up with each other, the roof conversion is unpredictable.

TIP: The Plate Height of a roof should be set equal to the Wall Height.

Lesson 6
Roofs

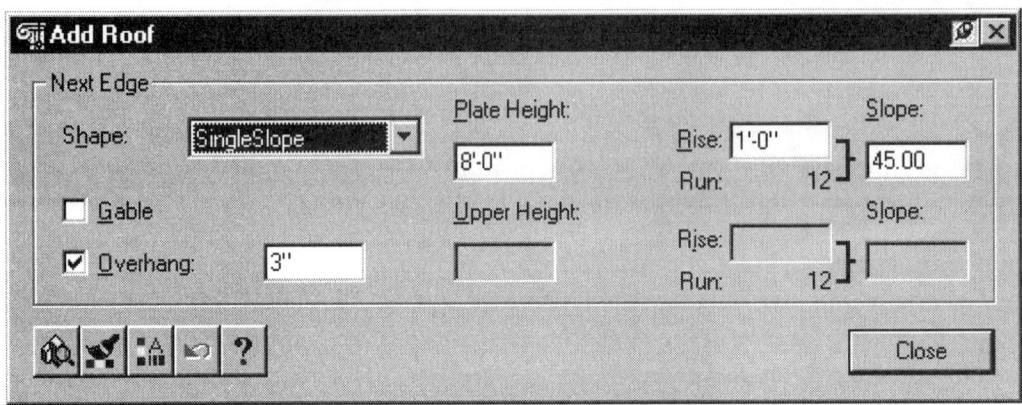

Shape – Select the Shape option on the command line by typing 'S'.	Single Slope – Extends a roof plane at an angle from the Plate Height.	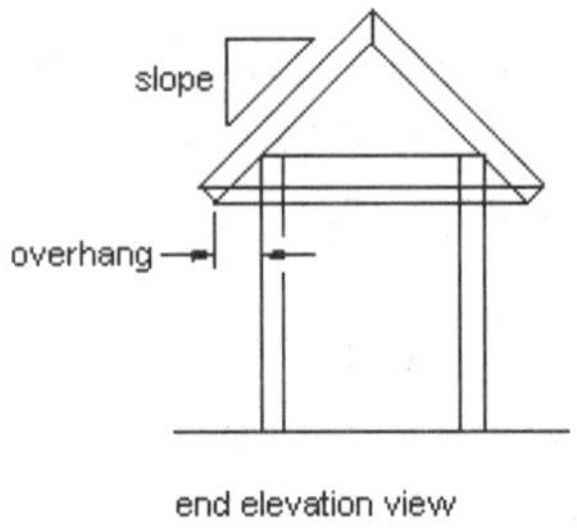 ![slope diagram with overhang, end elevation view]
	Double Slope – Includes a single slope and adds another slope, which begins at the intersection of the first slope and the height specified for the first slope.	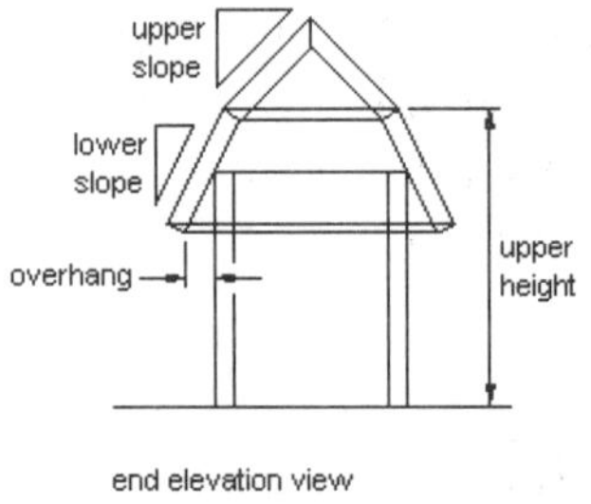 ![upper slope, lower slope, overhang, upper height, end elevation view]
Gable – Select the Gable option on the command line by typing 'G'.	If this is enabled, turns off the slope of the roof place. To create a gable edge, select Gable prior to identifying the first corner of the gable end. Turn off gable to continue to create the roof.	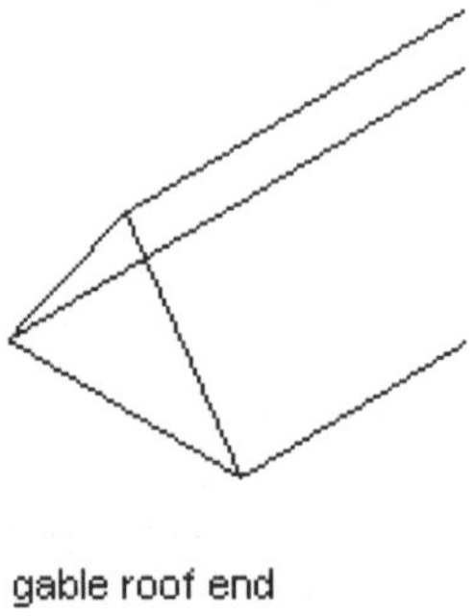 ![gable roof end]
Plate Height – Set the Plate Height on the command line by typing 'PH'.	Specify the top plate from which the roof plane is projected. The height is relative to the XY plane with a Z coordinate of 0.	

Rise– Set the Rise on the command line by typing 'PR'.	Sets the angle of the roof based on a run value of 12	A rise value of 5 creates a 5/12 roof, which forms a slope angle of 22.62 degrees.
Slope – Set the Slope on the command line by typing 'PS'.	Angle of the roof rise from the horizontal.	If slope angles are entered, then the rise will automatically be calculated.
Upper Height - Set the Upper Height on the command line by typing 'UH'.	This is only available if a Double Slope roof is being created. This is the height where the second slope will start.	
Rise (upper) - Set the Upper Rise on the command line by typing 'UR'.	This is only available if a Double Slope roof is being created. This is the slope angle for the second slope.	A rise value of 5 creates a 5/12 roof, which forms a slope angle of 22.62 degrees.
Slope (upper) - Set the Upper Slope on the command line by typing 'US'.	This is only available if a Double Slope roof is being created. Defines the slope angle for the second slope.	If an upper rise value is set, this is automatically calculated.
Overhang- To enable on the command line, type'O'. To set the value of the Overhang, type 'V'.	If enabled, extends the roofline down from the plate height by the value set.	
![icon]	The Floating Viewer opens a viewer window displaying a preview of the roof.	
![icon]	The match button allows you to select an existing roof to match its properties.	
![icon]	The properties button opens the Roof Properties dialog.	
![icon]	The Undo button allows you to undo the last roof operation. You can step back as many operations as you like up to the start.	
![icon]	Opens the Roof Help file.	

TIP: You can create a gable on a roof by gripping any ridgeline point and stretching it past the roof edge. You can not make a gable into a hip using grips.

Lesson 6
Roofs

Exercise 1:
Creating a Roof using Existing Walls

Drawing Name: Lesson 5.dwg saved from the previous lesson.
Estimated Time: 15 minutes

This exercise reinforces the following skills:

- Convert to Roof

MENU	Design->Roofs->Convert to Roof
Roofs Toolbar	
Command Line	RoofConvert

Switch to the Work-3D tab so you can watch how the roof is created.
Activate the right viewport, which shows the top view of your model.
Freeze the equipment and furnishings layers to save screen regen time.

Select the Convert to Roof command from the toolbar or menu.

Select all the exterior walls in the drawing. You can select using a window for the entire building. (This selection method is applicable for this command – do not assume it will work for all commands.)

Press Enter to continue.

Enter NO to retain the layout geometry.

6-4

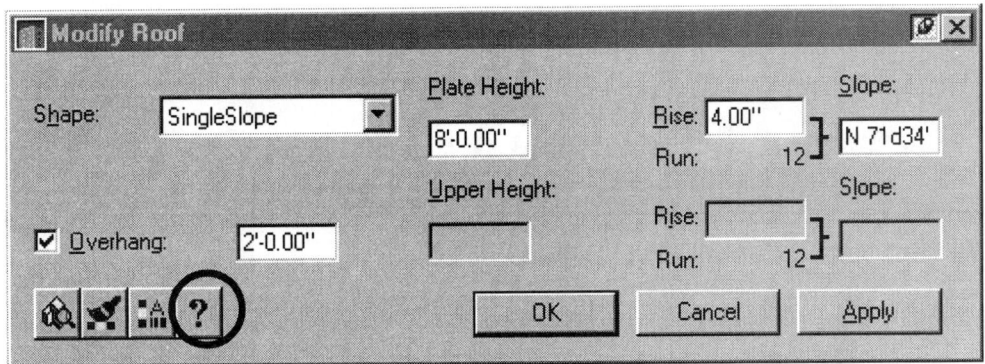

Set the Shape to SingleSlope.
Set the Plate Height to 8'.
Set the Overhang to 2'-0". Set the Rise to 4".
Select the Properties button.

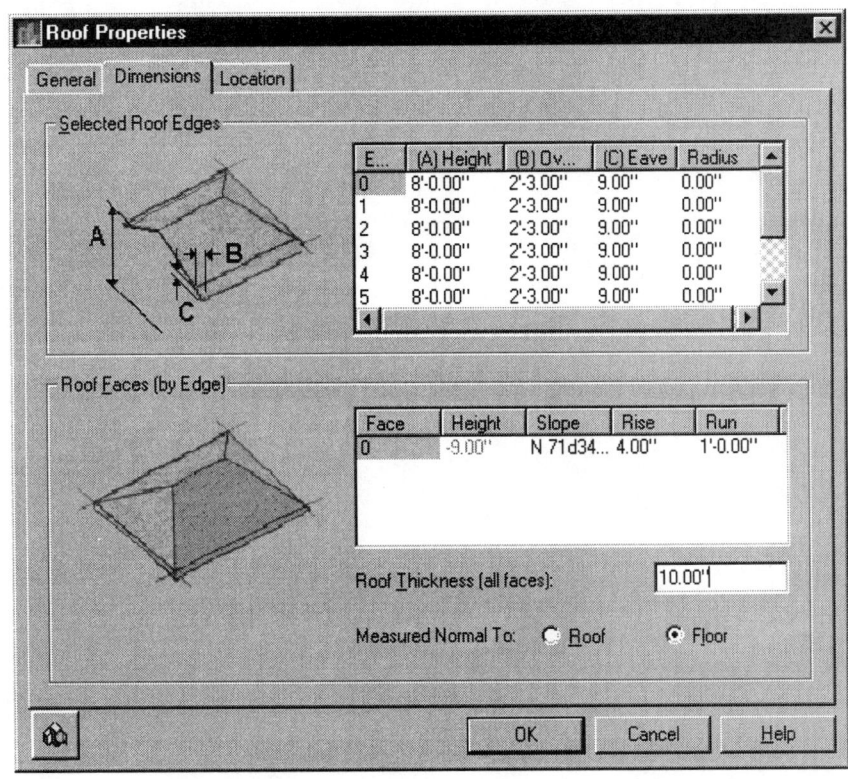

Select the Dimensions tab. Change the roof thickness to 10". This value corresponds to the framing material thickness. Change the Measured Normal toggle to Floor. ADT defaults to having the roof edge perpendicular to the roof, which on sloped roofs doesn't make sense. You can't hang a gutter on a tilted fascia.

Press 'OK'.

Lesson 6
Roofs

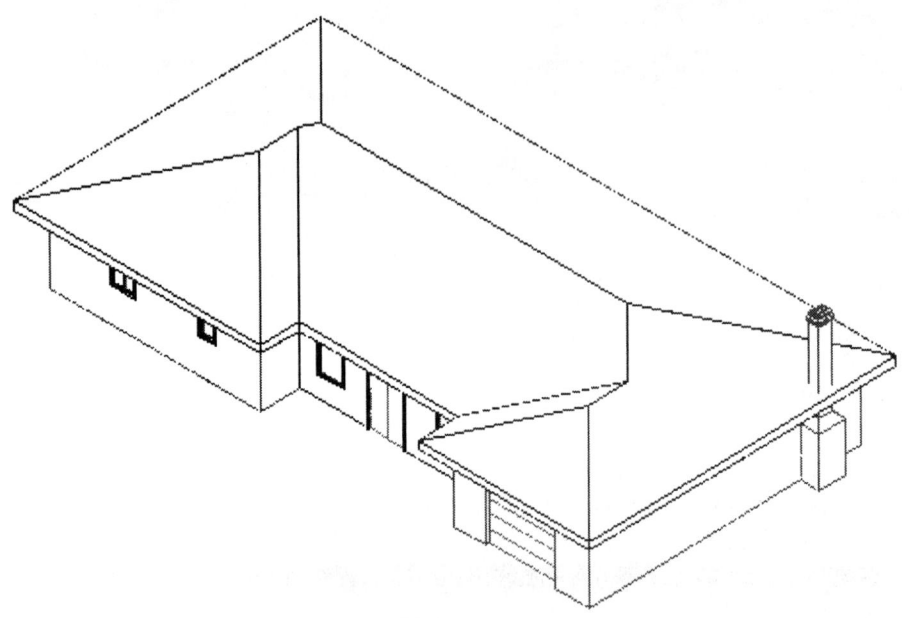

Save as Ex6-1.dwg.

Exercise 2:
Roof Slabs

Drawing Name: Lesson 6-1 dwg saved from the previous lesson.
Estimated Time: 30 minutes

This exercise reinforces the following skills:

- Convert to Roof
- Roof Slab Tools

Our roof works, but strictly speaking we can make it better with a couple of changes around the chimney. ADT allows us to cut holes through roofs (to allow for vents, chimneys and skylights), and to add other faces or subsidiary roofs such as dormers.

In order to edit a roof you have to convert it to Roof Slabs.

Menu	Design->Roofs->Convert to Roof Slabs
Toolbar	
Command Line	ROOFSLABCONVERT

Select the roof and enter.
At the "Erase layout geometry?" prompt type y, then enter.
At the next prompt hit enter to accept the Standard roof slab type.
The roof will change appearance slightly. It now consists of individual slabs on the A-Roof-Slab layer.

NOTE: The Roof Slab Hole command is on the Roof Slab Tools toolbar which is different from the Roof toolbar.

Activate the Roof Slab Hole command:

Menu	Design->Roof Slab Tools-> Roof Slab Hole
Toolbar	
Command Line	ROOFSLABHOLE

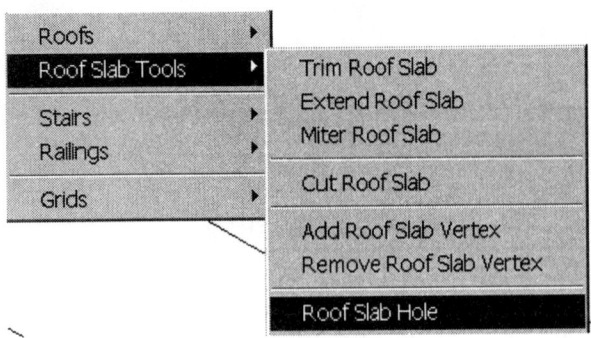

At the [Add/Remove] <Add>: prompt, hit enter to Add a new hole.
Select the roof face (slab) with the chimney.
At the next prompt pick the chimney. Be careful where you pick so you don't pick walls or furniture, which you can't see below the roof.
At the "Erase Layout Geometry?" prompt hit enter to accept the default No.
At the "Hole location" prompt type O for Outside and hit enter.

The roof slab will change appearance slightly. Now the fascia is angled and the slab appears to have dropped.

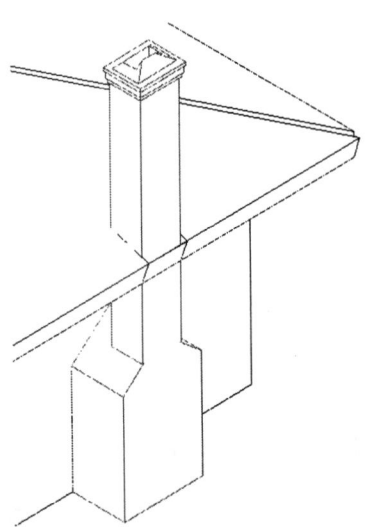

If you have difficulty selecting the chimney, switch to a top view. Draw a rectangle as an outline for the chimney and select that when prompted for a polyline. The ROOFSLABHOLE command will automatically project the hole properly.

Pick the roof slab, right slick and pick Roof Slab Properties.

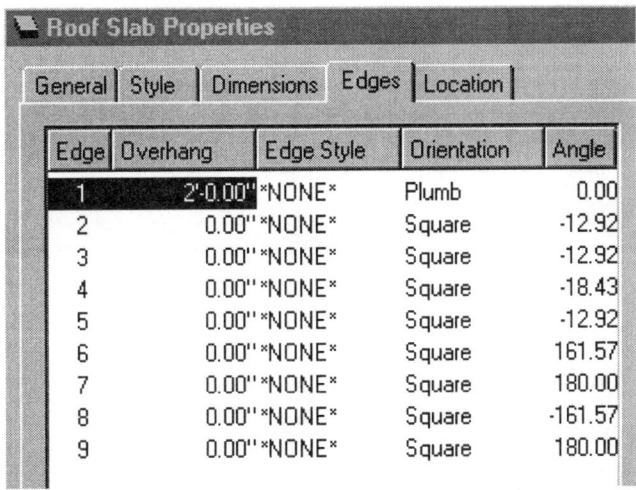

On the Edges tab note that the slab shows that one edge has a 2' overhang (as we specified) but no edge style.

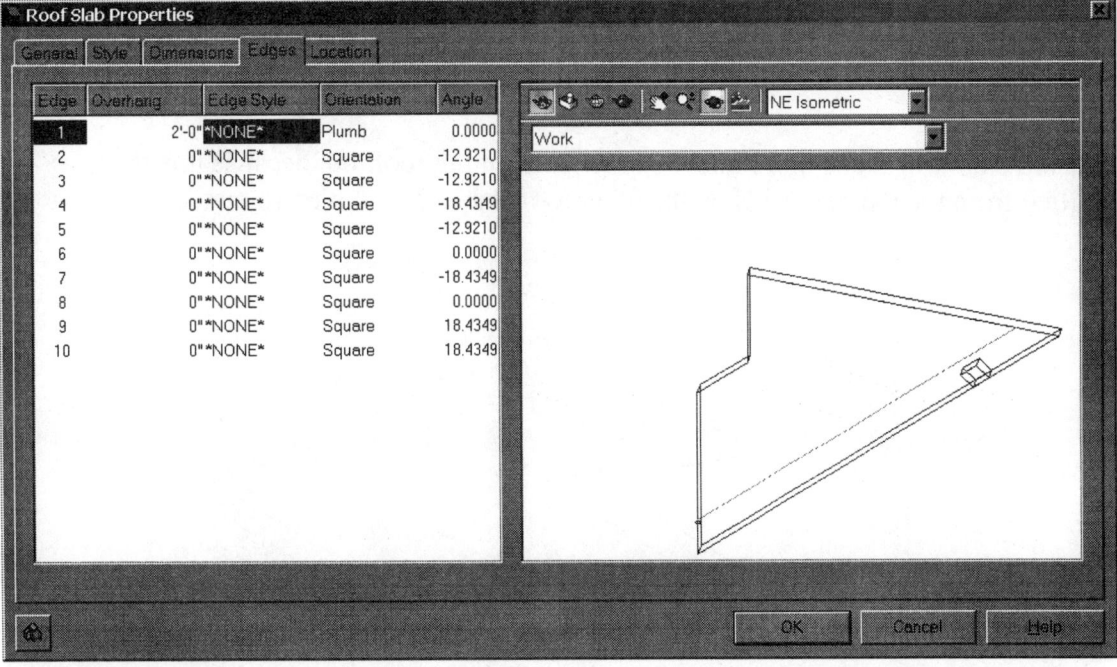

Slabs have edges, and edges can be defined by styles. In this case changing the style of the edge alone will not make the roof slab look correct, so we'll change the style of the slab. On the Style tab pick 10 – 1x8 Fascia, then pick OK.

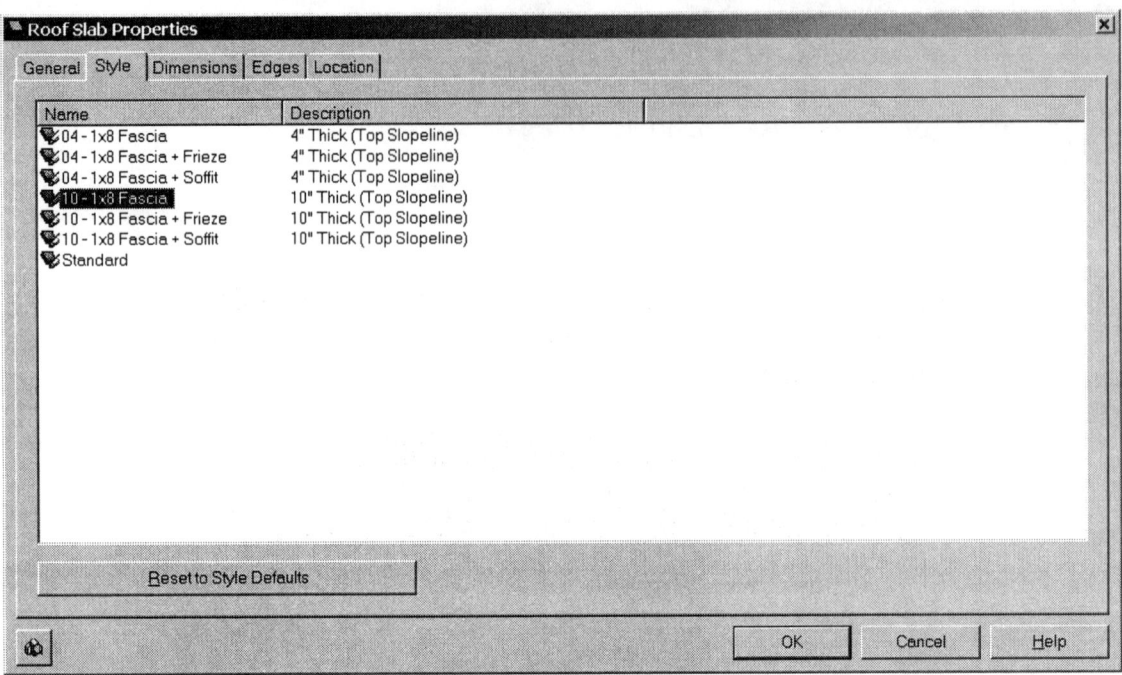

The roof slab now looks more like the other parts of the roof, but because this style is justified from the top, it sits below the other roof slabs. Move it up to match.

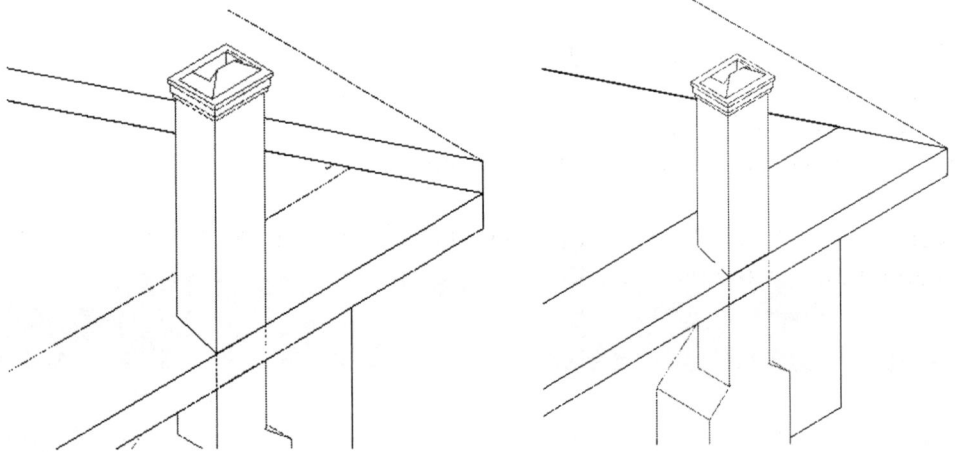

Turn off the layer of the chimney and you will see we have indeed cut a hole through the roof in the correct spot.

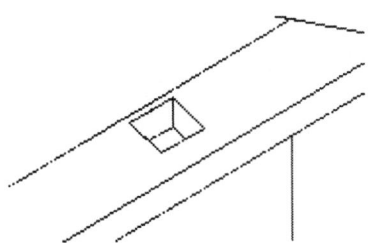

Our roof as it exists has one problem at the chimney. The "uphill" side of the chimney will catch water and snow. It needs a tiny little roof gable, called a cricket, to divert water away from the chimney.

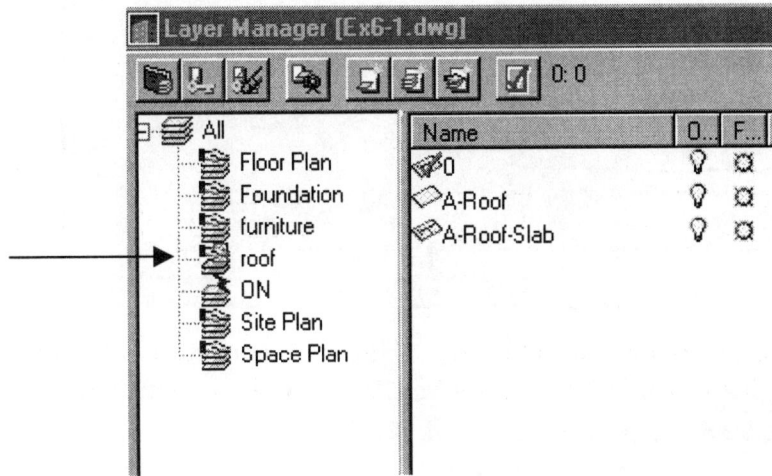

Create a Layer User Group called Roof.
Place the layers 0, A-Roof, and A-Roof-Slab in that group.
Make layer 0 the current layer.
Isolate the layer group so you only see those layers.

Draw a rectangle 1'6" wide by 36" long, using the SE corner of the roof hole as the start point and going away from the hole.

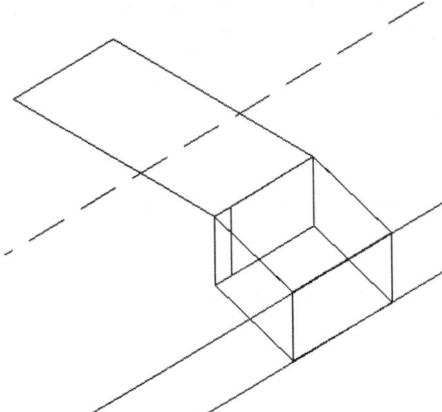

This view is wire frame with the roof layers and layer 0 isolated.

Lesson 6
Roofs

Activate the Convert to Roof tool you used before to create the main roof.

Pick the polyline (you can type "l" for "last" at the Select Objects prompt) and hit enter.
Hit enter to accept No for "Erase layout geometry?".
In the Modify Roof dialogue box,
Accept the plate height.
Make the rise 4"
Uncheck overhang to deactivate it.

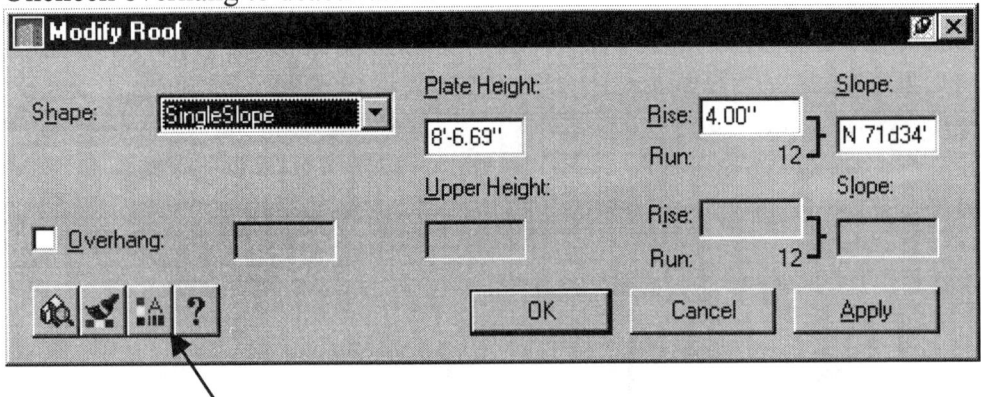

Pick the Properties button.

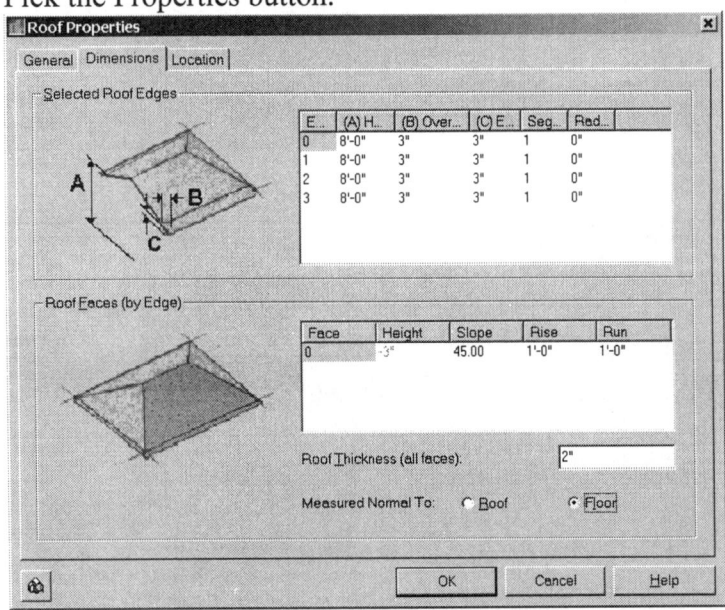

On the Roof Properties dialogue make the roof thickness for the cricket 2" and set its edges normal to the Floor.
Pick OK
Pick OK.

6-12

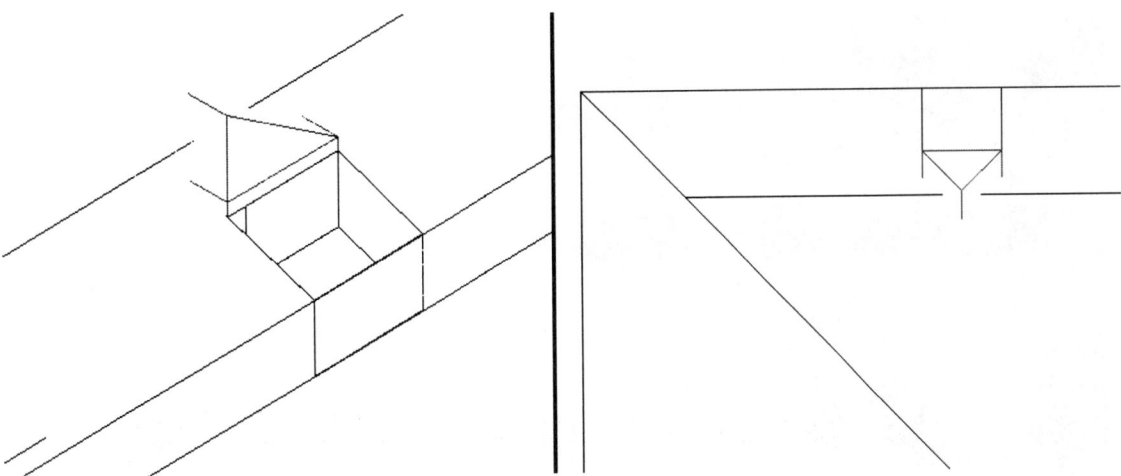

These are hidden line views.

The new roof is hipped, as was the first. Grip edit its ridge to make the exposed face a gable.

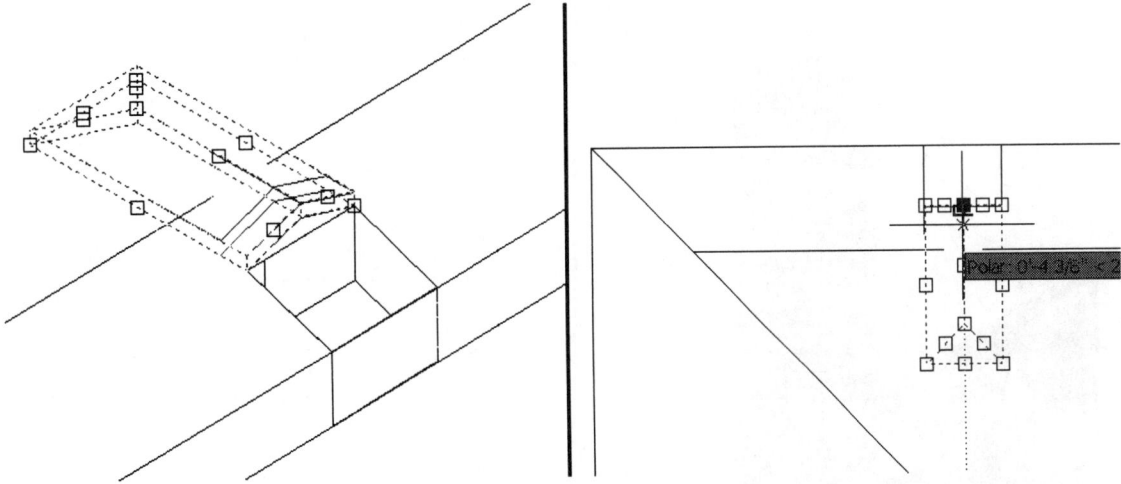

Move the new roof down 2" inches (@0,0,-2) so it lines up with the main roof.

Lesson 6
Roofs

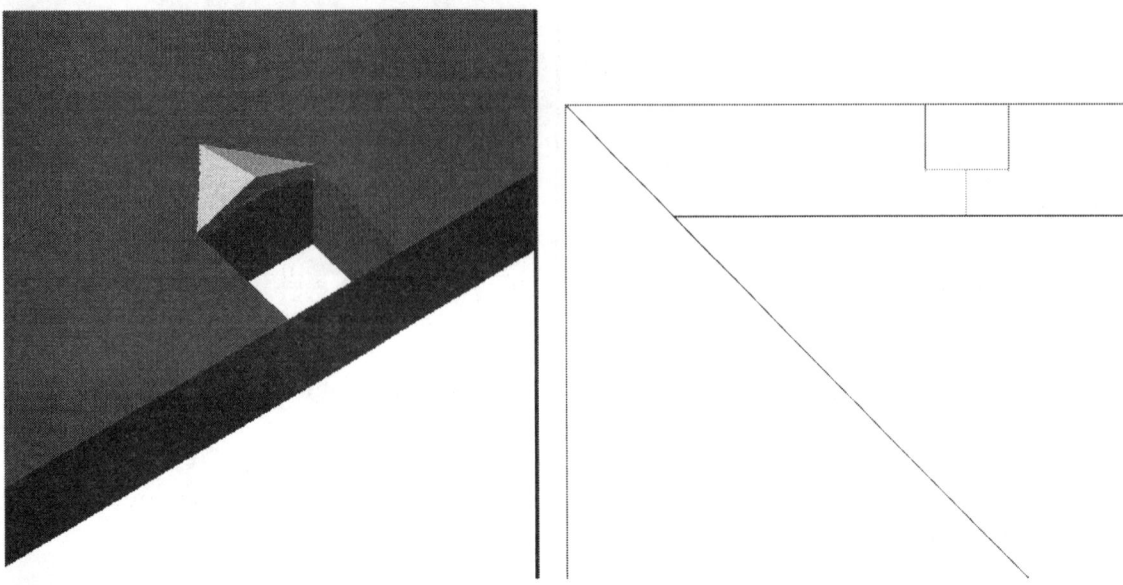

Turn on any layers you may have turned off for visibility, including the chimney. Here's a view from the other direction

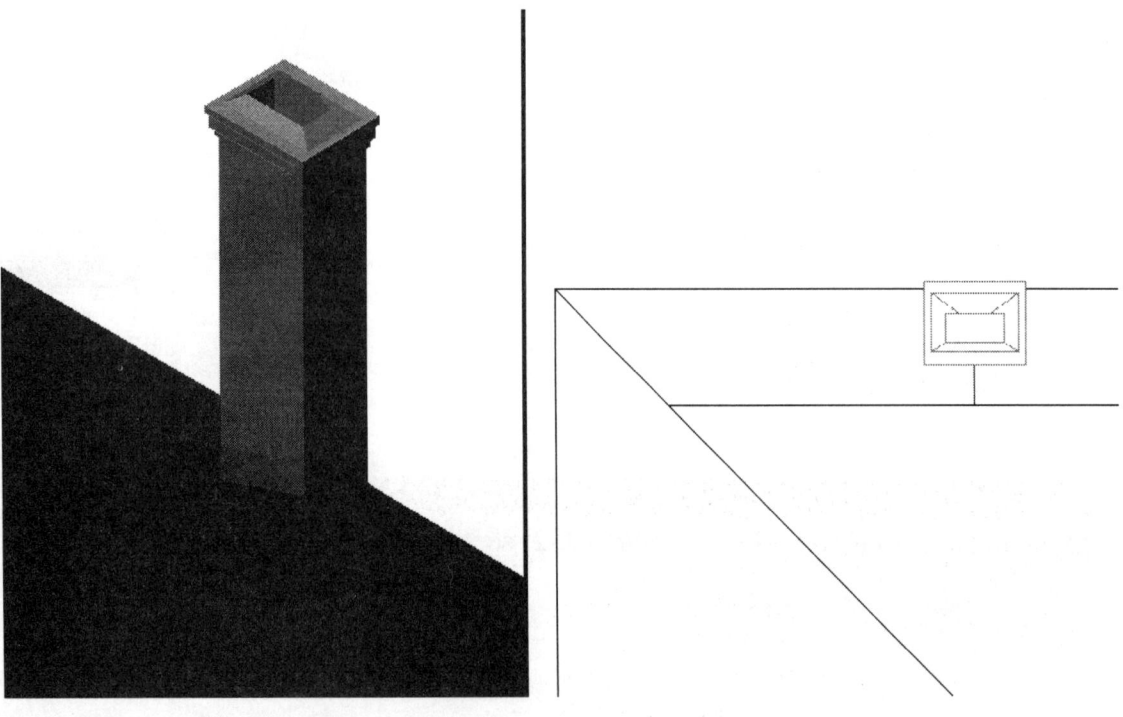

Save your file as Lesson 6.dwg.

QUIZ 3

True or False

1. Custom content can be located in any subdirectory and still function properly.
2. The sole purpose of the Space Planning process is to arrange furniture in a floor plan.
3. The Design Center only has 3D objects stored in the Content area because ADT is a strictly 3D software.
4. Appliances are automatically placed on the APPLIANCE layer.
5. When you place a wall cabinet, it is automatically placed at the specified height.

Multiple Choice

6. A residential structure is divided into:

    A. Four basic areas
    B. Three basic areas
    C. Two basic areas
    D. One basic area

7. Kitchen cabinets are located in the _____ subfolder.

    A. Casework
    B. Cabinets
    C. Bookcases
    D. Furniture

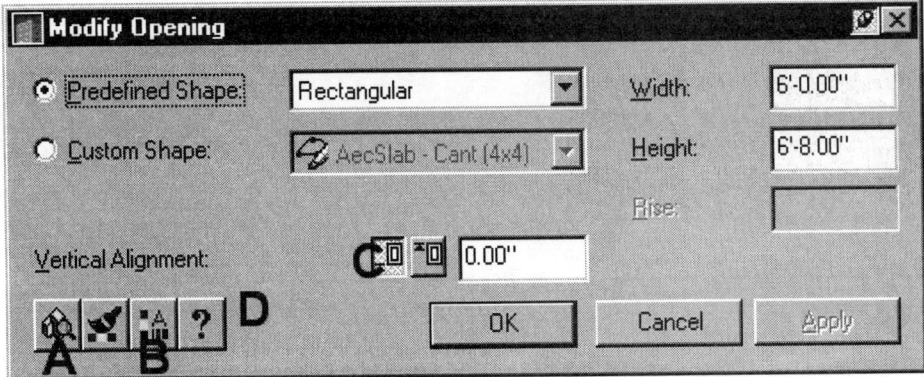

8. To modify the properties of an AEC object, select here.

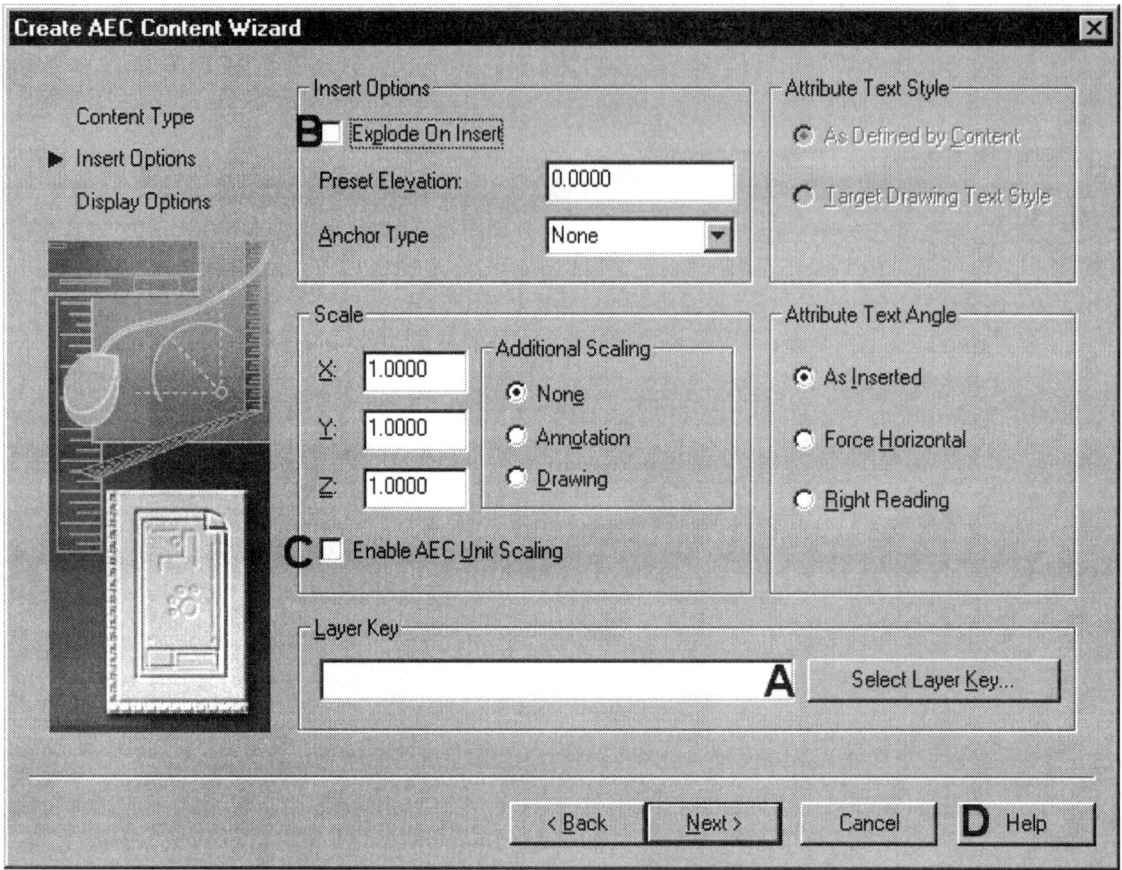

9. You want to set up a drawing so that it will be inserted onto the correct layer when dragged from the Design Center. You use 'Create AEC Content' from the Desktop Menu. Select the button to use to set the correct layer setting.

10. Select the area type that is NOT part of a private residence:

A. Bedrooms
B. Common Areas
C. Service Areas
D. Public Areas

ANSWERS:

1) F; 2) F; 3) F; 4) F; 5) T; 6) B; 7) A; 8) B; 9) A; 10) D

Lesson 7
Structural Members

Architectural documentation for residential construction will always include plans (top views) of each floor of a building, showing features and structural information of the floor platform itself. Walls are located on the plan but not shown in structural detail.

Wall sections and details are used to show:
- the elements within walls (exterior siding, sheathing, block, brick or wood studs, insulation, air cavities, interior sheathing, trim)
- how walls relate to floors, ceilings, roofs, eaves,
- openings within the walls (doors/windows with their associated sills and headers)
- how walls relate to openings in floors (stairs).

Stick-framed (stud) walls usually have their framing patterns determined by the carpenters on site. Once window and door openings are located on the plan, and stud spacing is specified by the designer (or the local building code), the specific arrangement of vertical members is usually left to the fabricators and not drafted, except where specific structural details require explanation.

The structural members in framed floors that have to hold themselves and/or other walls and floors up are usually drafted as framing plans. Designers must specify the size and spacing of joists or trusses, beams and columns. Plans show the orientation and relation of members, locate openings through the floor and show support information for openings and other specific conditions.

In the next exercises we shall create a floor framing plan. Since the ground floor of our one-story lesson house has already been defined as a concrete slab, we'll assume that the ground level slopes down at the rear of the house and create a wood deck at the sliding door to the family room. The deck will need a railing for safety.

Autodesk Architectural Desktop includes a Structural Member Catalog that allows you to easily access industry-standard structural shapes. To create most standard column, brace, and beam styles, you can access the Structural Member Catalog, select a structural member shape, and create a style that contains the shape that you selected. The shape, similar to an AEC profile, is a 2D cross-section of a structural member. When you create a structural member with a style that you created from the Structural Member Catalog, you define the path to extrude the shape along.

You can create your own structural shapes that you can add to existing structural members, or use to create new structural members. The design rules in a structural member style allow you to add these custom shapes to a structural member, as well as create custom structural members from more than one shape.

Lesson 7
Structural Members

All the columns, braces, and beams that you create are sub-types of a single Structural Member object type. The styles that you create for columns, braces, and beams have the same Structural Member Styles style type as well. When you change the display or style of a structural member, use the Structural Member object in the Display Manager and the Structural Member Styles style type in the Style Manager.

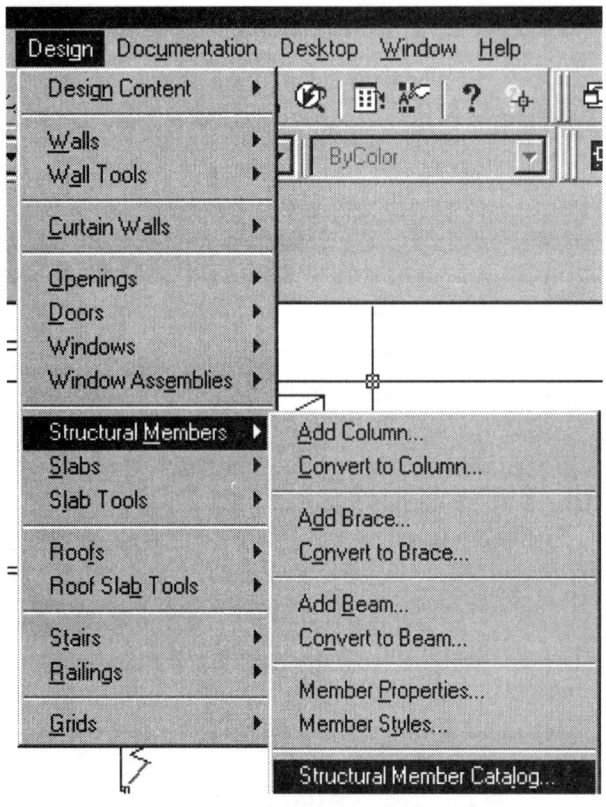

Framing is created using Structural Members.

These can be accessed through the menu or through the Structural Members toolbar.

If you are operating with the AIA layering system as your current layer standard, when you create members or convert AutoCAD entities to structural members using the menu picks or toolbars, Architectural Desktop assigns the new members to layers: A-Cols, A-Cols-Brce or A-Beam, respectively. If Generic Architectural Desktop is your standard, the layers used are A_Columns, A_Beams and A_Braces. If your layer standard is Current Layer, new entities come in on the current layer, as in plain vanilla AutoCAD.

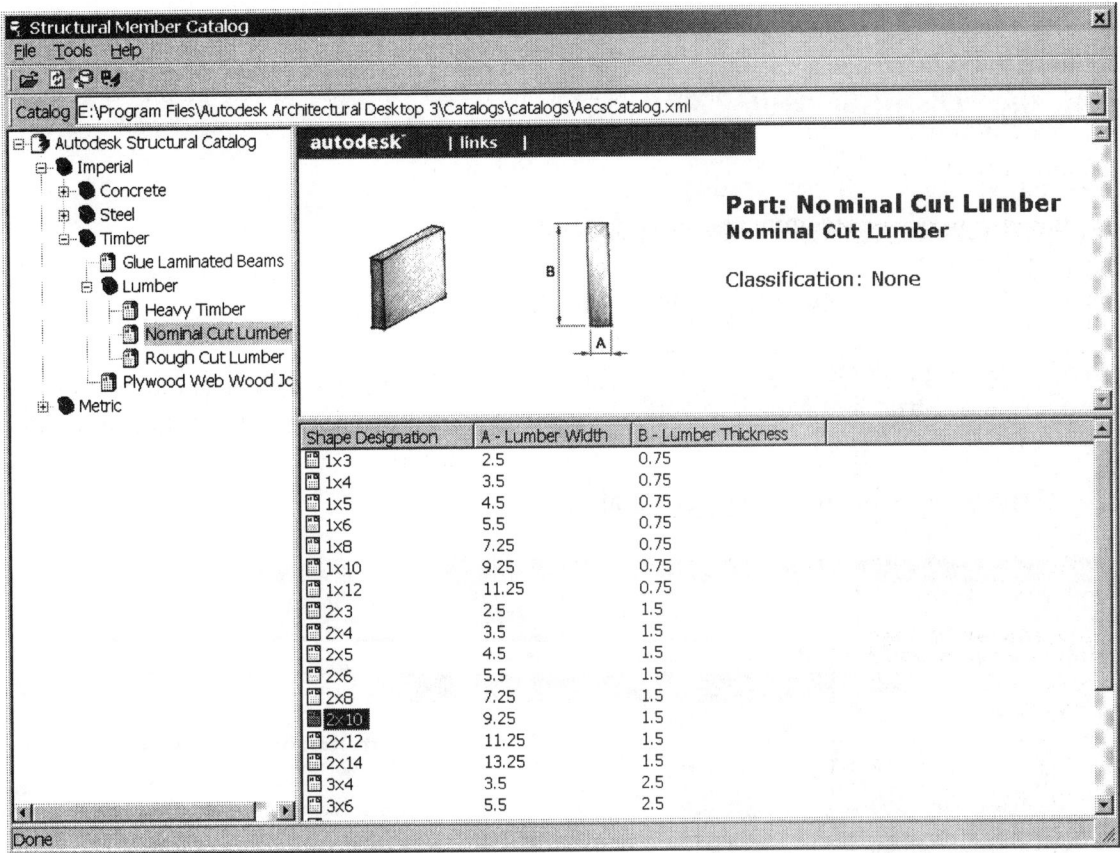

The Structural Member Catalog includes specifications for standard structural shapes. You can choose shapes from the Structural Member Catalog, and generate styles for structural members that you create in your drawings.

The left pane of the Structural Member Catalog contains a hierarchical tree view. Several industry standard catalogs are organized in the tree, first by imperial or metric units, and then by material.

📂	Open a catalog file - The default is the catalog that comes with ADT, but you can create your own custom catalog. The default catalog is located in the following directory path: \\Program Files\Autodesk Architectural Desktop R3\Catalogs\catalogs.
🔄	Refresh Data
🔍	Locate Catalog item based on an existing member – allows you to select a member in a drawing and then locates it in your catalog
💾	Generate Member Style – allows you to create a style to be used

Lesson 7
Structural Members

Exercise 1:
Creating Member Styles

Drawing Name: Lesson 6.dwg
Estimated Time: 15 minutes

This exercise reinforces the following skills:

- Creating Member Styles
- Use of Structural Members tools

Activate the Structural Member Catalog.

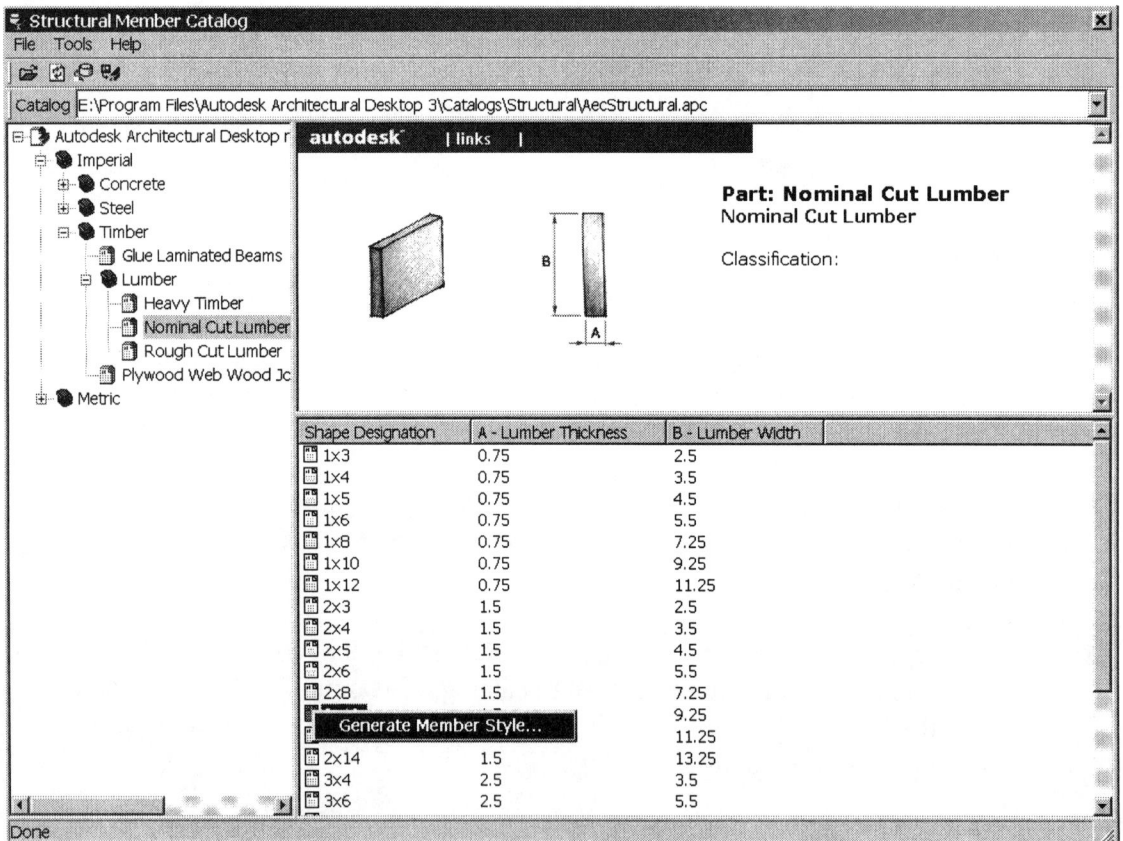

Locate the 2x10 shape designation under Imperial->Timber->Lumber->Nominal Cut Lumber. Highlight, right click and select Generate Member Style.

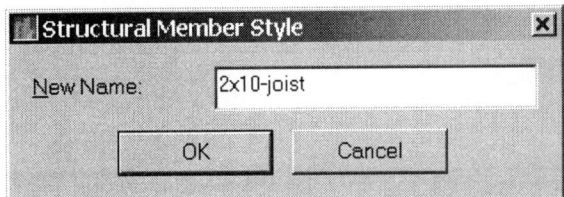

In the Structural Member Style dialog box, type '2x10-joist' - the name for your style, and click OK.

A style that contains the catalog shape that you selected is created. You can view the style in the Style Manager, create a new structural member from the style, or apply the style to an existing member.

When you add a structural member to your drawing, the shape inside the style that you created defines the shape of the member. You define the length, justification, roll or rise, and start and end offsets of the structural member when you draw it.

You cannot use the following special characters in your style names:

- less-than and greater-than symbols (< >)
- forward slashes and backslashes (/ \)
- quotation marks (")
- colons (:)
- semicolons (;)
- question marks (?)
- commas (,)
- asterisks (*)
- vertical bars (|)
- equal signs (=)
- backquotes (`)

Save the file as Ex7-1.dwg.

Exercise 2:
Adding Members

Drawing Name: Ex7-1.dwg
Estimated Time: 30 minutes

This exercise reinforces the following skills:

- Layer Standards
- ADT Layer Manager – User groups
- Adding Structural Members
- Use of Structural Members tools

Open the Ex7-1 drawing.

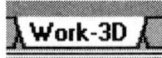

Switch to the Work-3D tab.

Make sure you are in Paper Space (use the MODEL/PAPER toggle at the right end of the status bar at the bottom of the graphics screen),

Zoom Extents to maximize your view of the two viewports. In the left viewport, which defaults to an isometric view, use the Named Views dialogue or NW Isometric View icon to switch to a Northwest Isometric view, then zoom to the double sliding doors into the family room. In the right viewport, turn off SNAP and GRID (using the status bar toggles), and Zoom Extents.

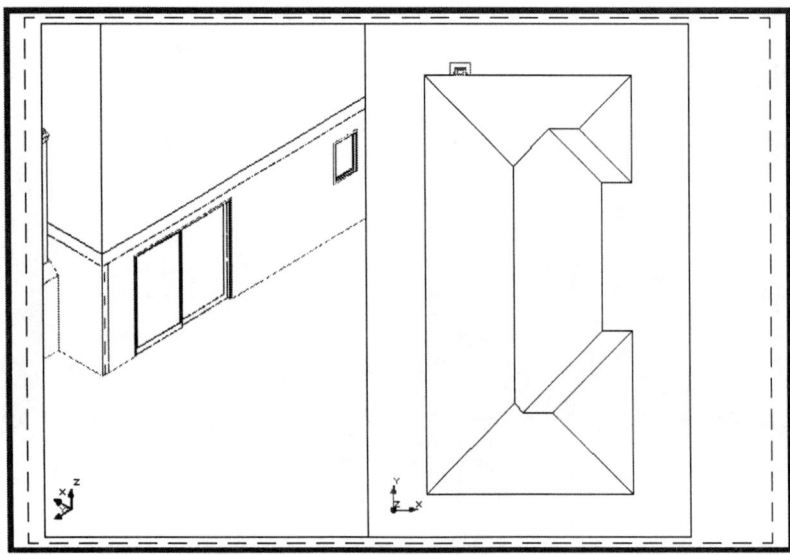

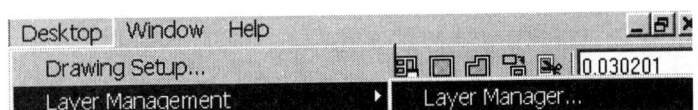

Activate the ADT Layer Manager by picking Desktop>Layer Management/Layer Manager:

Create a new user group and name it Deck Framing:

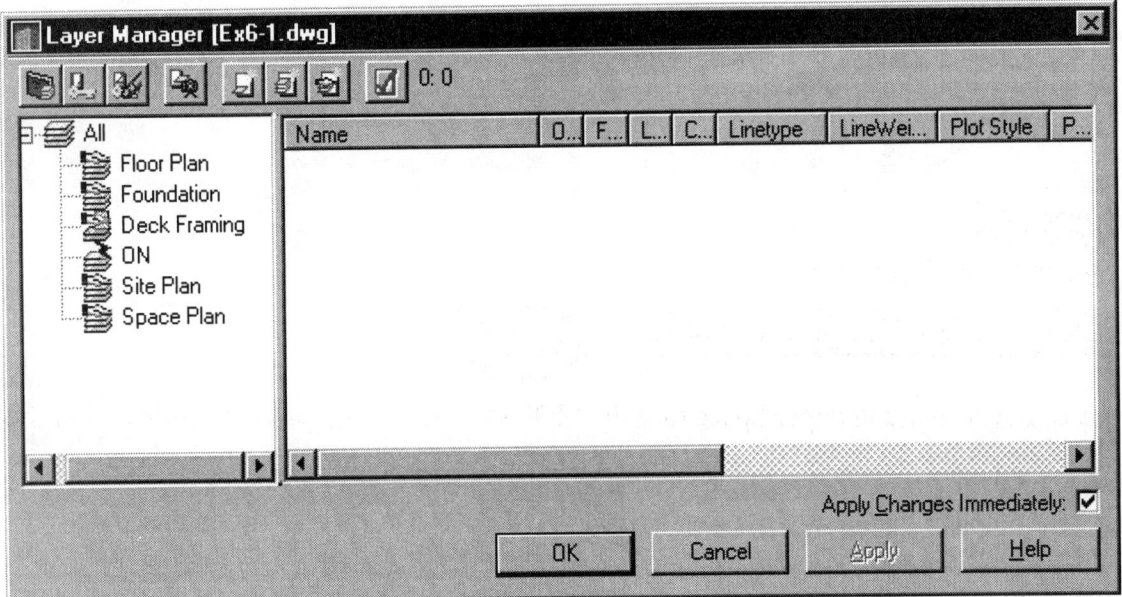

Double click on layer 0 to make it current. The icon to the left of the layer name will turn green.

Lesson 7
Structural Members

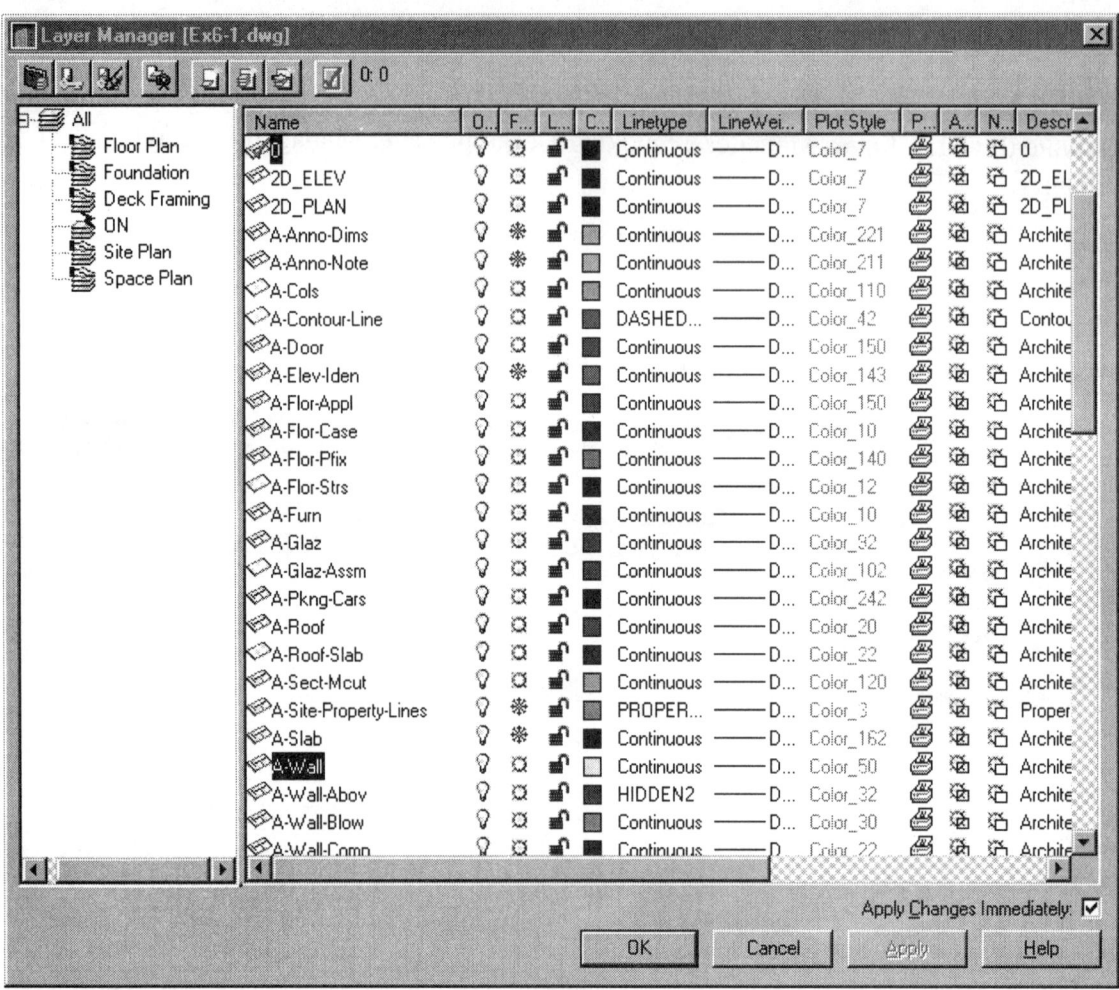

Pick on All in the tree view to expand the drawing layers.

Drag layers 0 and A-Wall into the Deck Framing user group.

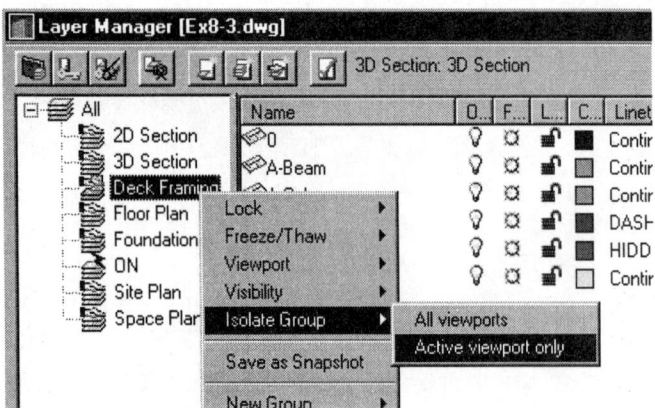

Pick the Deck Framing group name, right click and pick Isolate Group>Active Viewport only. Click Okay:

Lesson 7
Structural Members

We will be adding a wood framed deck, 14' x 9', to the back of the house. For purposes of this exercise we will assume that the ground level is 12" below the slab at the back of the house and falls away so that grade level below the edge of the deck away from the house is 6' below floor level: -6'-0" a.f.f. (above finish floor) in architectural notation. We will place the top of the deck floorboards even with the top of the floor slab.

We will use support and rim joists as in standard wood floor framing, and they will all be at the same level, rather than joists crossing a support beam below. In practice this means the use of metal hangers, which will not be drawn. Once the floor system is drawn we will add support columns at the outside rim joist and braces at the columns.

Add the first joist, known as a ledger, which will be fastened to the side of the floor slab:

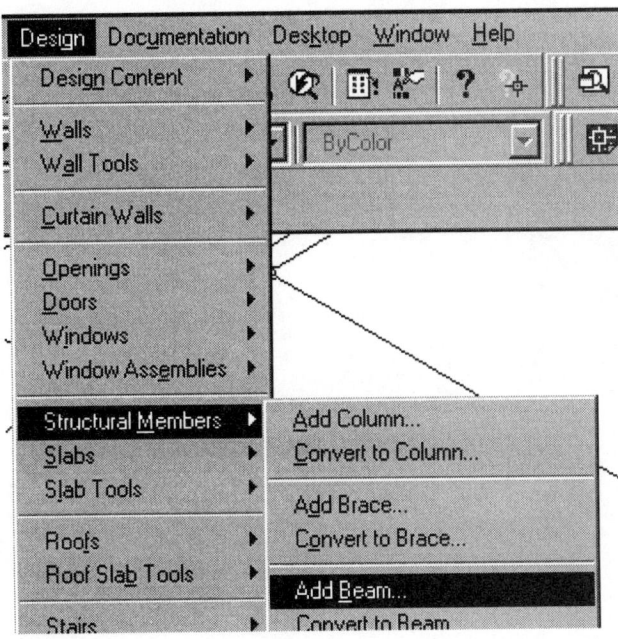

MENU	Design->Structural Members->Add Beam
	Add Beam
Command Line	BEAMADD

Use the BeamAdd tool.

Lesson 7
Structural Members

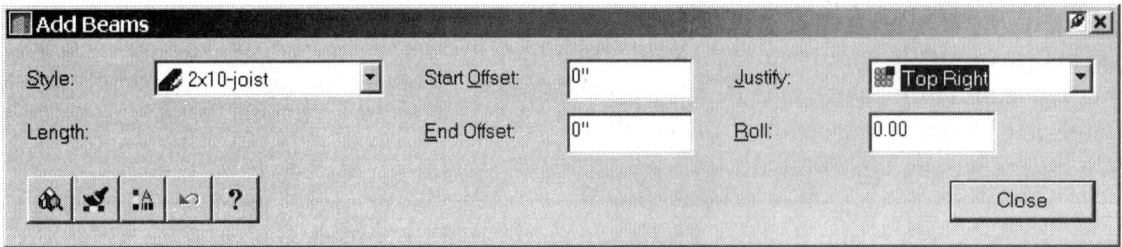

Set the Style to 2x10-joist.
Set Justify to Top Right.
Set the Roll to 0.
Set the Start Offset to 0"

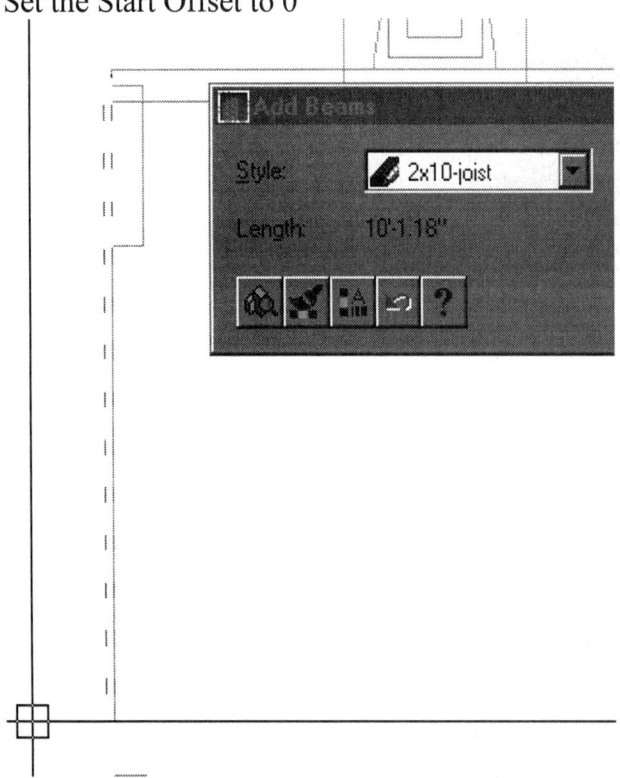

Use the From Osnap, pick the upper left corner of the house wall, and type "@1.5<-90" for the Offset; Press ENTER. Pull the cursor down at a 270° angle, and type in 13'9" for the length. DO NOT EXIT THE COMMAND.

```
Command: _AecsBeamAdd
Start point or [STyle/STArt offset/ENd offset/Justify/Roll/Match]: _from Base
point: <Offset>: @1.5<-90

End point or [STyle/STArt offset/ENd offset/Justify/Roll/Match]: 13'-9"

End point or [STyle/STArt offset/ENd offset/Justify/Roll/Match/Undo]:
```

Change the Justification to Top Left. Pull the cursor to the left (180°) and enter 9' for the length. Hit enter to terminate the command:

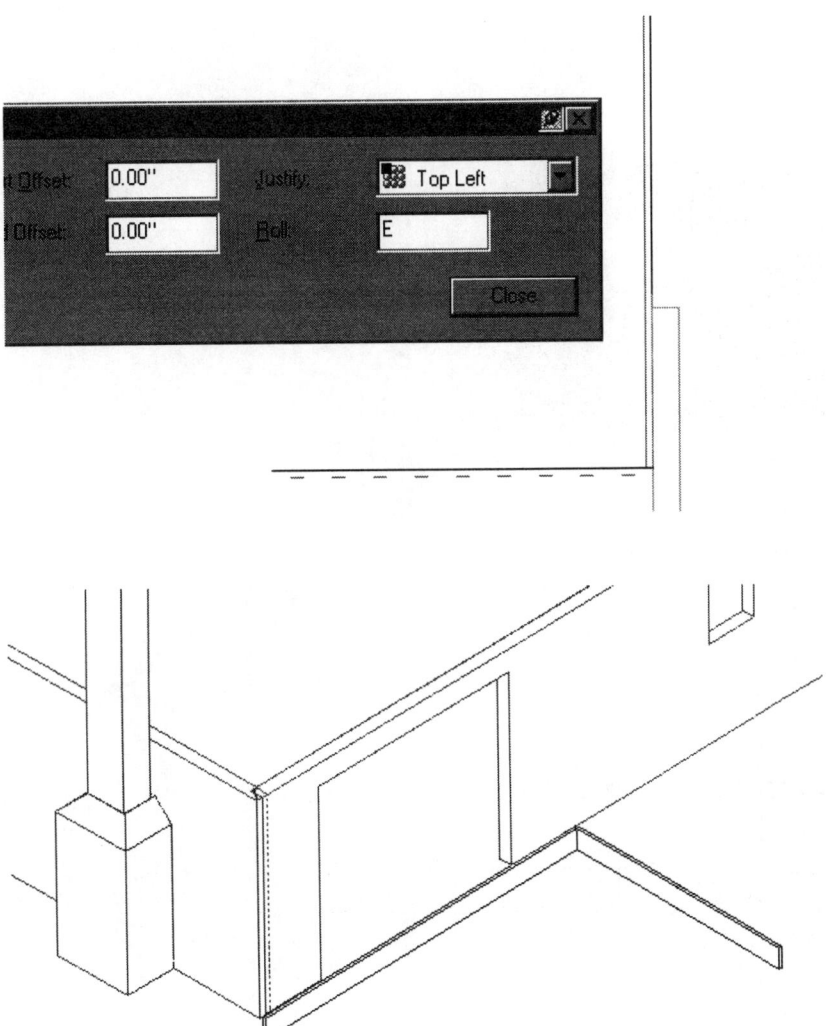

Mirror the first joist. (use the Midpoint Osnap with Polar or Ortho)

Lesson 7
Structural Members

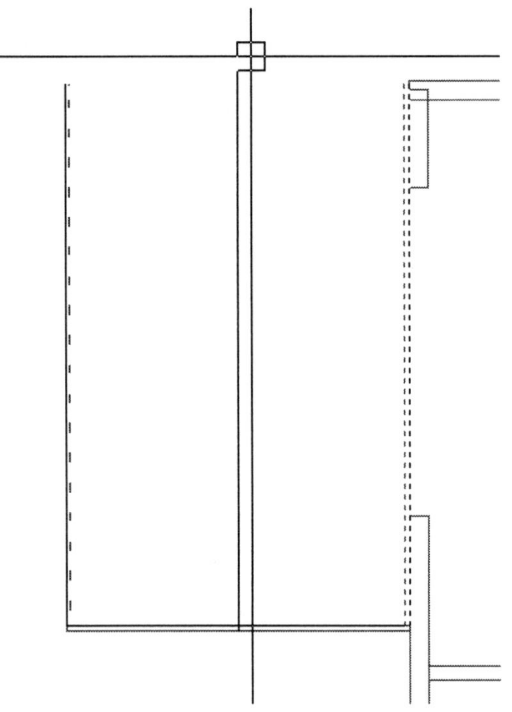

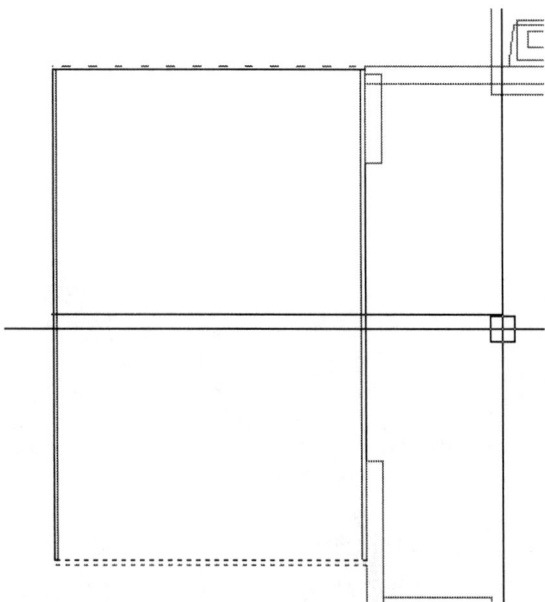

Mirror the second from bottom to top (use the Midpoint Osnap with Polar or Ortho) to finish the outside of the frame:

Lesson 7
Structural Members

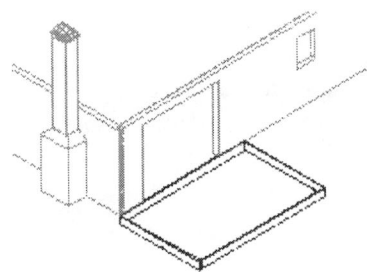

Copy the left (outside) joist 1.5" to the right to form a beam to support the rest of the joists we are about to create.

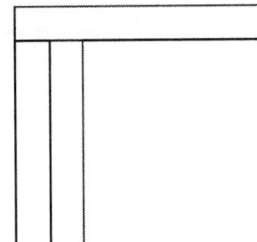

Activate BEAMADD and create another beam inside the left and right joists along the bottom joist.

Move the new joist 13.75" up, so that the center of it is 1'-4" from the bottom edge of the bottom horizontal joist.

TIP: Starting a 16" spacing from one edge to centers minimizes waste when using flooring (boards or sheets of sub floor) in nominal 4' increments. The same patterning usually holds for wall studs and drywall.

Use a rectangular array to copy the last joist 10 times at a 16" distance to finish the floor joist framing:

7-13

Lesson 7
Structural Members

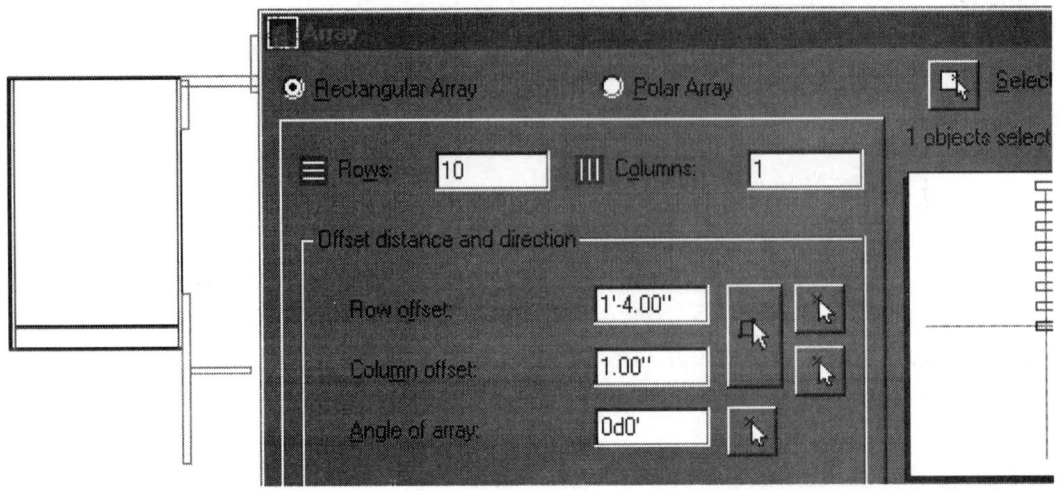

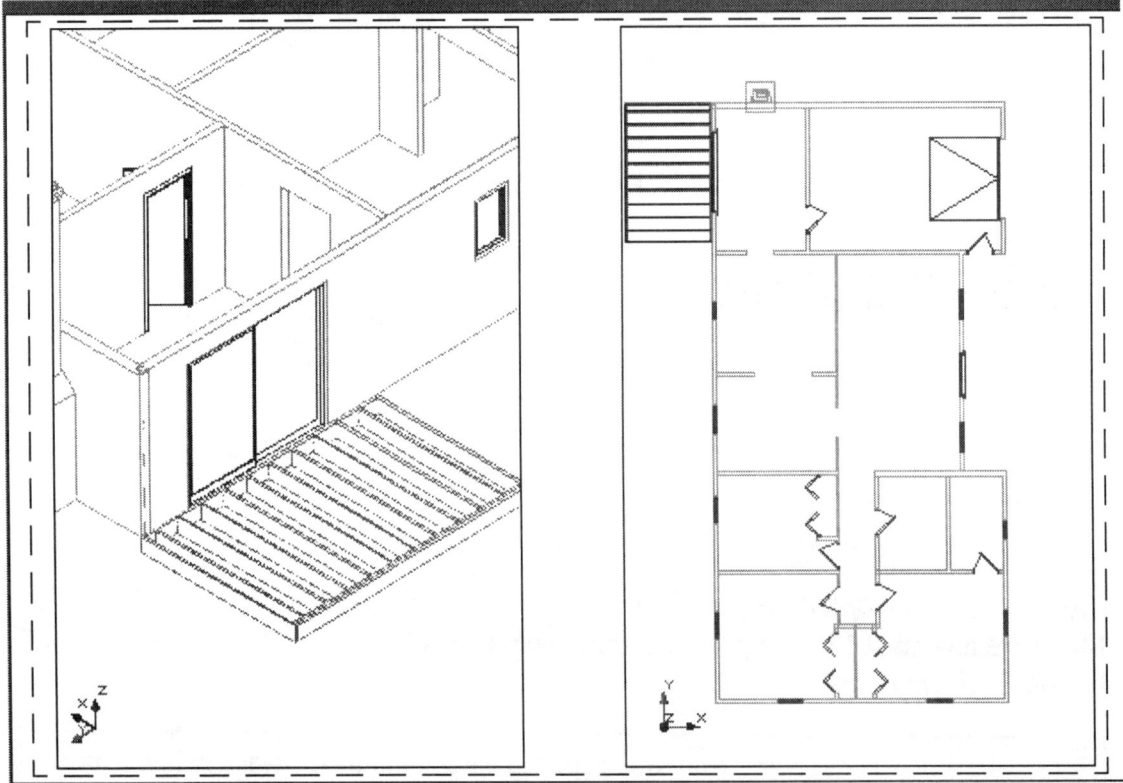

Open the Structural Member Catalog.

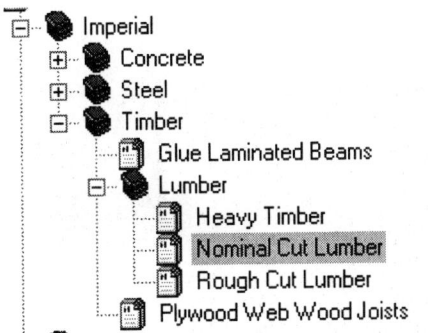

Go to Imperial>Lumber>Nominal Cut Lumber.

Right click on Shape Designation 4x4 and pick Generate Member Style.
Create a style named "4x4 post."

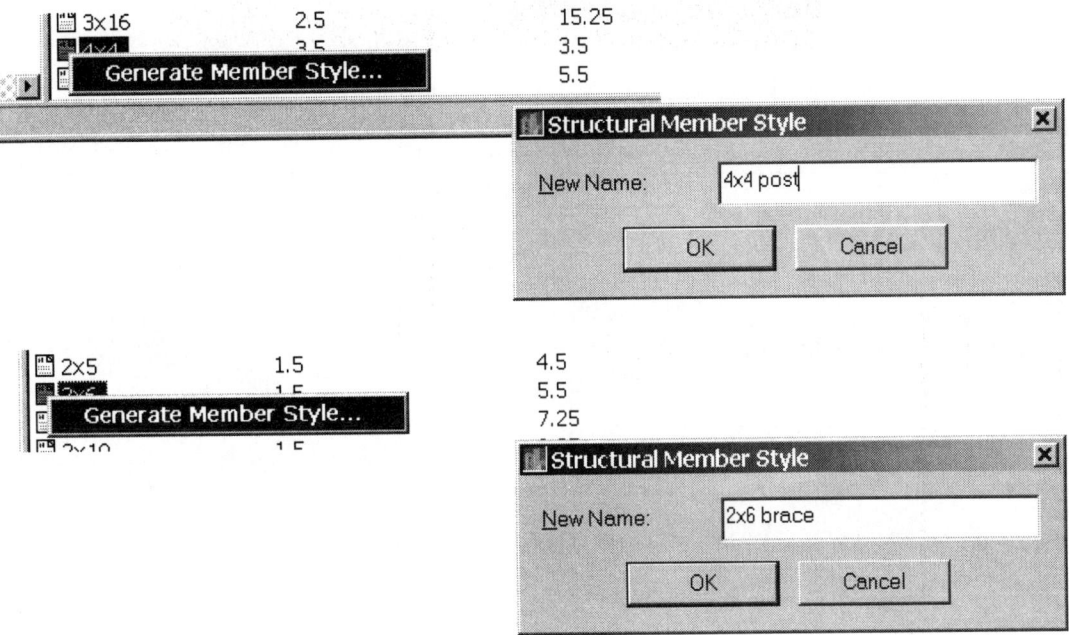

Right click on Shape Designation 2x6 and create a member style named "2x6 brace:"
Close the catalog.

Lesson 7
Structural Members

Adding Columns

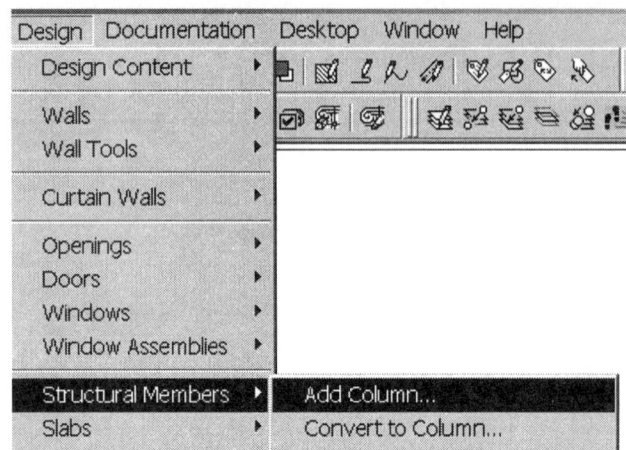

MENU	Design->Structural Members->Add Column
	Add Column
Command Line	COLUMNADD

Activate the ColumnAdd tool.

7-16

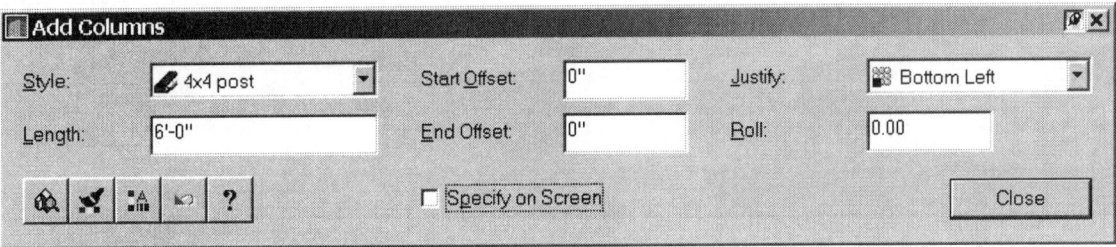

Set the Style to 4x4 post.
Set Justify to Bottom Left.
Set the Length to 6'.
Disable the 'Specify on Screen.'
Set the Roll to 0.

At the "Select grid or RETURN:" prompt, hit Enter.

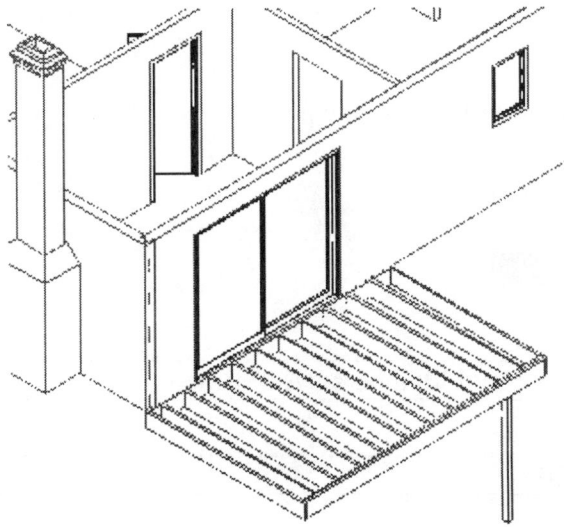

In the right viewport, use the From Osnap, pick the lower left corner of the joist frame, and enter "@0,24,-6'10' to place the post so the left face aligns with the outside of the doubled joist, the bottom face is 2' from the bottom edge, and the top is even with the bottom of the joists.

Lesson 7
Structural Members

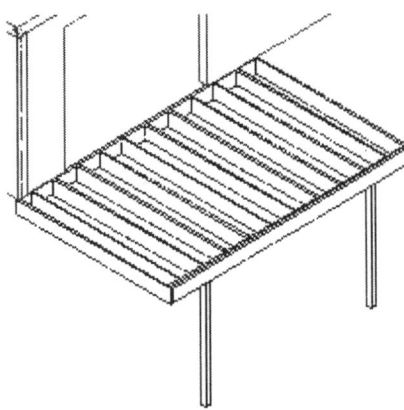

Mirror the post to place another column 2' in from the top of the frame:

Zoom in close to the base of one of the columns in the left viewport.

Menu	Design->Structural Members->Add Brace
Structural Members Toolbar	
Command Line	BRACEADD

Add braces to the posts.

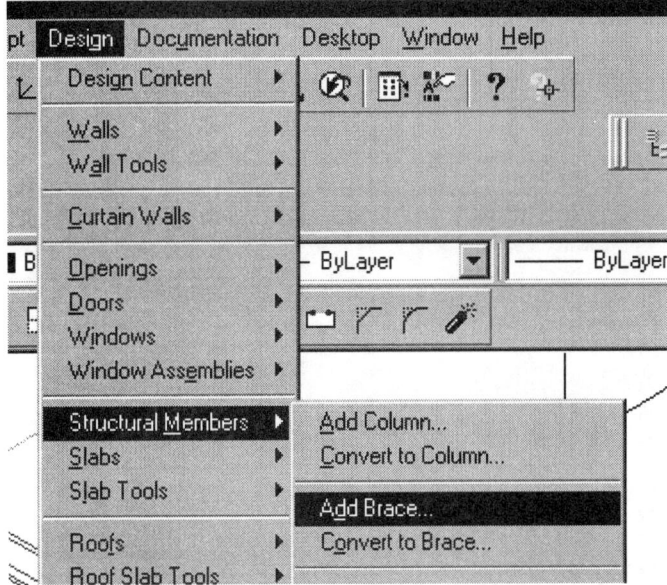

Use the BraceAdd tool.

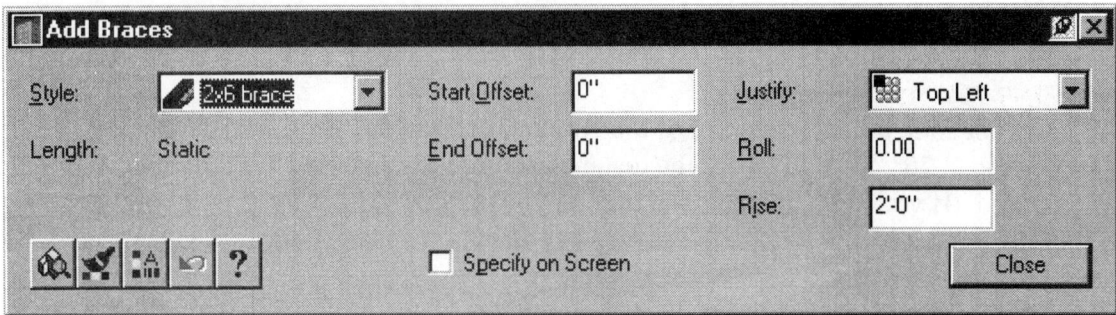

Set the Style to 2x6 brace.
Set Justify to Top Left.
Disable 'Specify on Screen'.
Set the Rise to 2'-0".
Set the Roll to 0.

Lesson 7
Structural Members

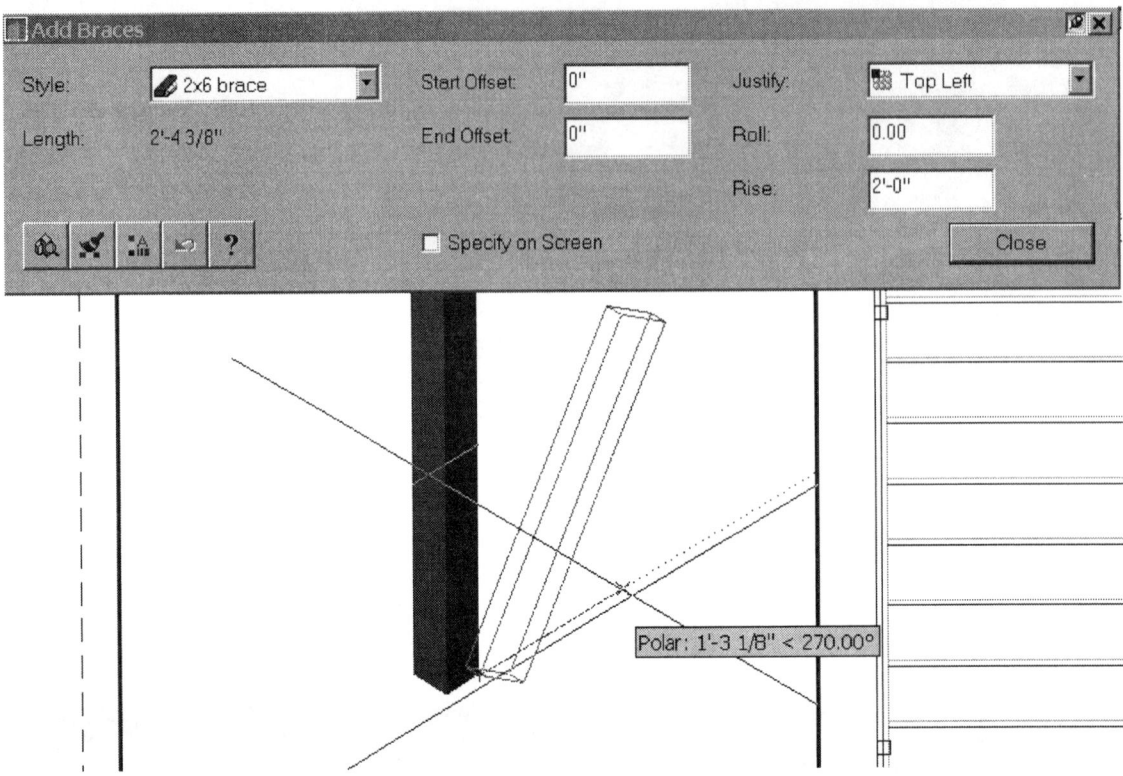

Pick the outside bottom endpoint of the column, pull the cursor to the right so the Polar angle reads 270°, type in 24 for the distance and return, then return again to end the command. If you can't get the polar angle, type in @24<270 at the prompt and press ENTER.

In order to show the brace as it will be cut and placed, we need to add trim planes.

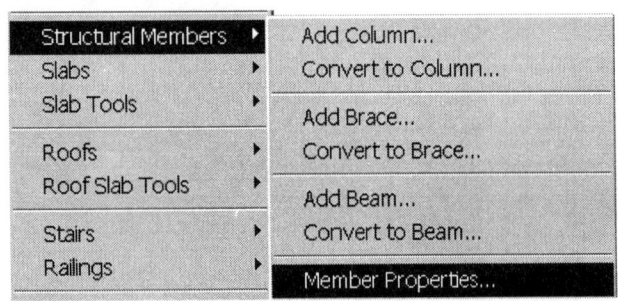

MENU	Design->Structural Members->Member Properties
	Member Properties
Command Line	MEMBERPROPS

Pick the MemberProps tool.

Select the brace.

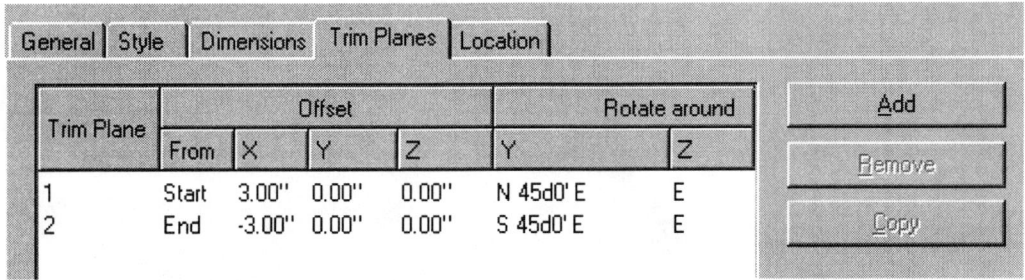

Select the Trim Planes tab of the Structural Member Properties dialogue.
Select the 'Add' button.
Add two trim planes with the settings shown.
Note that you need to set the first trim plane From Start and the second trim plane From End.

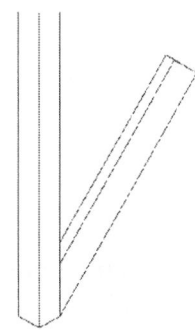

Move the resulting brace shape so that the endpoint aligns with the endpoint of the post.

Switch to a Left View.

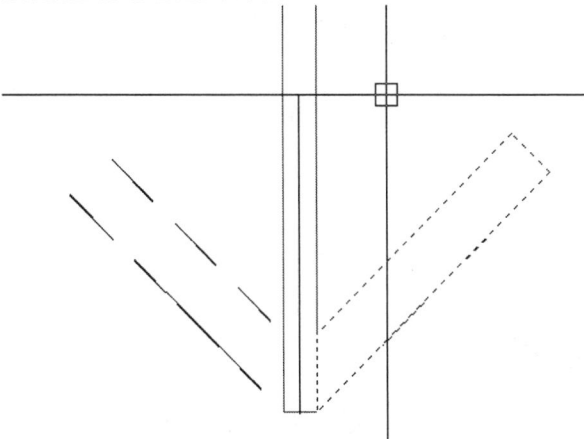

Mirror it to the other side of the post.

Move the two braces up so their tops touch the underside of the joists. Use the Endpoint and Perpendicular Osnaps. Copy the two braces from one post to the other:

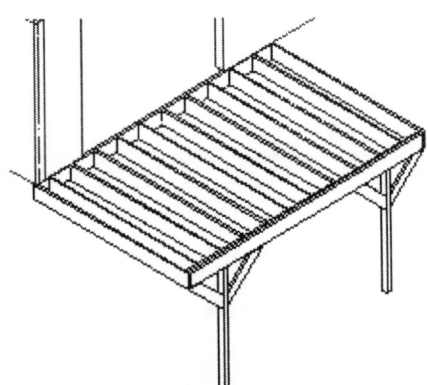

Save as Ex 7-2.dwg

Exercise 3:
Add Floorboards

Drawing Name: Ex7-2.dwg
Estimated Time: 15 minutes

This exercise reinforces the following skills:

- Structural Members
- Use of Structural Members tools

The top of the joist frame we created in the last exercise is sitting even with the top of the floor slab. It needs to be lowered to allow for nominal 2x6 deck boards.

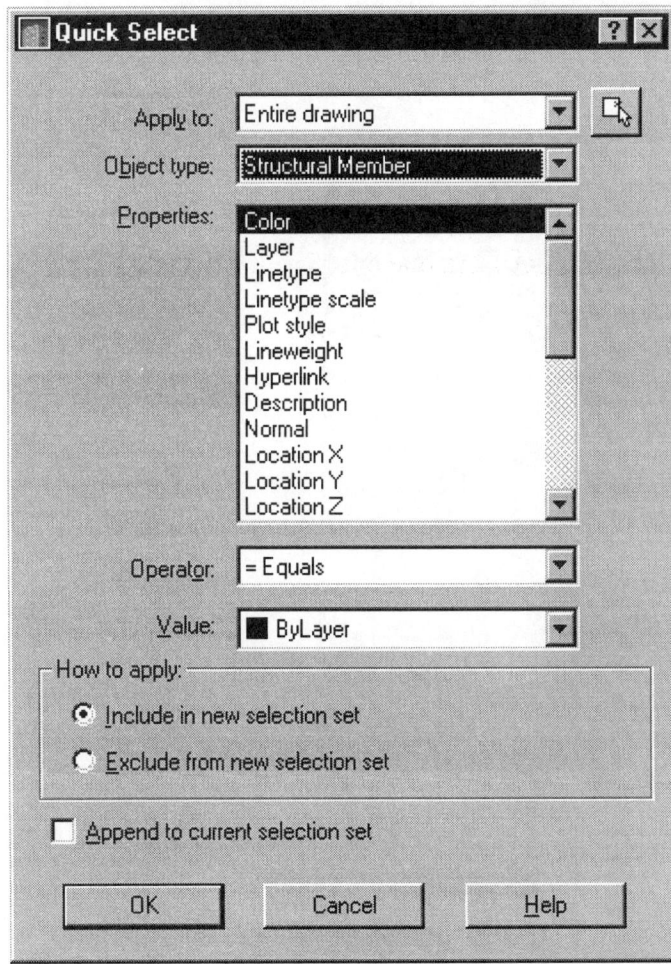

Activate QUICK SELECT.
Select all the Structural Members.

Select the MOVE tool.
When prompted for the base point, select the left end point of one of the joists.
When prompted for the destination point, type @0,0,-1.5.

Create a new layer for the deck floorboards.

Activate the ADT Layer Manager by picking Desktop>Layer Management/Layer Manager:

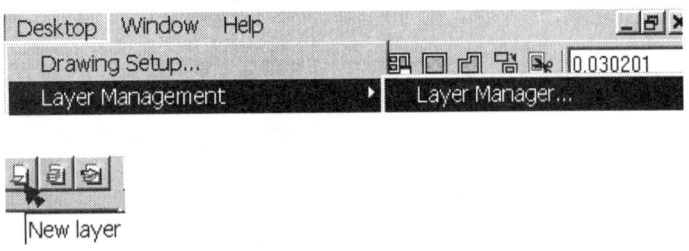

Pick the New Layer icon to create a layer.

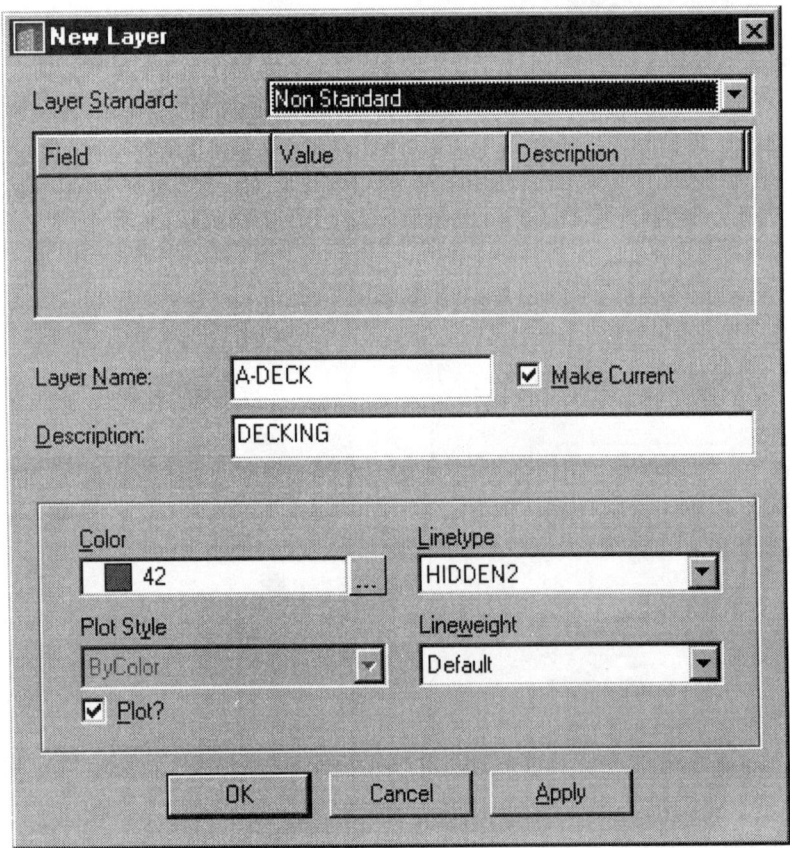

Make the Layer Standard Non Standard
Name the layer A-Deck
Make it current
Make it Hidden2 linetype

7-24

Select Color 42.

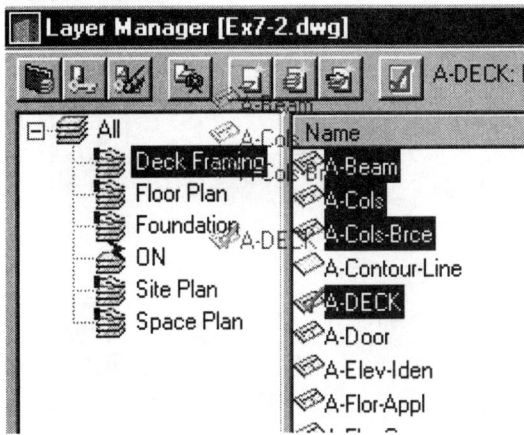

Drag the new layer into the Deck Framing user group. Add layers A-Beams, A-Cols, and A-Cols-Brce to the group:

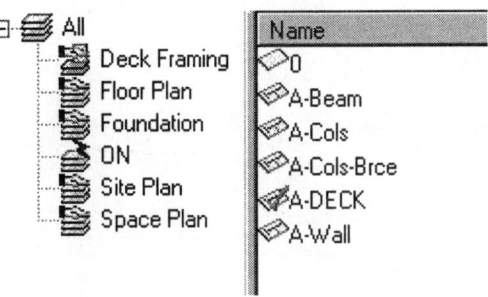

If you look at the Deck Framing group, it should show the list you see above.

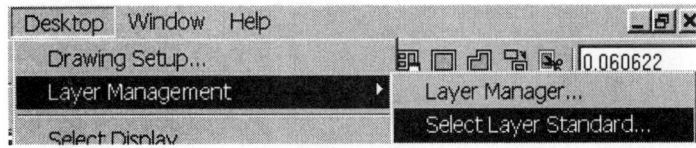

Activate the Layer Standards dialogue box by picking Desktop>Layer Management>Select Layer Standard.

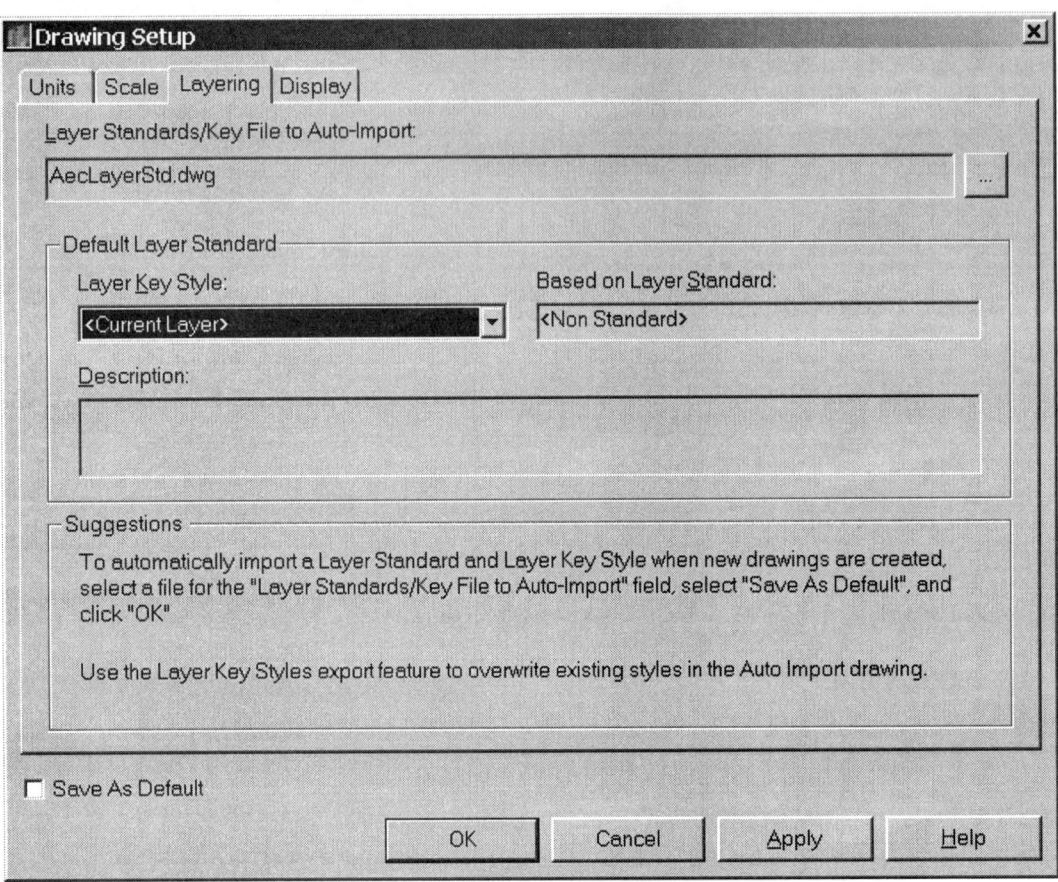

Pick the <Current Layer> Layer Key style from the drop down list.
Press 'Apply' and click OK:

Create deck boards as beams.

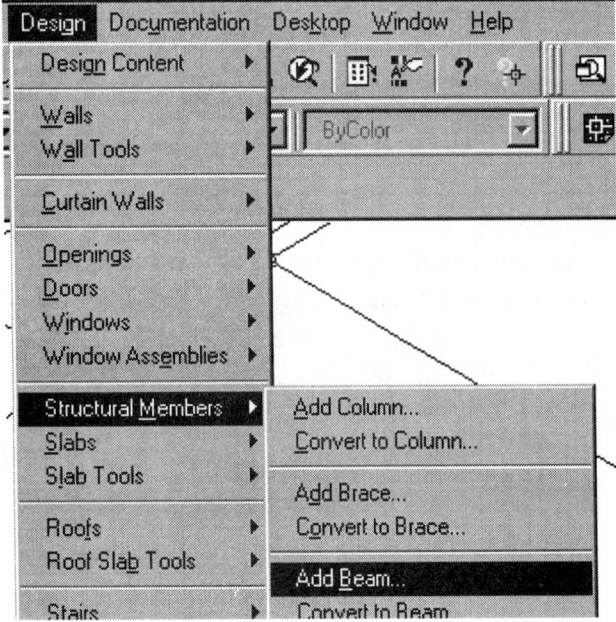

MENU	Design->Structural Members->Add Beam
	Add Beam
Command Line	BEAMADD

Use the BeamAdd tool.

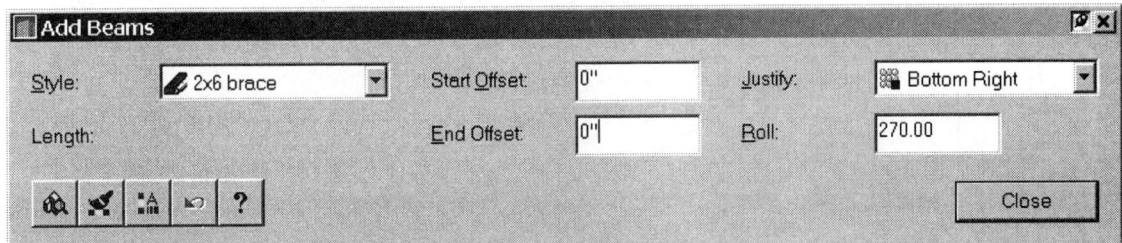

Set the Justification to Bottom Right, so the board sits on top of the joists.
Set the Roll to 270°, so that it lies flat rather than vertical.

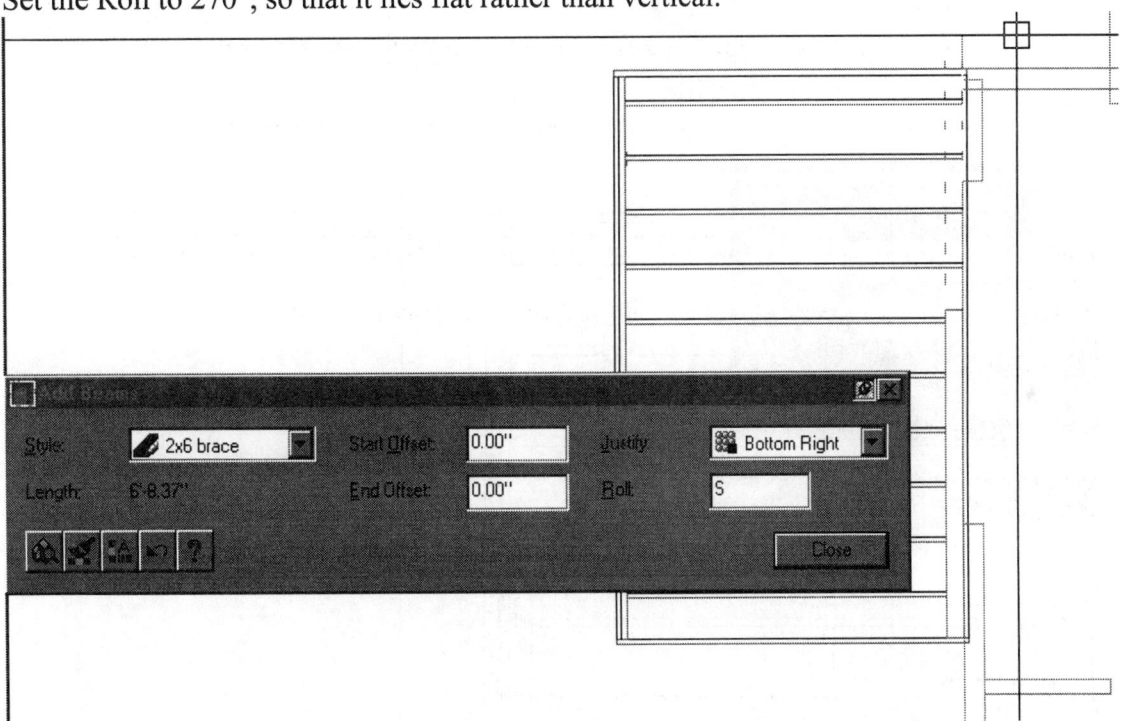

Pick the lower right corner of the joist at the house wall, drag the cursor up (90°), and give the board a length of 8'. Drag the cursor up at 90° and add a second board of length 6':

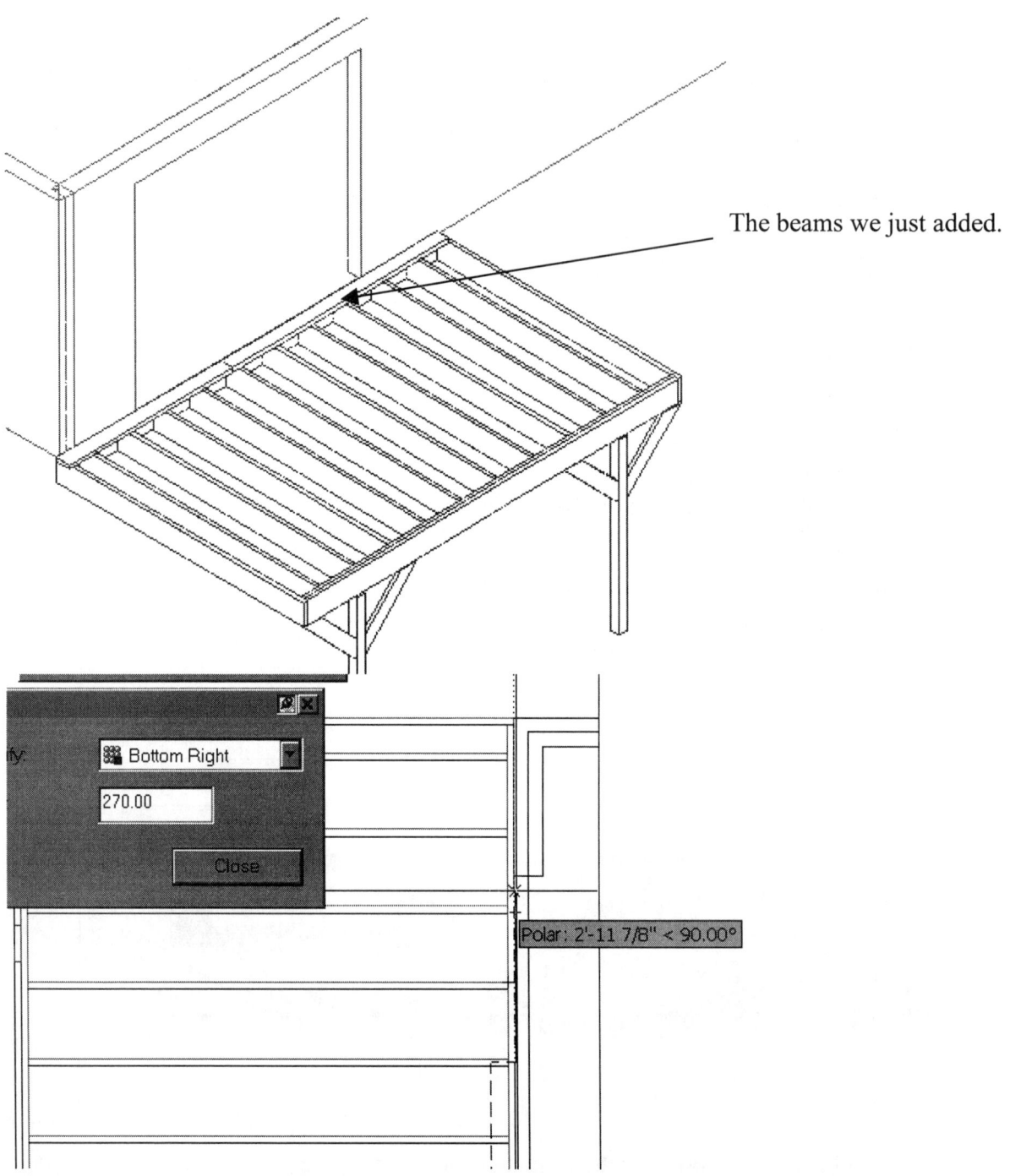

The beams we just added.

Enter twice to terminate and restart the command so you can pick a new starting point. Pick the lower left corner of the first deck board you created, pull the cursor up at 90°, enter a distance of 4', then pull the cursor up at 90° and enter a distance of 10'. (Although not strictly necessary, we are providing suggested lengths of deck boards to minimize wastage.) Deck boards are laid with a nominal ½" space between them to allow for drainage and board warping, so move the last two deck boards .5 to the left.

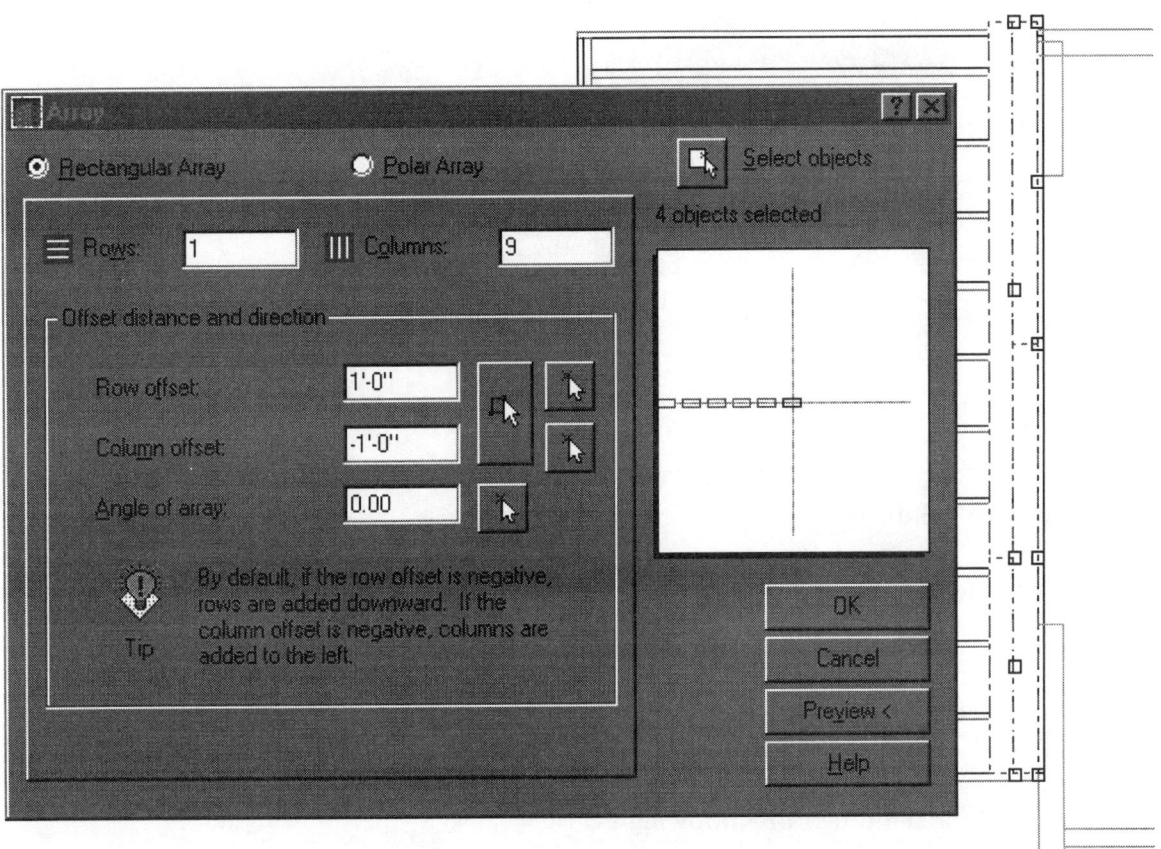

Array the 4 deck boards to fill the deck: 1 row, 9 columns with a column offset of –1'0".

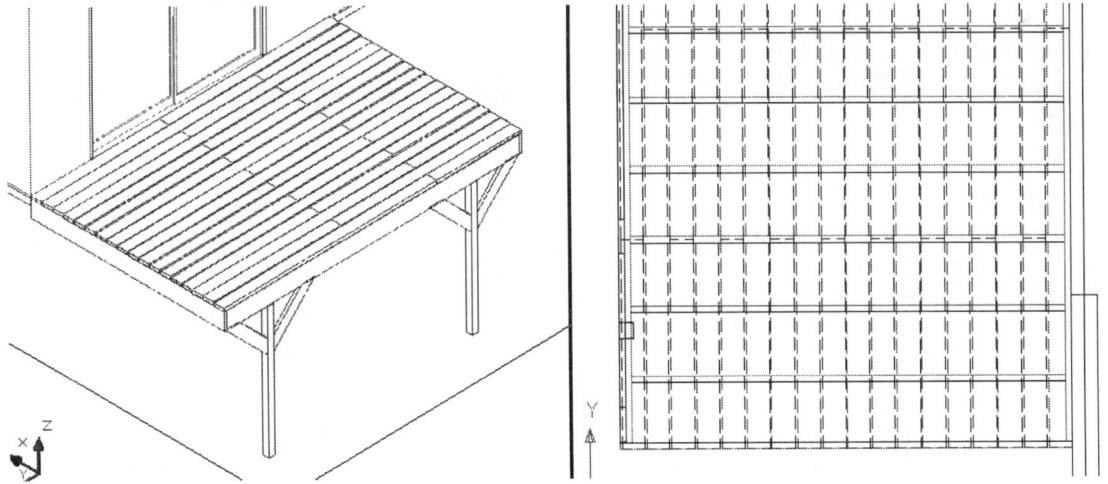

A hidden line view of the 3D model shows the board spacing and end lines.

Lesson 7
Structural Members

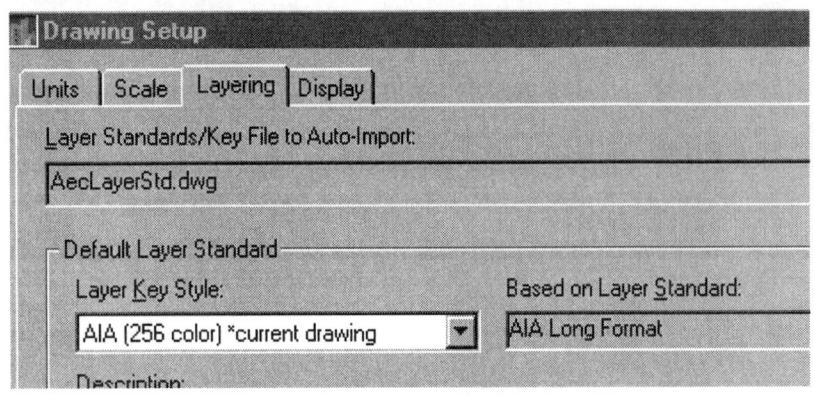

Activate the Layer Standards dialogue and change to the AIA (256 color) Layer Key Style.

Save as Ex7-3.dwg.

Exercise 4:
Add Railing

Drawing Name: Ex7-3.dwg
Estimated Time: 15 minutes

This exercise reinforces the following skills:

- Railings
- Railing Styles

Next, we create a railing for the deck.

MENU	Design->Railings->Add Railing
	Add Beam
Command Line	RAILINGADD

Use the RailingAdd tool.

Lesson 7
Structural Members

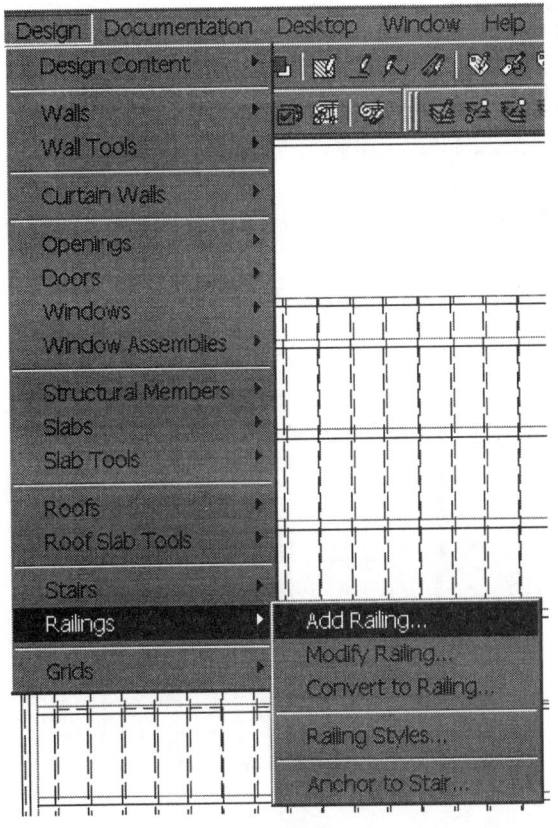

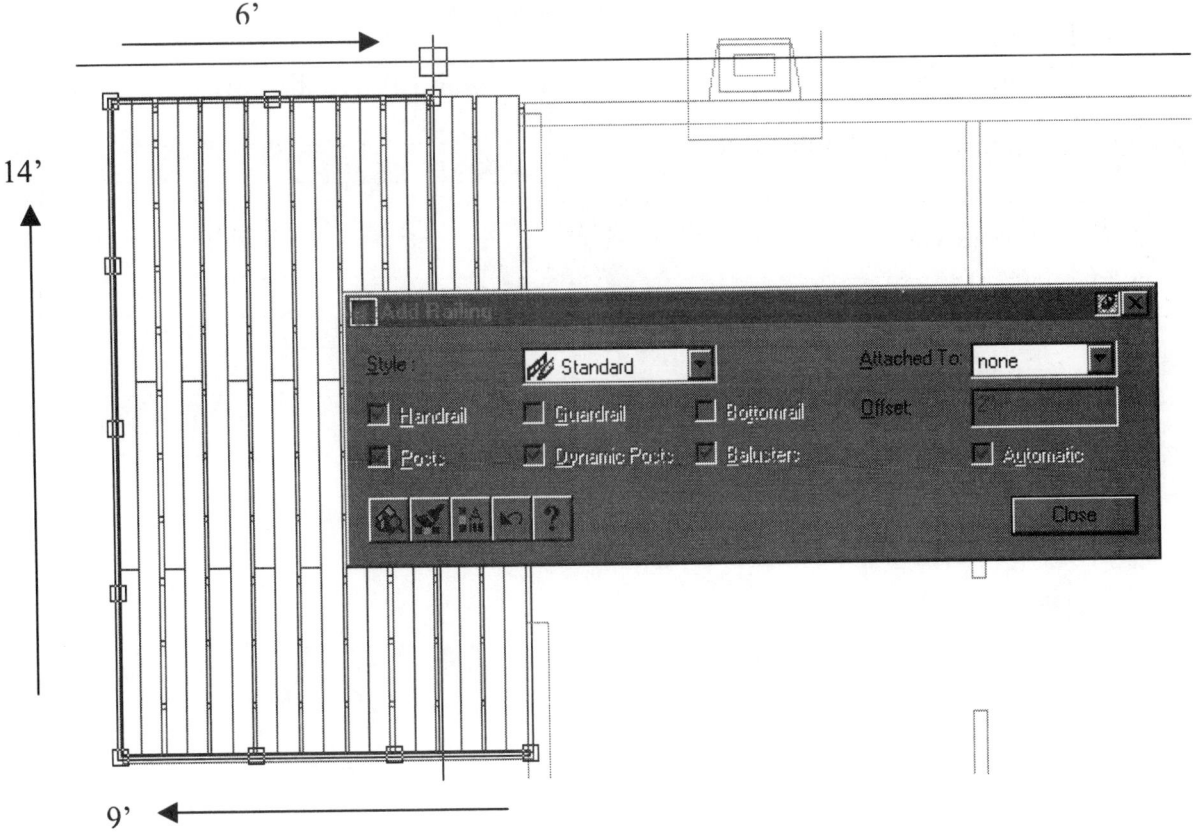

7-31

Pick the lower right corner of the deck for the start point, and create endpoints at 9' to the left (180°), then 14' up (90°), and 6' to the right (0°). This leaves a 3' opening at the corner of the house.

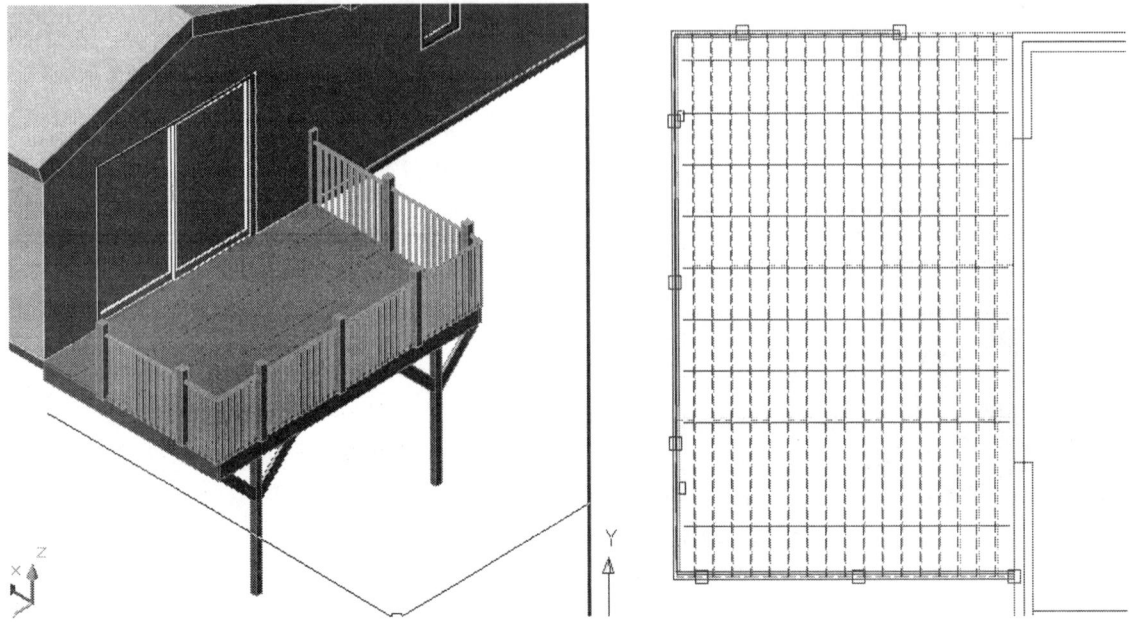

The Standard railing style does not put posts at corners, nor does it contain a bottom rail. Create a new Railing Style to change the appearance of the railing.

MENU	Design->Railings->Railing Styles
![icon]	Railing Styles
Command Line	RAILINGSTYLE

Use the RailingStyle tool to open the Style Manager. Pick the Standard Style, right click and pick New.

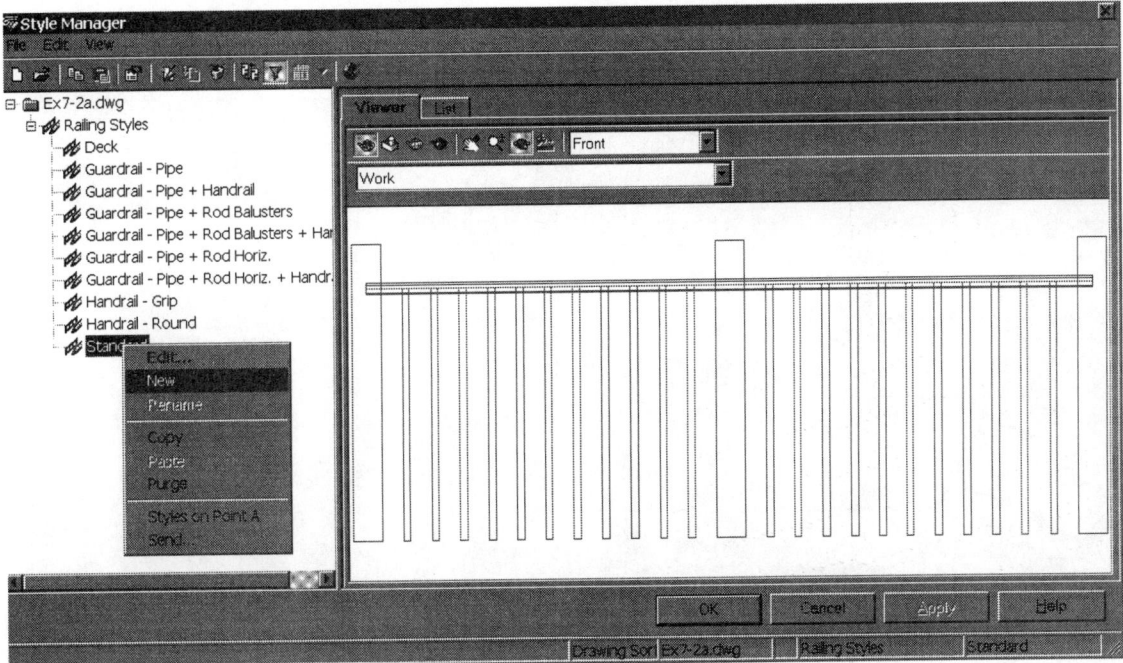

Name the new railing style Deck.

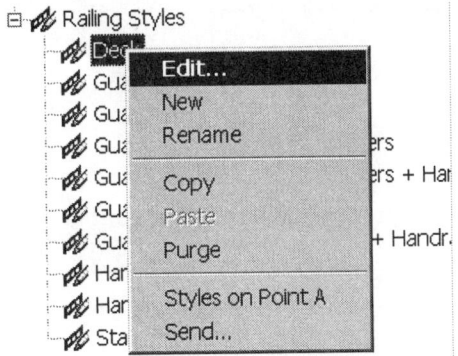

Pick the new style, right click and pick Edit.

Lesson 7
Structural Members

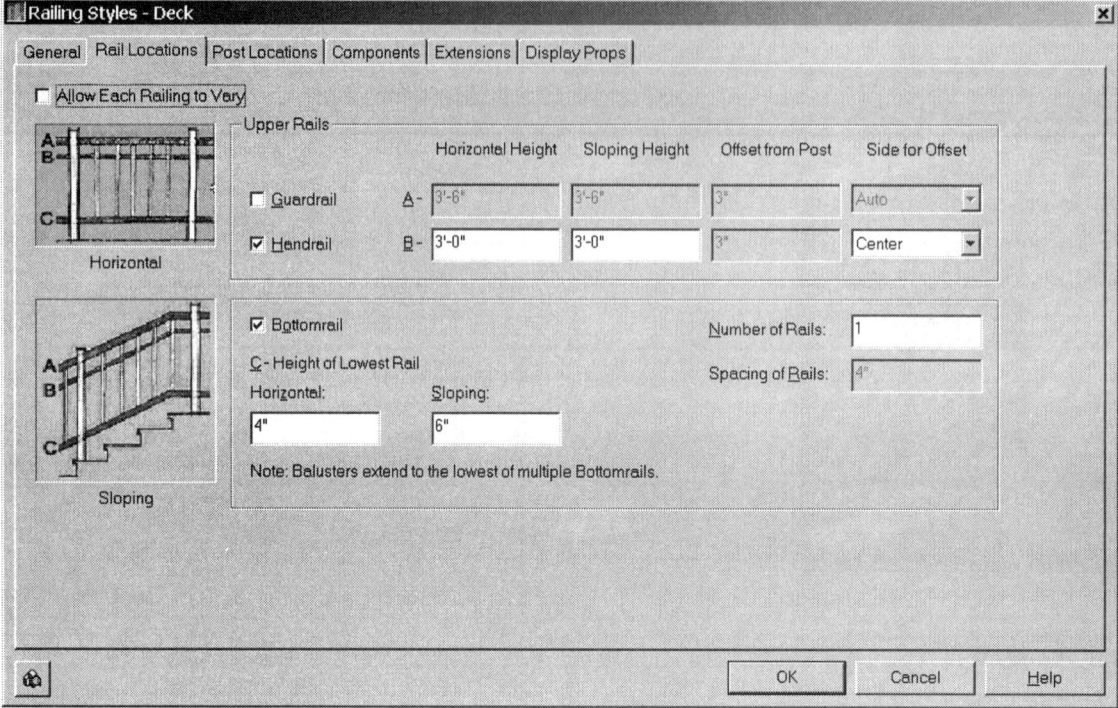

Select the Rail Locations tab.
Enable Handrail for the Upper Rails.
Set the Horizontal Height to 3'.
Set the Sloping Height to 3'.
Set the Side for Offset to Center.
Enable the Bottom Rail. (This adds a bottom rail)
Set the Horizontal Height to 4'".
Set the Sloping Height to 6'".
Set the Number of Rails to 1.

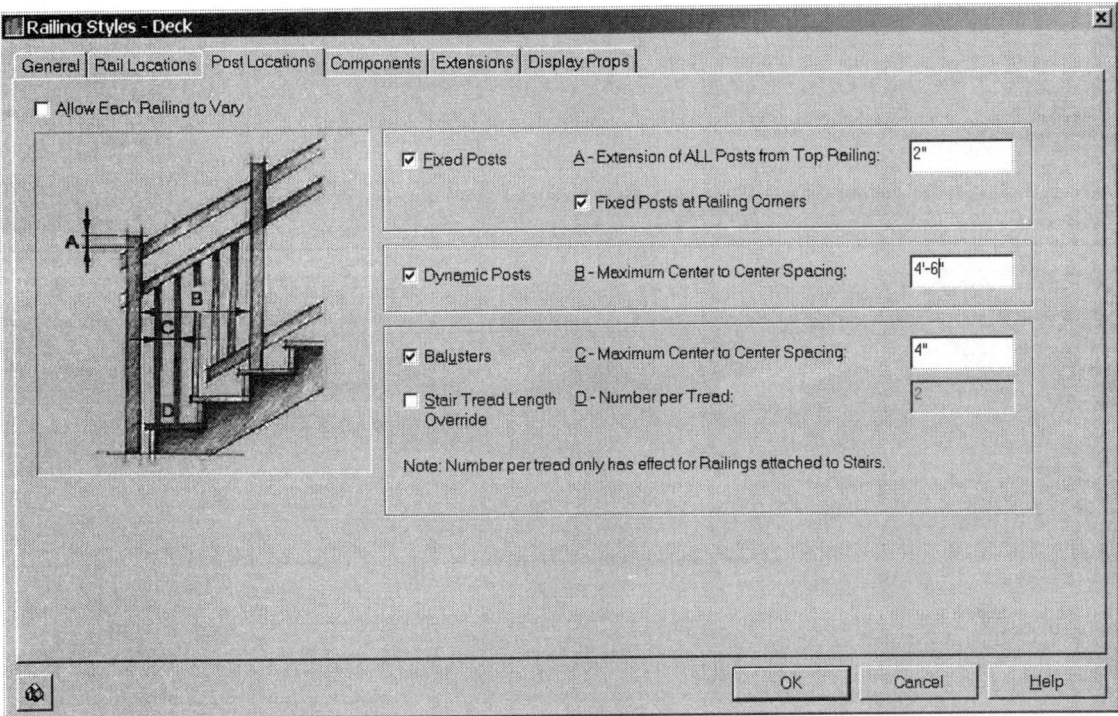

Select the Post Locations tab.
Disable 'Allow Each Railing to Vary'.
Enable Fixed Posts.
Enable Fixed Posts at Railing Corners. (This adds a post to each corner)
Set the Extensions of ALL Posts from Top Railing to 2".
Enable Dynamic Posts.
Set the Maximum Center to Center Spacing to 4'-6".
Enable Balusters.
Set the Maximum Center to Center Spacing to 4".
Disable the Stair Tread Length Override.

Lesson 7
Structural Members

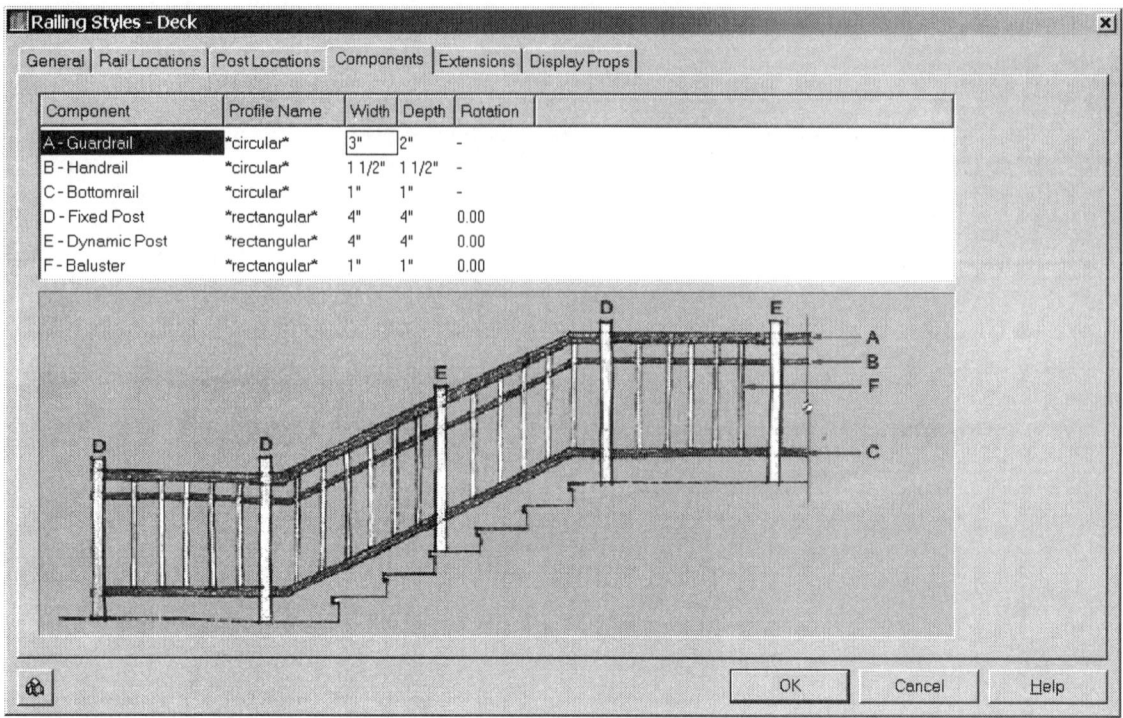

Select the Components tab.
Change the Width of the Guardrail to 3".
All the other settings are unchanged.

Press 'OK' to close the dialog box.

Press 'Apply' and 'OK'.

Method 1:

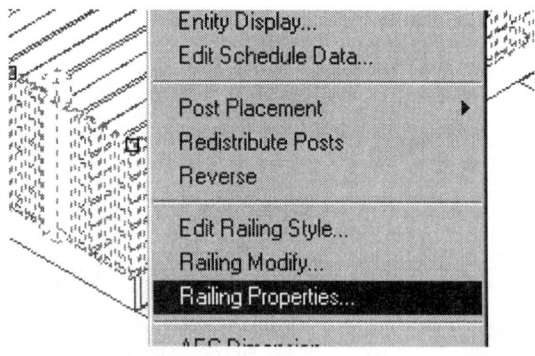

Select the Railing.
Right click and select 'Railing Properties.'

Lesson 7
Structural Members

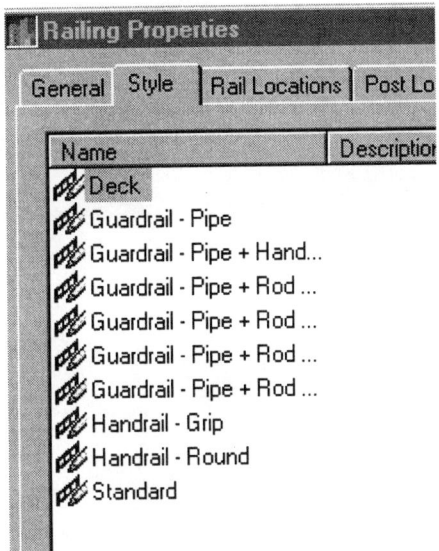

Select the Style tab.
Highlight Deck.
Press 'OK'.

Method 2:

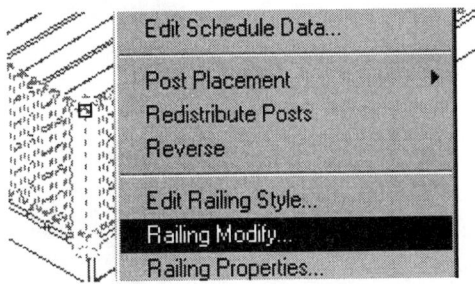

Pick the Railing.
Right click and select Railing Modify.

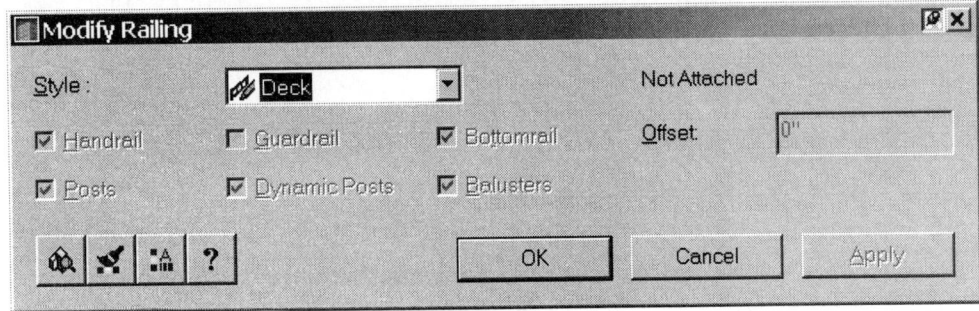

Change the Style to Deck and pick OK.

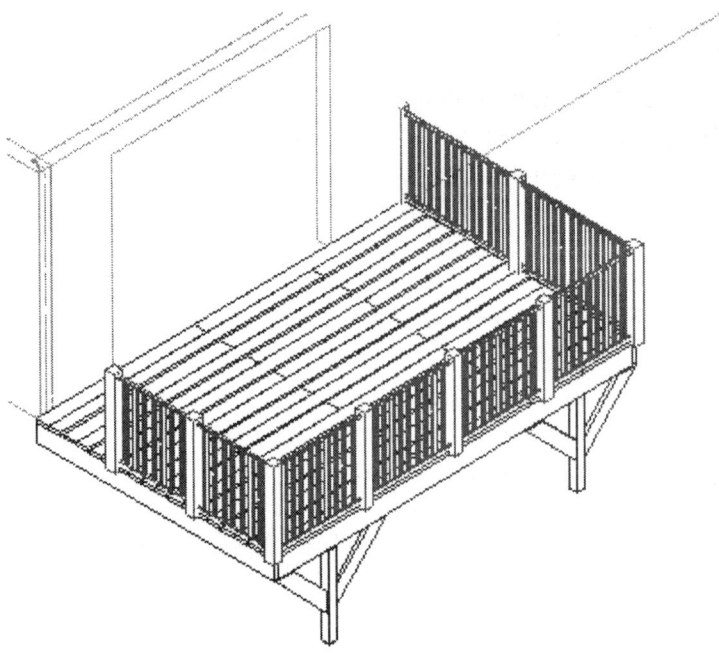

Save your drawing as Ex 7-3.

Stairs

A house may have main stairs (from the first floor to the second floor) and/or a set of service stairs. Main stairs are usually constructed using pre-fabricated parts and are generally of better quality than service stairs. Service stairs are built on location. They are generally constructed of construction lumber.

There are six general types of stairs commonly used in residential construction. They are straight-run, L stairs, double-L stairs, U stairs, winder stairs and spiral stairs.

Straight run stairs are the most common. They are the least expensive to build, but they require a long open space.

Common terms associated with stairs include:

Balusters:	vertical members that support the handrail on open stairs
Enclosed stairs:	stairs that have a wall on both sides (also known as closed, housed, or box stairs). These can be stairs leading down to a basement or cellar.
Headroom:	The shortest clear vertical distance measured from the nosing of the tread and the ceiling.
Housed stringer:	A stringer that has been routed or grooved to accommodate the treads and risers.
Landing:	The floor area at either end of the stairs; also the area between a set of stairs, such as in an L stairs.
Newel:	The main posts of the handrail at the top and bottom or at points where the stairs change direction.
Nosing:	The rounded projection of the tread which extends past the face of the riser.

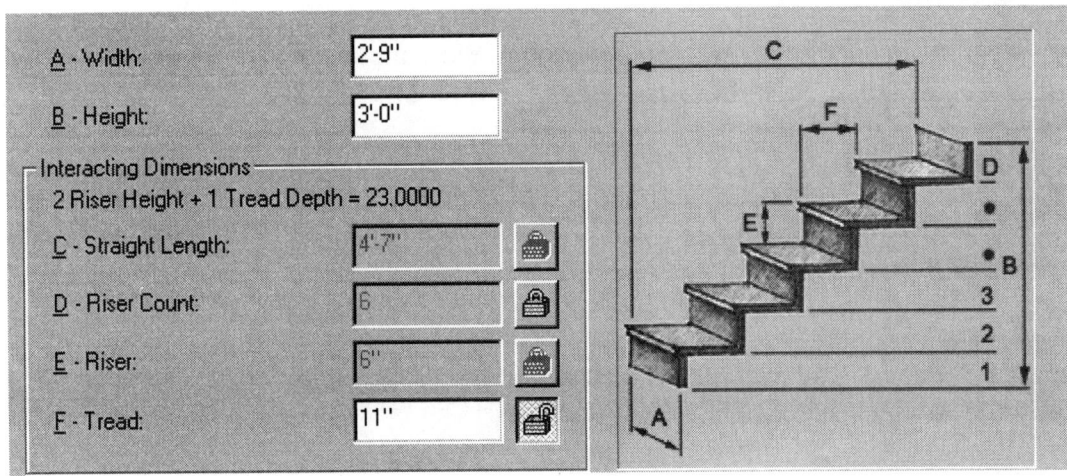

In the dialog box, the letter E indicates the riser.
The tread (where the foot is placed) is indicated by the letter F.
ADT automatically computes the Straight Length C (open space) required to place the stairs based on the Width A and the Height B.

We will be adding service stairs to our deck.

Lesson 7
Structural Members

Exercise 5:
Add Stairs

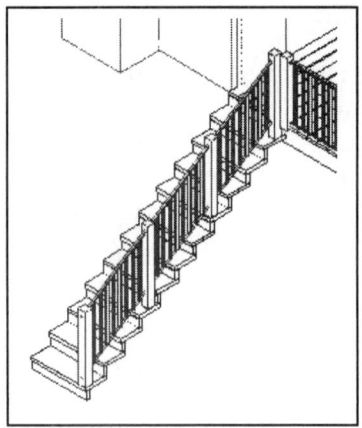

Drawing Name: Ex7-3.dwg
Estimated Time: 15 minutes

This exercise reinforces the following skills:

- Add Stairs
- Add Railing

Menu	Design->Stairs->Add Stair
Stairs Toolbar	
Command Line	StairAdd

Stairs run from the foot of the stair to the top of the stair, so before we can place our stairs, we need to reset our UCS so the Z depth is at the foot of the stairs.

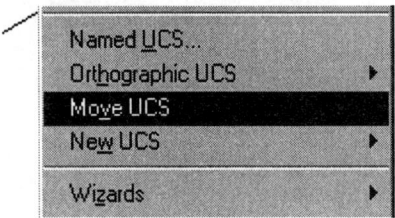

Under Tools, locate Move UCS.

7-40

Lesson 7
Structural Members

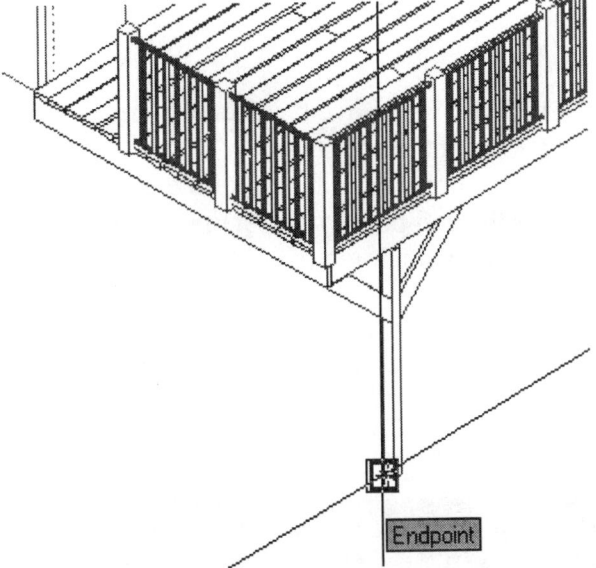

Select one of the endpoint corners of the columns placed to support the decking.

Start the AddStair command.

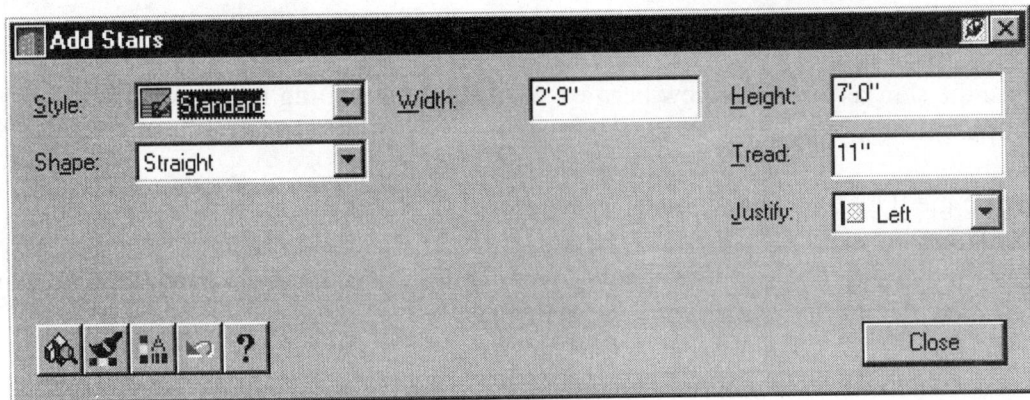

Set the Style to Standard
Set the Shape to Straight.
Set the Width to 2'-9".
Set the Height to 7'.
Set the Tread to 11".
Set Justify to Left.

Lesson 7
Structural Members

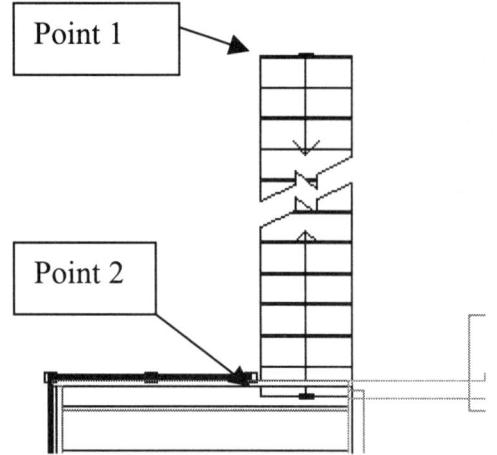

Pick a point above the deck for the start point of the stairs.
Pick a point near the top of the deck for the end point of the stairs.
Close the dialog box..

Switch to a NW view.

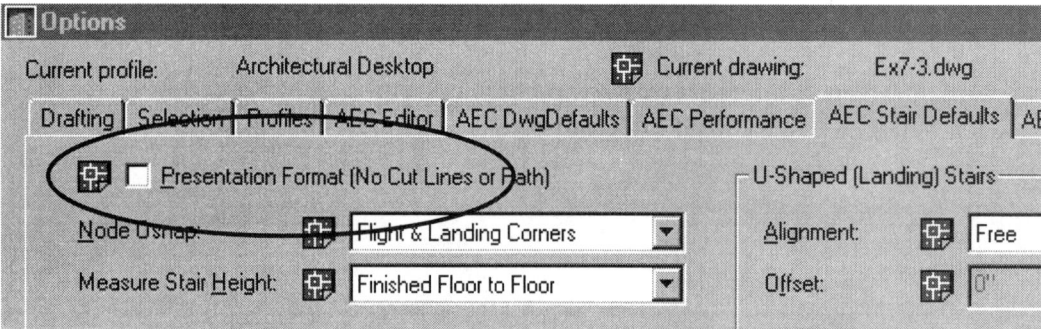

We see the stair as a broken view because of the Options setting in our AEC Stair Defaults.

Bring up the 'Options' Dialog box. Select the AEC Stair Defaults tab. Enable the Presentation Format.
Press 'Apply' and 'OK'. Notice that the stairs appearance does not change, even if you do a Regen. Erase the stairs and place them again.

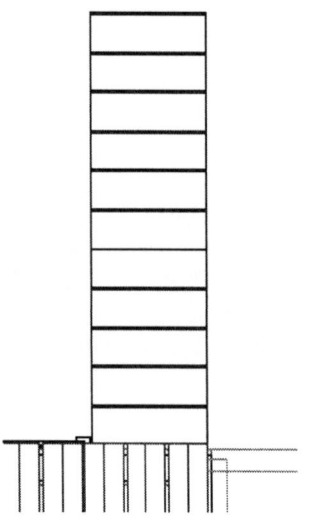

Now they appear without the break mark and arrows.

7-42

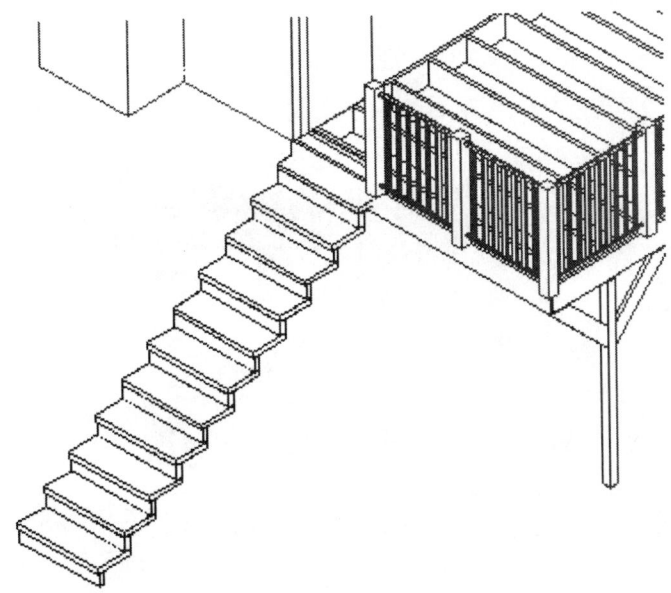

The stairs are placed, but they need to be shifted slightly to fit with the deck.
Use the MOVE tool to position the stairs properly.

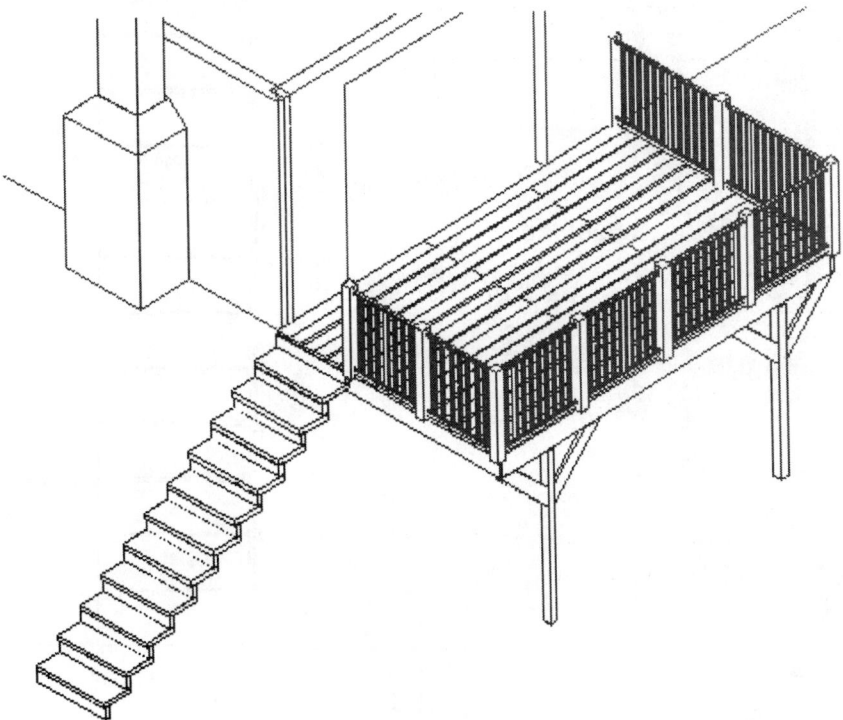

Safety regulations require a handrail down the stairs.
Switch back to a plan/top view to place the rail.

Lesson 7
Structural Members

RAILINGADD

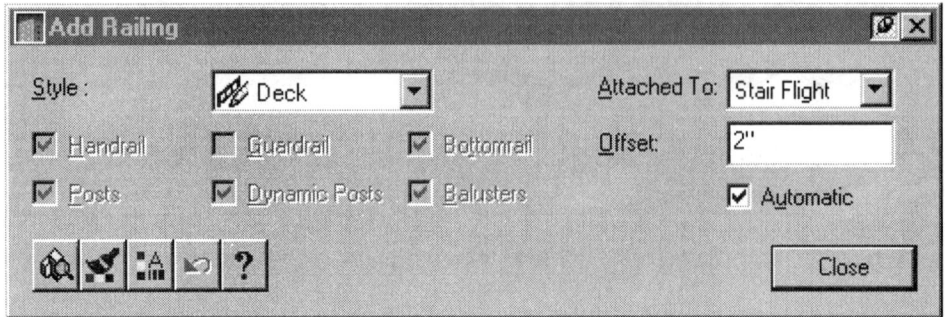

Set the Style to Deck.
Set Attached to: Stair Flight.
Set Offset to 2".

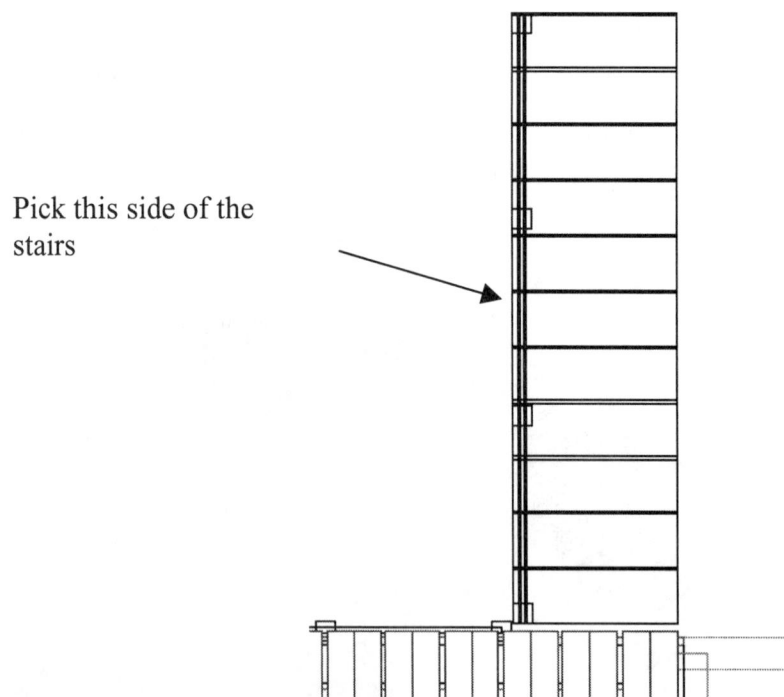

Pick this side of the stairs

You will be prompted to select the stairs.
Pick the left side of the stair to place railing on the left side.

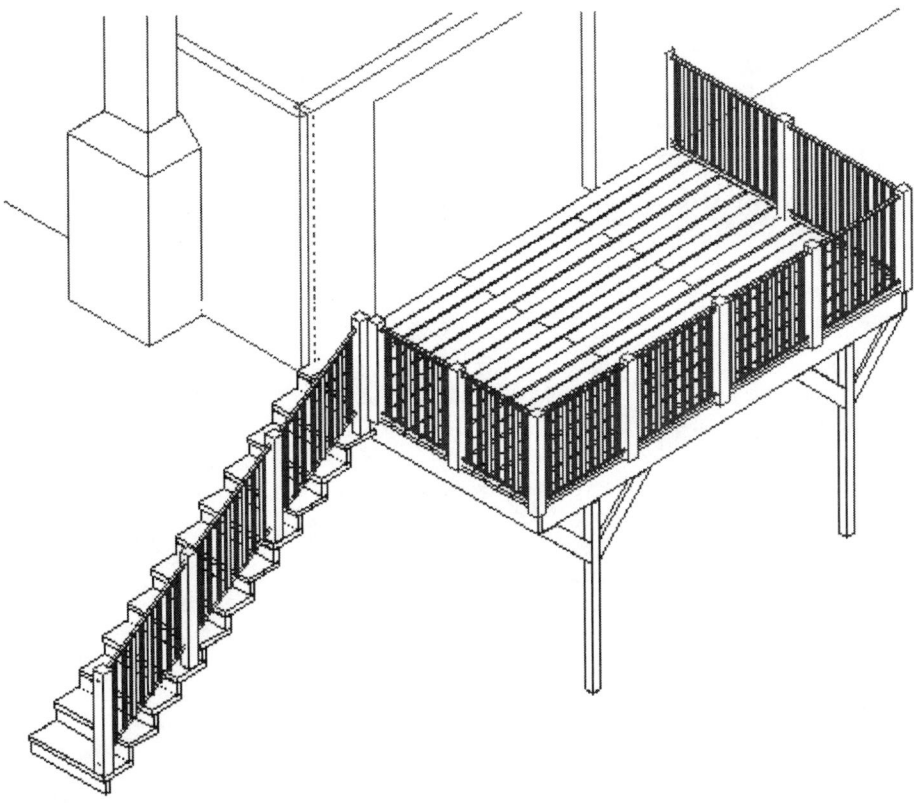

Save as Lesson 7.dwg

Notes:

Lesson 8
Layouts

Before we can create our construction drawings, we need to create layouts or views to present the design. Architectural Desktop is similar to AutoCAD in that the user can work in Model and Paper Space. Model Space is where we create the 3D model of our house. Paper Space is where we create or setup various views that can be used in our construction drawings. In each view, we can control what we see by turning off layers, zooming, panning, etc.

To understand paper space and model space, imagine a cardboard box. Inside the cardboard box is Model Space. This is where your 3D model is located. On each side of the cardboard box, tear out a small rectangular window. The windows are your viewports. You can look through the viewports to see your model. To reach inside the window so you can move your model around or modify it, you double-click inside the viewport. If your hand is not reaching through any of the windows and you are just looking from the outside, then you are in Paper Space.

You can create an elevation in your current drawing by first drawing an elevation line and mark, and then creating a 2D or 3D elevation based on that line. You can control the size and shape of the elevation that is generated. Unless you explode the elevation that you create, the elevation remains linked to the building model that you used to create it. Because of this link between the elevation and the building model, any changes to the building model can be made in the elevation as well.

When you create a 2D elevation, the elevation is created with hidden and overlapping lines removed. You can edit the 2D elevation that you created by changing its display properties. The 2D Section/Elevation style allows you to add your own display components to the display representation of the elevation, and create rules that assign different parts of the elevation to different display components. You can control the visibility, layer, color, linetype, lineweight, and linetype scale of each component. You can also use the line work editing commands to assign individual lines in your 2D elevation to display components, and merge geometry into your 2D elevation.

After you create a 2D elevation, you can use the AutoCAD BHATCH and AutoCAD DIMLINEAR commands to hatch and dimension the 2D elevation.

Lesson 8
Layouts

Exercise 1:
Creating Elevation Views

Drawing Name: Lesson 7.dwg saved from the previous lesson.
Estimated Time: 15 minutes

This exercise reinforces the following skills:

- VPORTs
- Locking and Unlocking Display

Activate the Plot-SEC layout tab. (You may have to use the scroll tools in order to access it.)

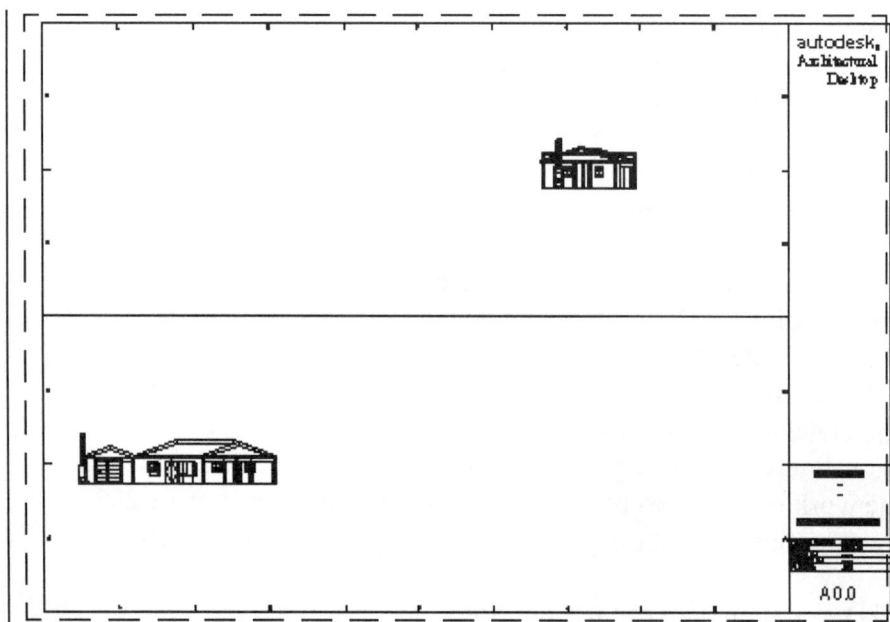

You should see two viewports. The top viewport shows a side elevation view. The bottom viewport shows a front elevation view.

Lesson 8
Layouts

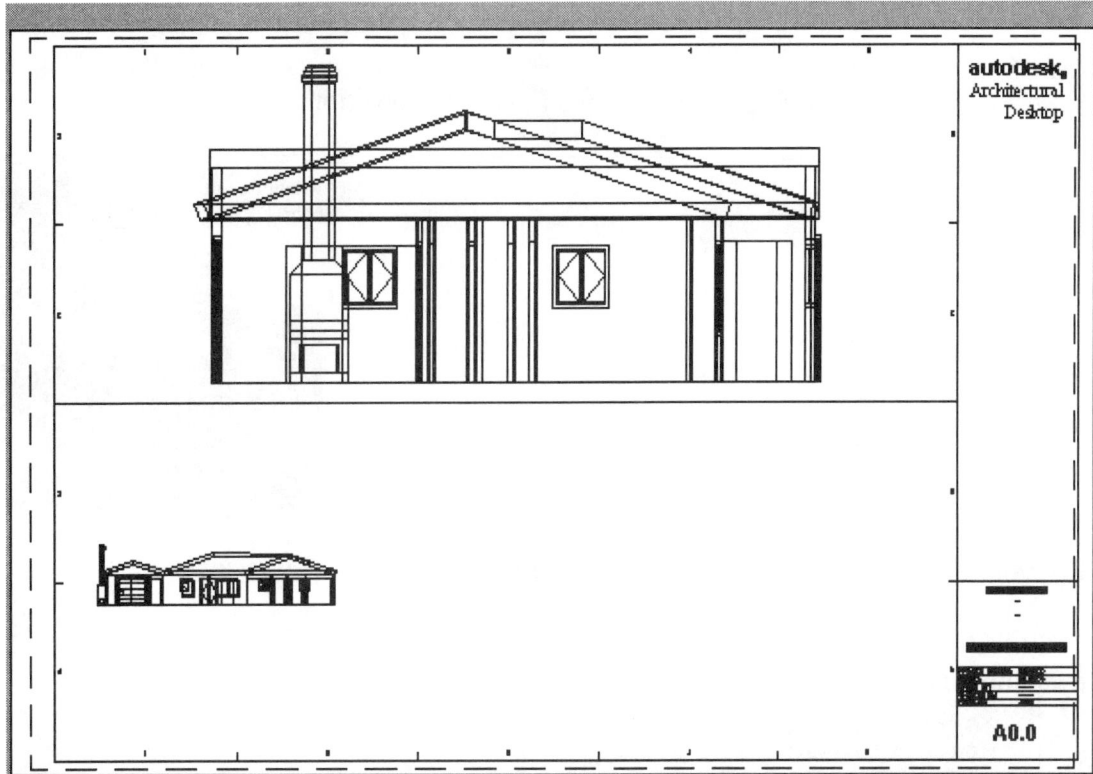

Plotted pages always have the scale identified, and it should be a standard architectural scale unless a perspective view is used.

Use the AutoCAD Viewports toolbar to set the viewing scale to ½" = 1'-0" and pan to locate the house appropriately."

Pick the top viewport.

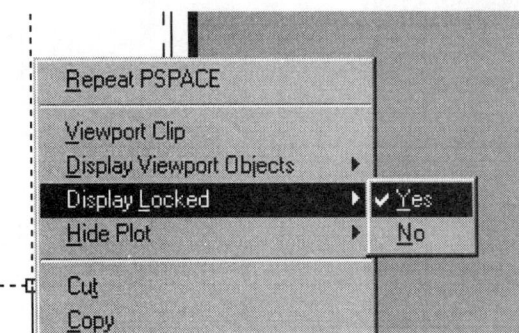

Right click and set the 'Display Locked" to No.

Lesson 8
Layouts

Set the Scale to ½" = 1'.
Double click inside the viewport and pan the model into position.

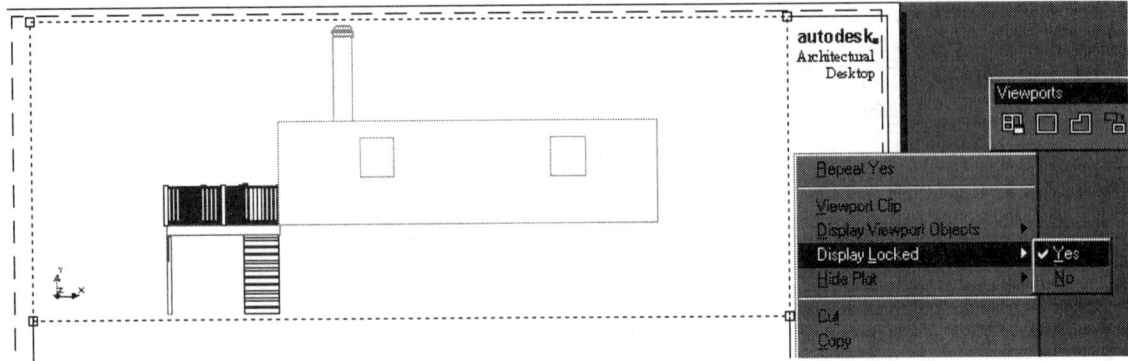

Once your display looks OK, select the Viewport and lock the display again.

Repeat process for the second viewport.

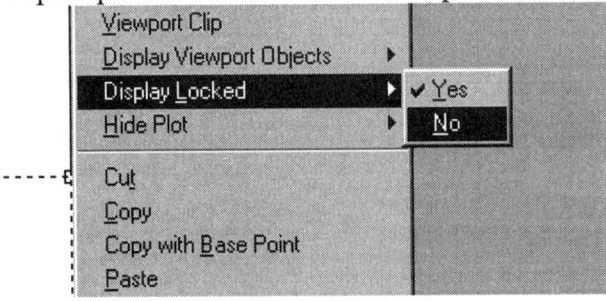

Select the bottom viewport.
Right click and set the Display Locked to No.

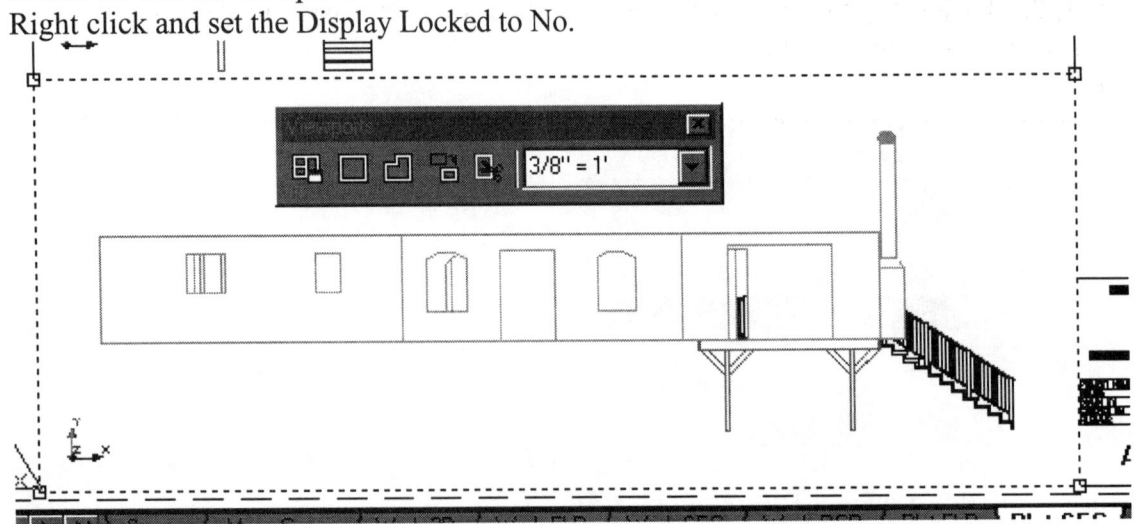

Set the scale to 3/8" = 1'.
Activate the viewport and pan to get the model located appropriately.
Right click and set the Display Locked to Yes.

8-4

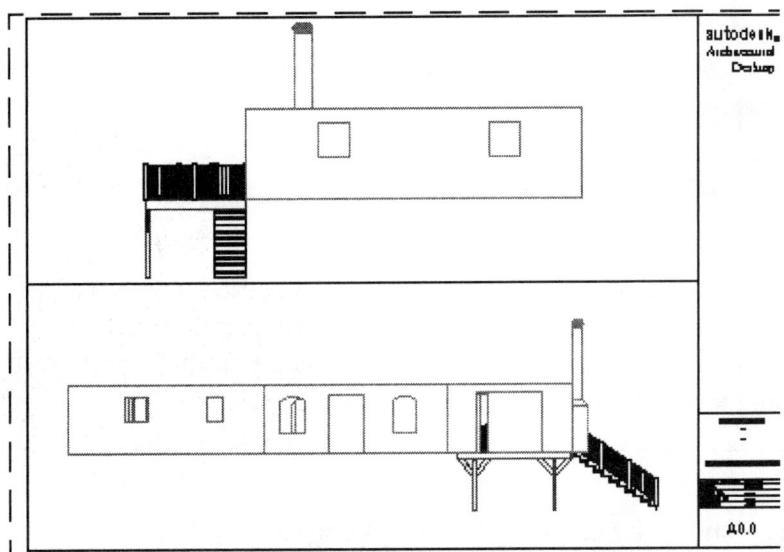

Save the file as Ex8-1.dwg

 TIP: To turn off the UCSICON, activate the viewport. Type UCSICON and OFF at the command line. You can only turn the UCSICON OFF/ON in model space.

 TIP: By locking the Display you ensure your model view will not accidentally shift if you activate the viewport.

Sections

You can draw a section line in your drawing by specifying a start point, an endpoint, a length, and a height for the section. You can specify additional points between the start and endpoint to create jogs in your section. The section line acts as a cutting plane when you create the section, slicing a section from the building model.

>
> **TIP:** If you draw your section line by choosing Documentation ⬀Sections ->Add Section Line, the section line includes two section marks. You can also create a Section line and mark by choosing Documentation-> Documentation Content-> Section Marks.

Section marks are placed to create a cutting plane line and a section identifier. The section identifier includes the section number and the sheet number where the section is located. ADT comes with four standard section marks: two allow you to define a section identifier and two allow you to define a section identifier and a sheet number. You can also create custom section marks and define them using Create AEC Content.

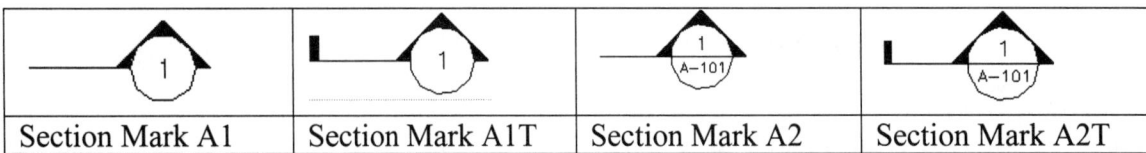

| Section Mark A1 | Section Mark A1T | Section Mark A2 | Section Mark A2T |

Menu	Documentation->Documentation Content->Section Marks
Toolbar	
Command Line	ANNOSECTIONMARKADD

When you place a section mark, you are prompted 'Add AEC Object?' A yes response to this prompt will use the section line as a section object for the development of a section. This allows you to generate a section drawing.

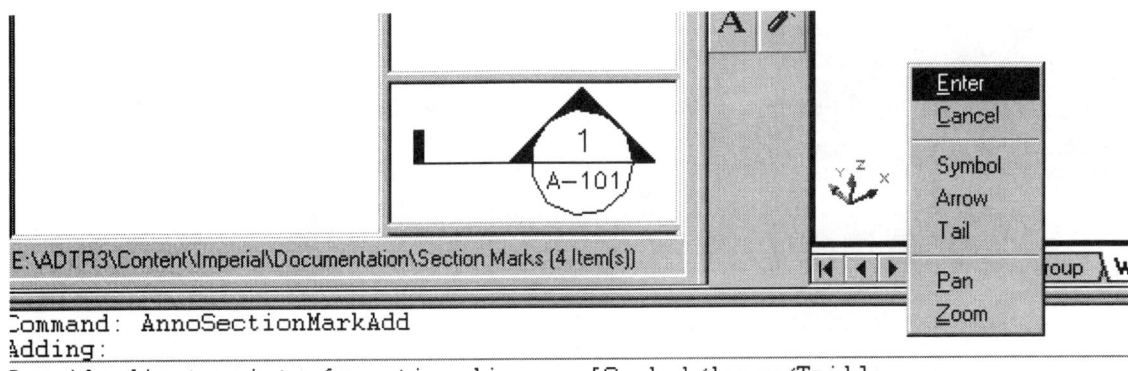

The command AnnoSectionMarkAdd has several options that allow the user to specify the symbol, the arrow, and whether or not the section mark has a tail. You can specify a custom symbol or a custom arrow as long as these drawing files reside in the Content path.

TIP: ADT only allows the user to specify a single path for Content. You can create a custom subdirectory under the Content path in order to protect your custom files.

The section line and mark should be placed on the A-Sect-Iden layer and is greenish-blue in color.

It will not automatically be placed onto the correct layer unless you drag and drop from the Design Center and have your layer standards set up properly.

Exercise 2:
Creating Section Views

Drawing Name: Ex8-1.dwg saved from the previous lesson.
Estimated Time: 15 minutes

This exercise reinforces the following skills:

- Creating a Layout
- Creating Section Lines
- Creating Section Marks
- Using AEC Content

Lesson 8
Layouts

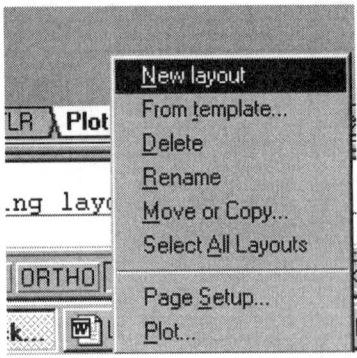

Right click on any of the existing layout names. Select 'New Layout' from the popup menu.

A new tab will appear at the end of all the existing tabs. The new tab is called Layout1.

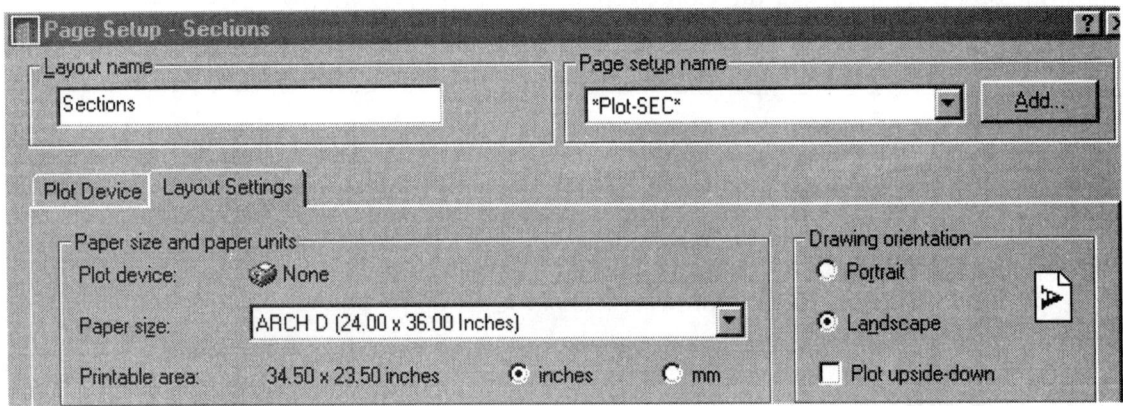

Left clicking (selecting) the Layout1 tab will bring up the Page Setup dialog box.
This allows you to set the page size, scale, printer, etc. for the new layout.
In the Layout name edit box, change the name from Layout1 to Sections.
In the Page Setup name drop down, select "Plot-SEC".
This automatically sets a paper size of ARCH D (24.00x 36.00 Inches)

TIP: This dialog box comes up automatically the first time you select a newly created layout. To modify any of the settings after the initial setup, you can select the tab, right click, and select 'Page Setup'.

Lesson 8
Layouts

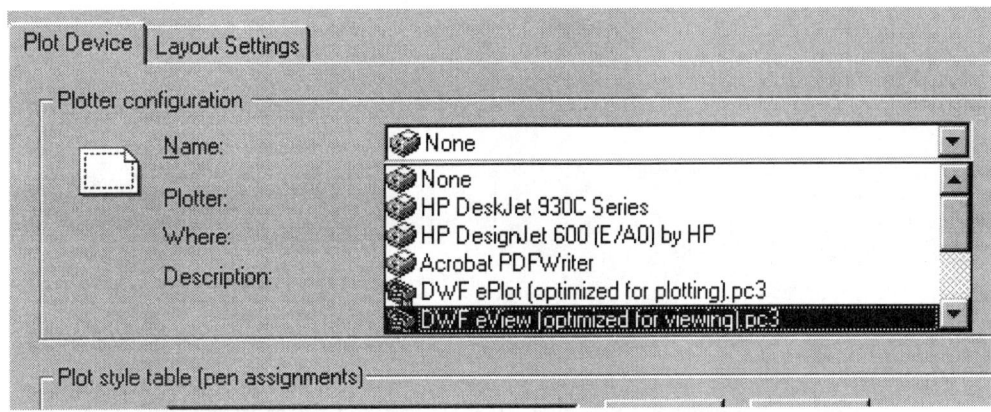

Select the Plot Device tab, and set the Plotter to DWF eView.
This allows you to view your plot no matter what type of plotter/printer you have on your system.

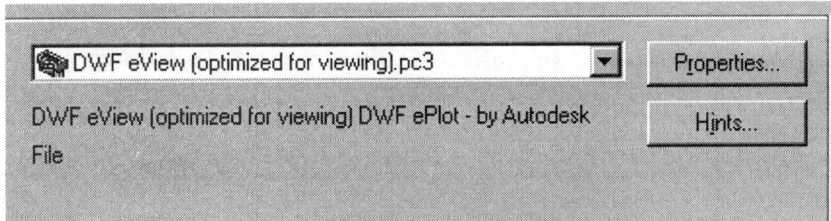

Select the 'Properties' button.

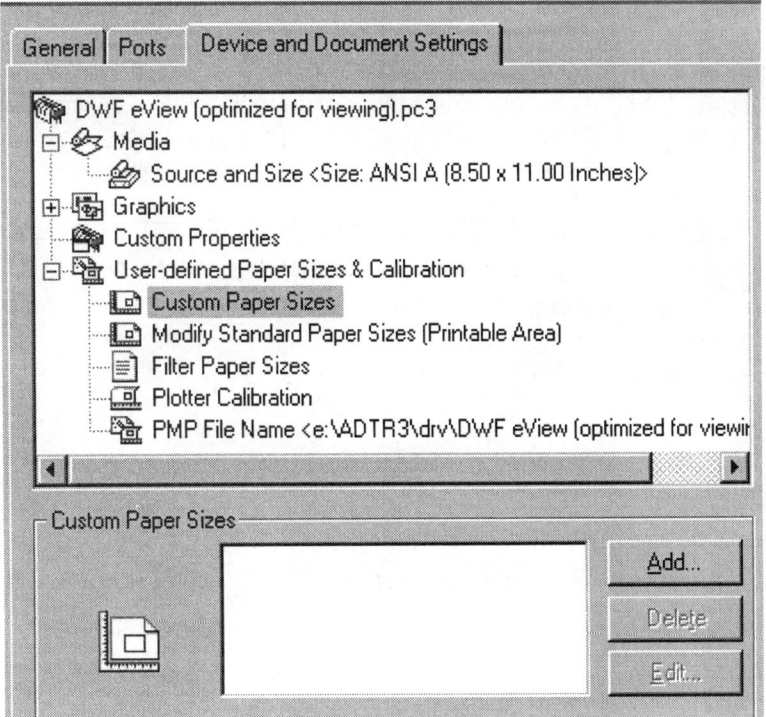

Lesson 8
Layouts

Select the Device and Document Settings tab.
Highlight Custom Paper Sizes.
Press the 'Add' button.

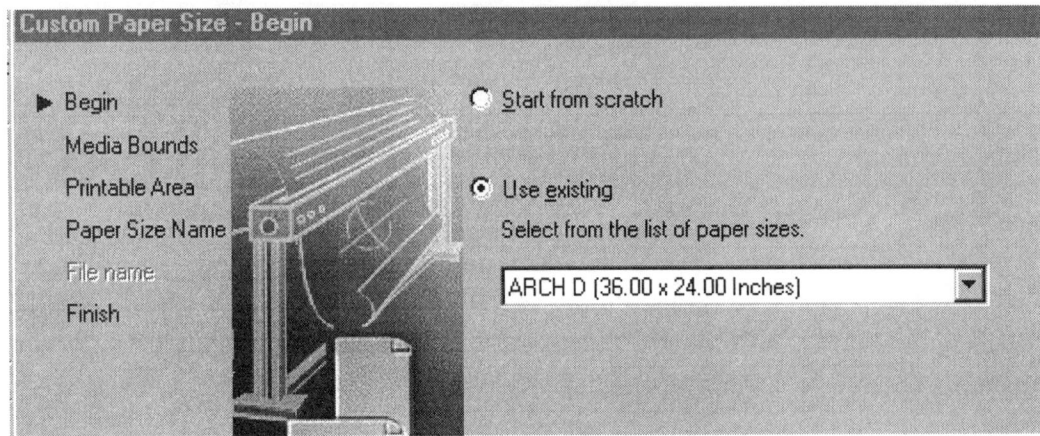

Enable the 'Use Existing' button and locate the ARCH D size from the drop down list.
Press 'Next'.

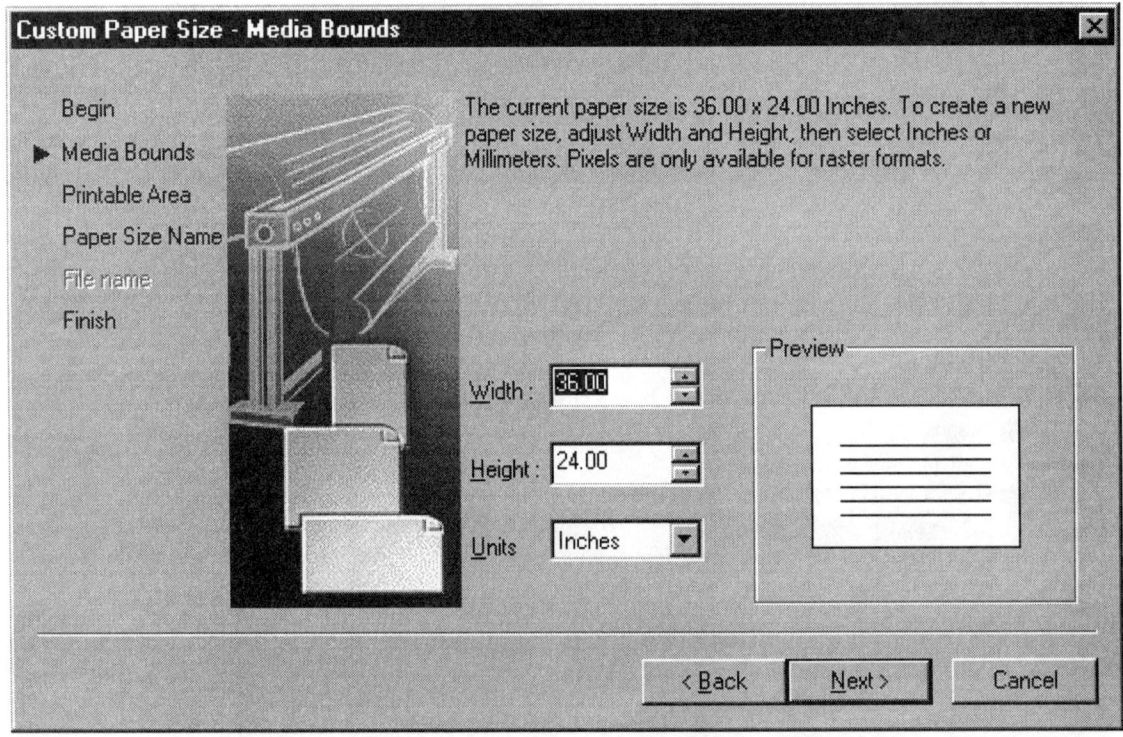

You do not need to alter any of the defaults for the Media Bounds dialog box.
Press 'Next'.

8-10

Lesson 8
Layouts

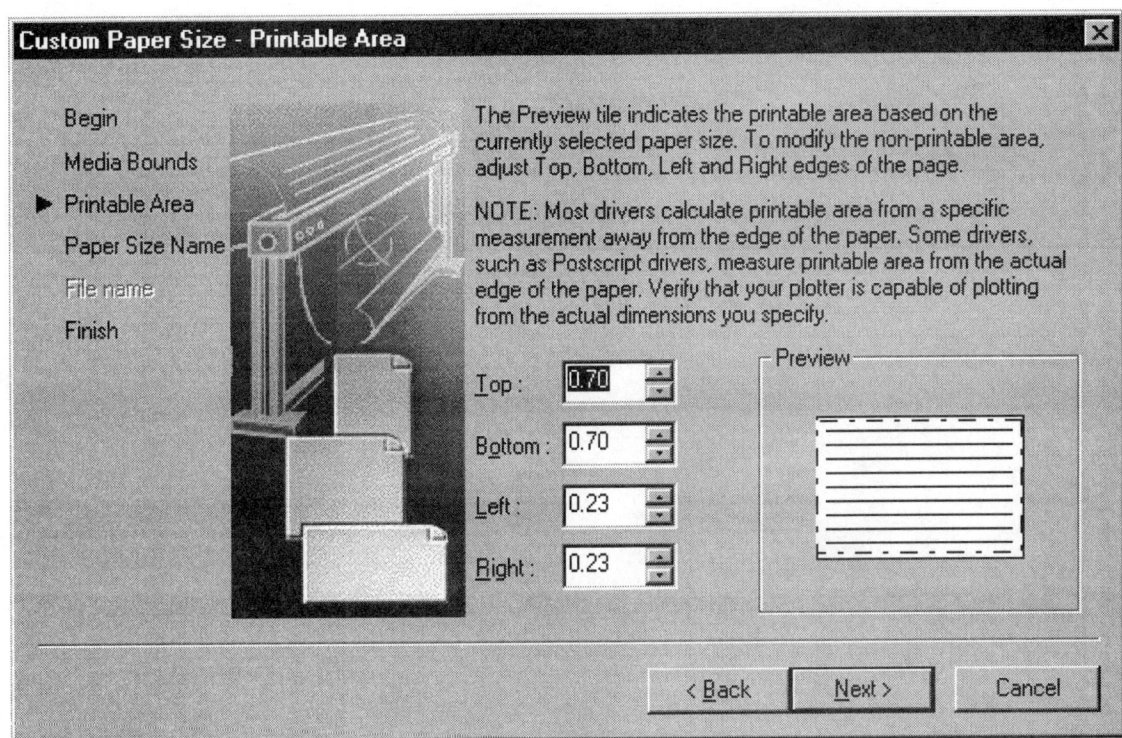

You do not need to alter any of the defaults for the Printable Area dialog box. Press 'Next'.

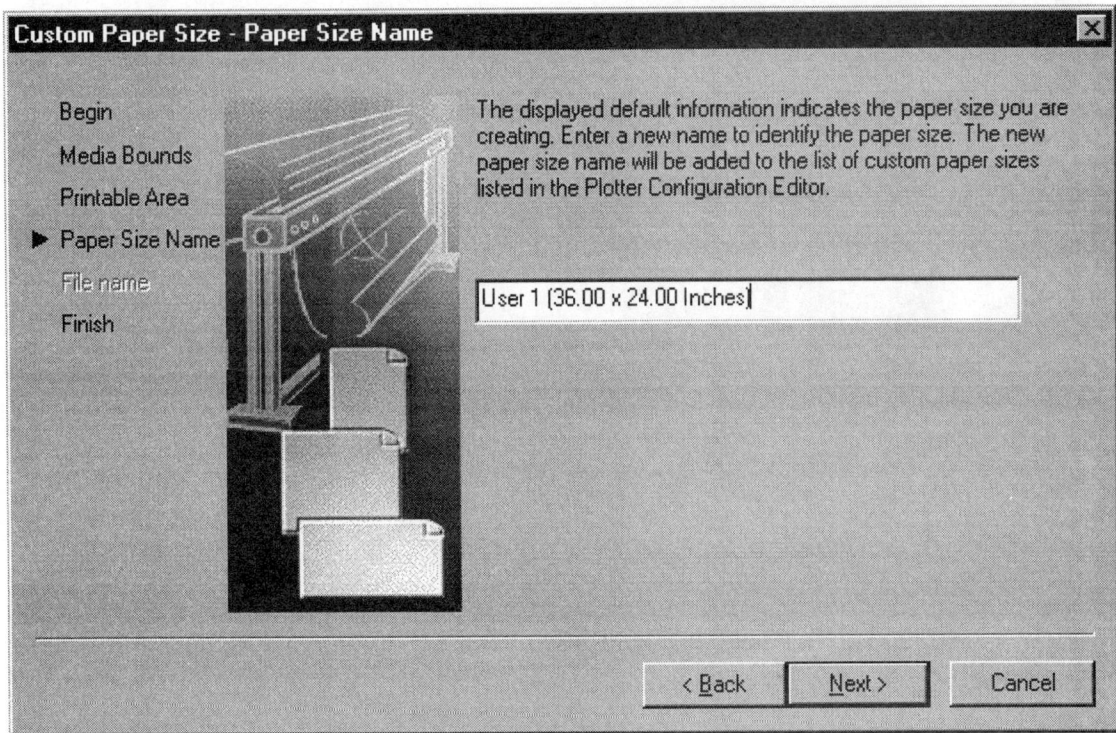

In the Paper Size Name, edit box you could assign your paper settings a special name, but you do not need to alter the default. Press 'Next'.

Lesson 8
Layouts

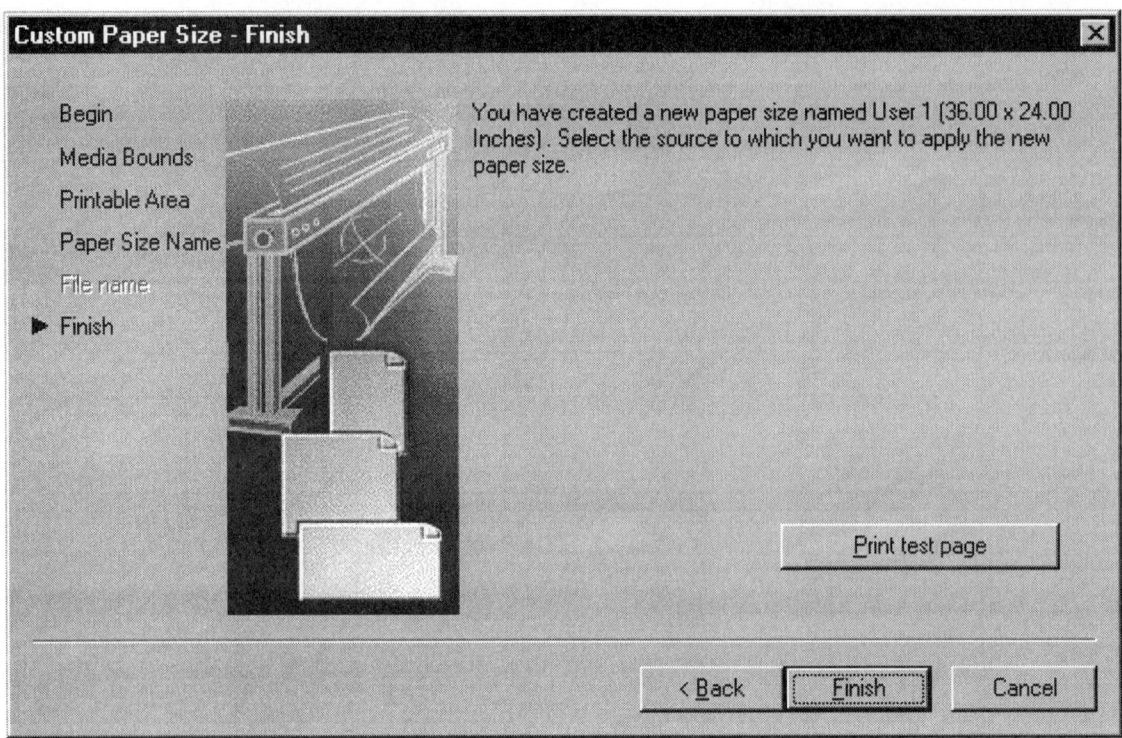

Press 'Finish'.

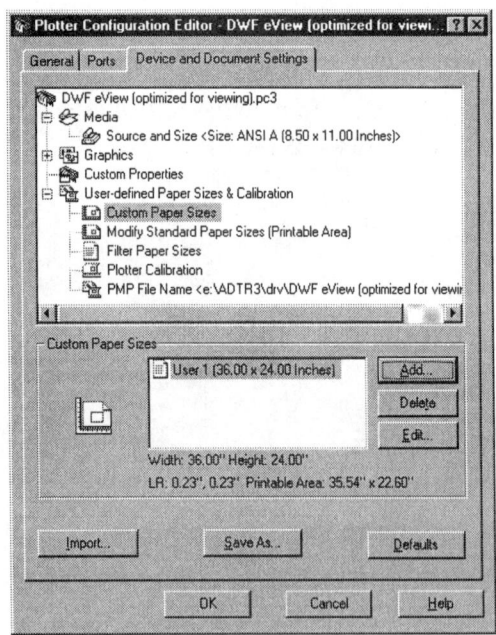

Your Custom Paper Size appears in the dialog box.
Press 'OK'.

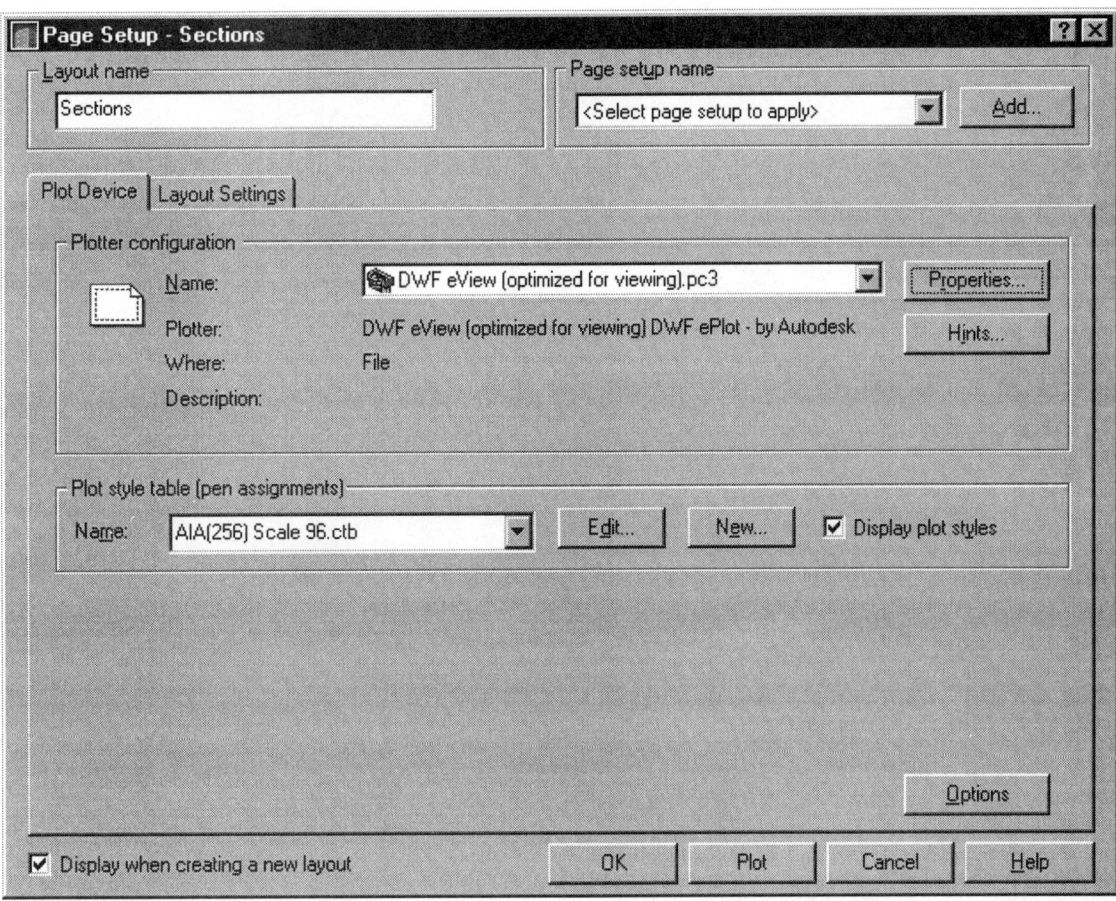

Note that if you have the DWF eView set up, then the Page setup name is not required. If you have the Page setup name specified, you don't need the DWF eView.
Press 'OK' to exit.

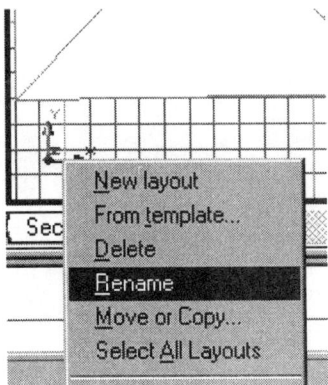

You can also rename a layout by selecting the tab, right click and select 'Rename'.

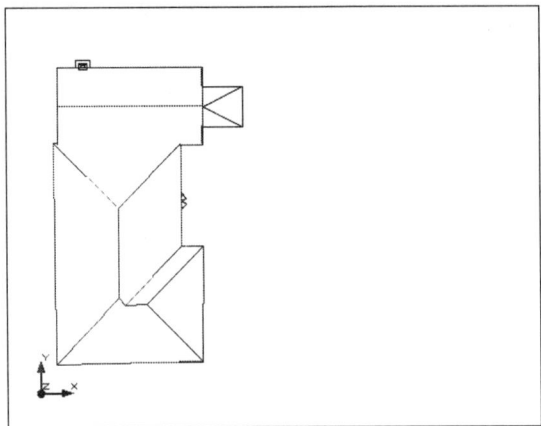

Select the Sections tab. You should have a single viewport.

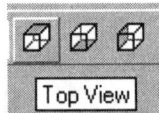

Select the Top View icon.

When prompted to select a viewport, pick the viewport in your graphics area.
It will automatically switch to a top view.

If you set the Model tab to top view then all new viewports will default to that view.

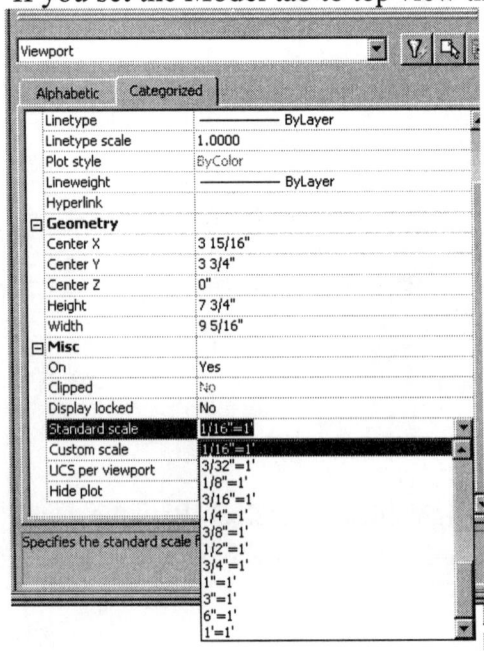

Pick the viewport. Right click and select 'Properties'.

Locate the Standard Scale field.
1/16" = 1'-0 will fit well on a nominal D sheet.

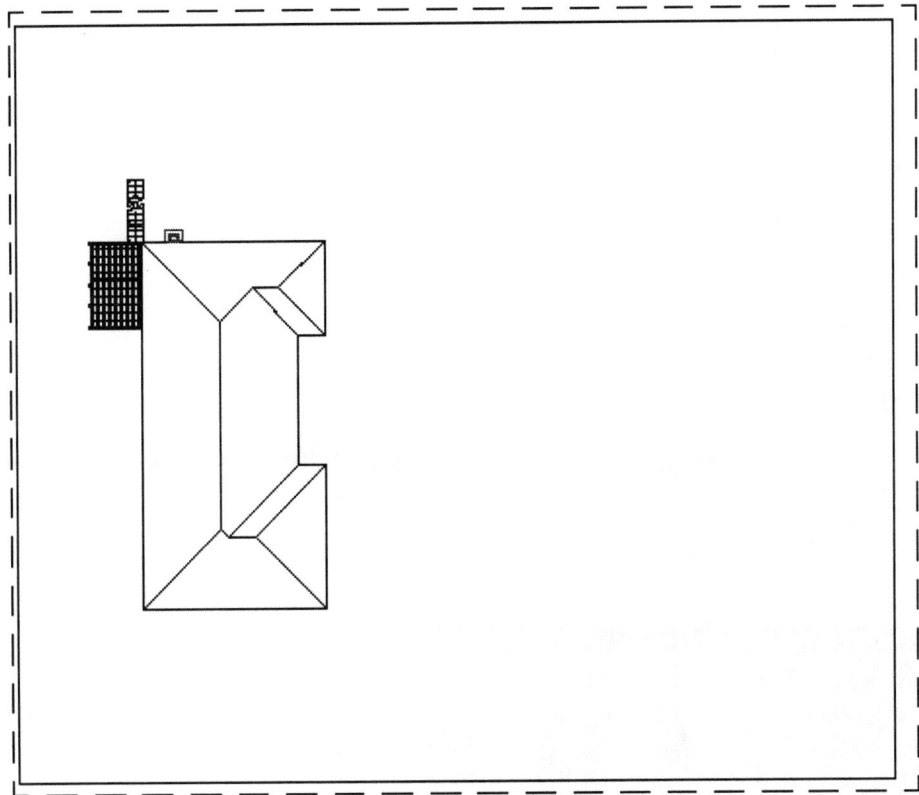

Activate MODEL Space.
Use PAN to move the model to the left side of the viewport.
Turn the UCSICON off
Lock the Display to ensure the scale stays at 1/16" = 1'.

MENU	Documentation>Sections>Add Section Line
Sections Toolbar	
Command Line	BLDGSECTIONLINEADD

Select the Add Section Line tool.

Lesson 8
Layouts

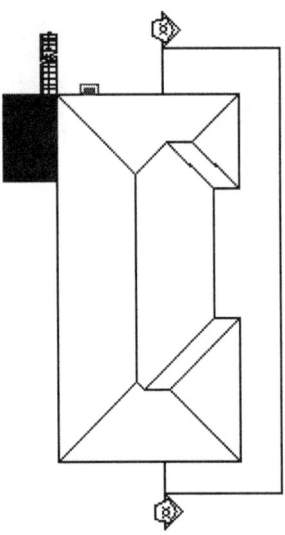

Place the section line as shown, going from top to bottom. Setting ORTHO ON will help create a straight line.
Set the length as the default.
Set the height as 20'.

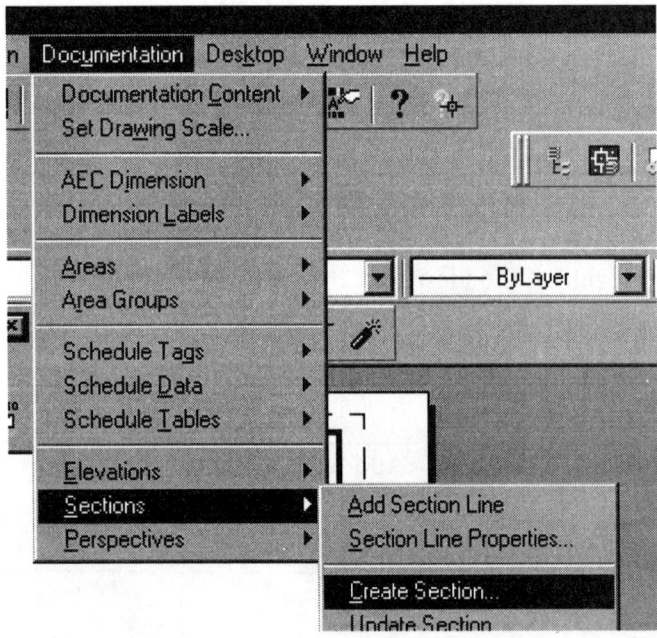

Menu	Documentation->Sections->Create Section
Sections Toolbar	
Command Line	BldgSectionLineGenerate

Lesson 8
Layouts

Select the 'Create Section' tool.
Select the section line you just created.

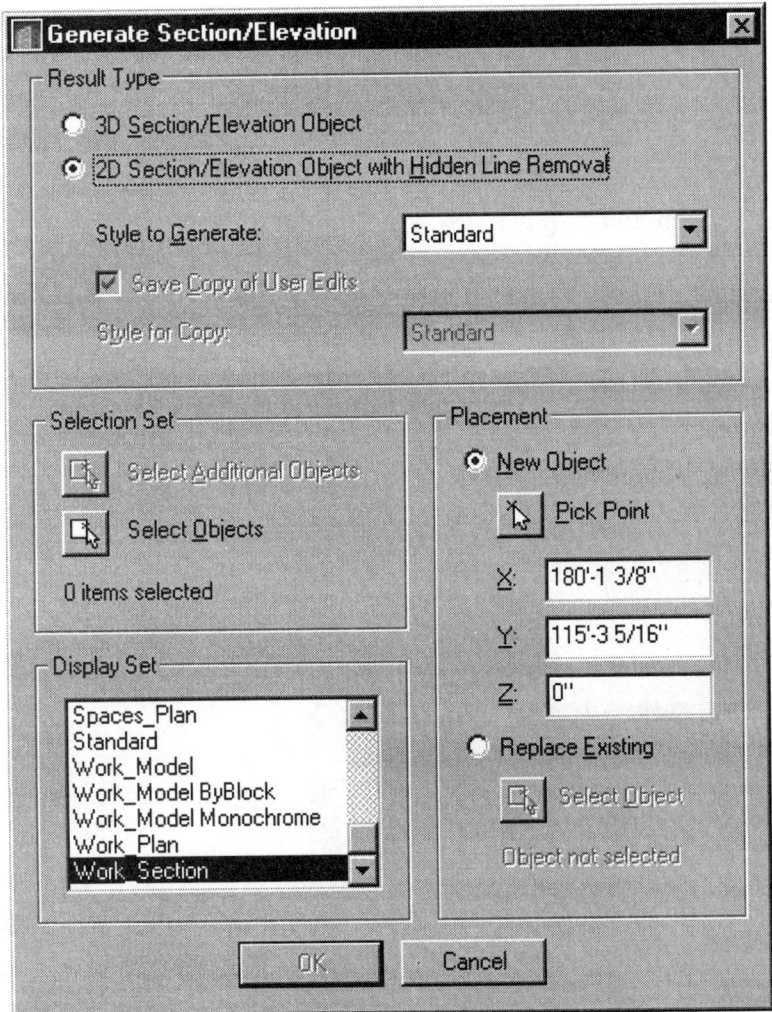

Set the Result Type to 2D Section

For the Selection Set, window around the entire building. Do not include the section line and viewport.

Set the Display Set to Work_Section.

Pick a point to the right of your plan view to place your elevation. You may need to enlarge your viewport or shift your views.

8-17

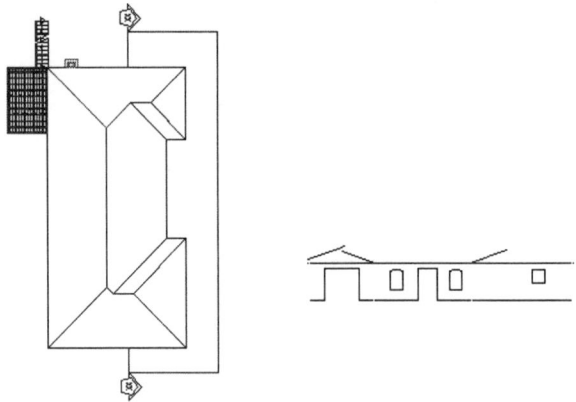

You can change the way the section line displays.

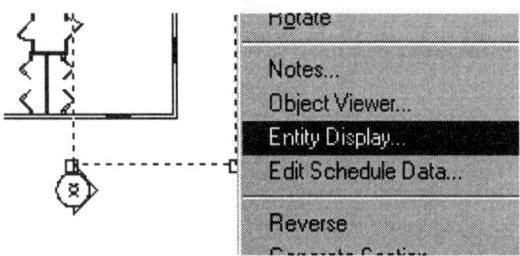

Select the Section Line.
Right click and select 'Entity Display'.

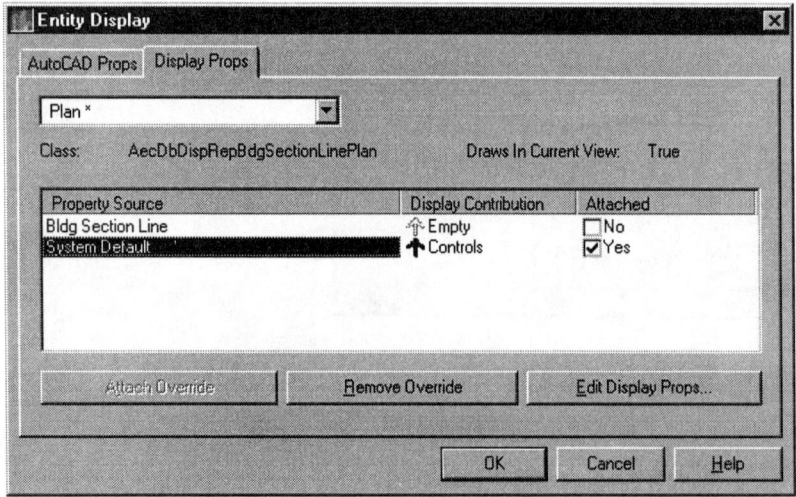

Select the 'Edit Display Props...' button.

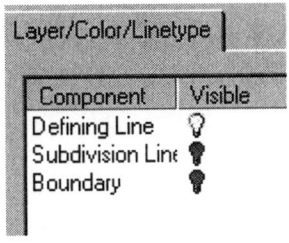

Set the Subdivision Line and Boundary so Visibility is turned OFF.

Press 'OK' twice to exit the dialog box.

You need to REGEN to see your changes take effect.

Select the Section Annotation Block.
Right click and select 'Multi-View Block Properties'.

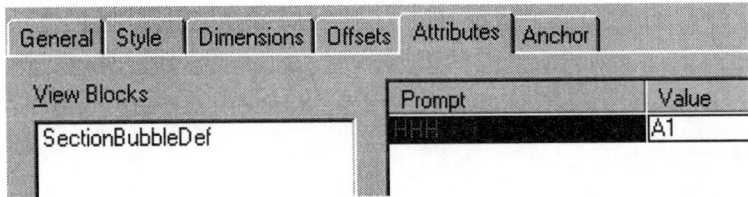

Select the Attributes tab.
Change the Value to 'A1'.
Press 'OK'.
Repeat for the other block.

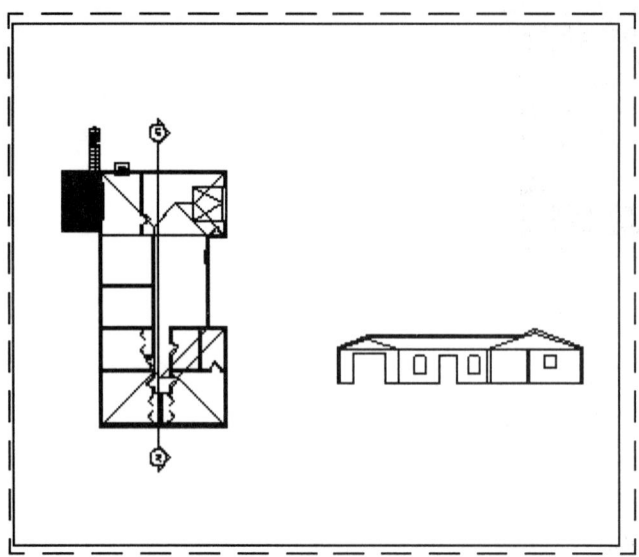

Create a Layer User Group for your 2D Section.

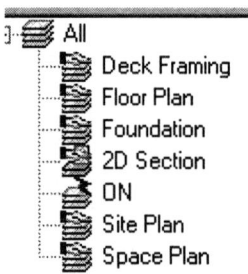

Call it 2D Section.

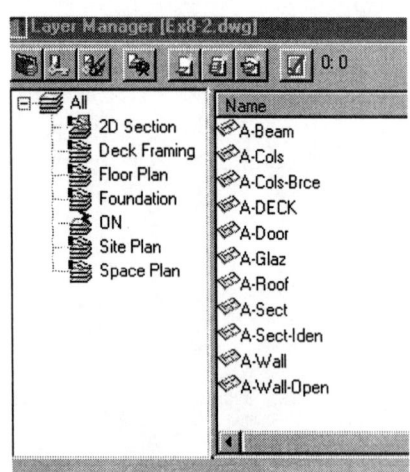

Place the layers shown into the 2D Section group.

Lesson 8
Layouts

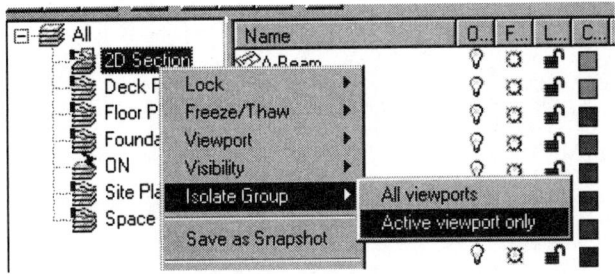

Select 'Isolate Group->Active Viewport only'.

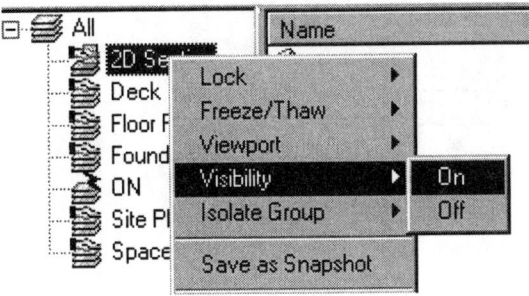

Set Visibility 'ON'.
Save your drawing as Ex8-2.

Exercise 3:
Creating 3D Section Views

Drawing Name: Ex8-2.dwg saved from the previous lesson.
Estimated Time: 15 minutes

This exercise reinforces the following skills:

- Creating a Layout
- Creating Section Lines
- Creating Section Marks
- Using AEC Content

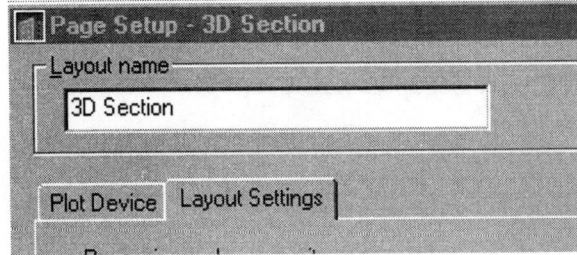

Create a new layout and name it 3D Section.

Lesson 8
Layouts

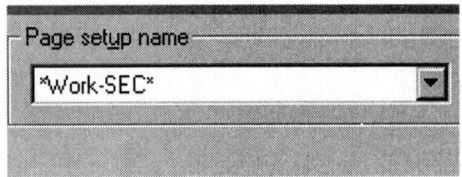

Set the Page setup name to Work-SEC.
Press 'OK'.
Click on it to make it current.

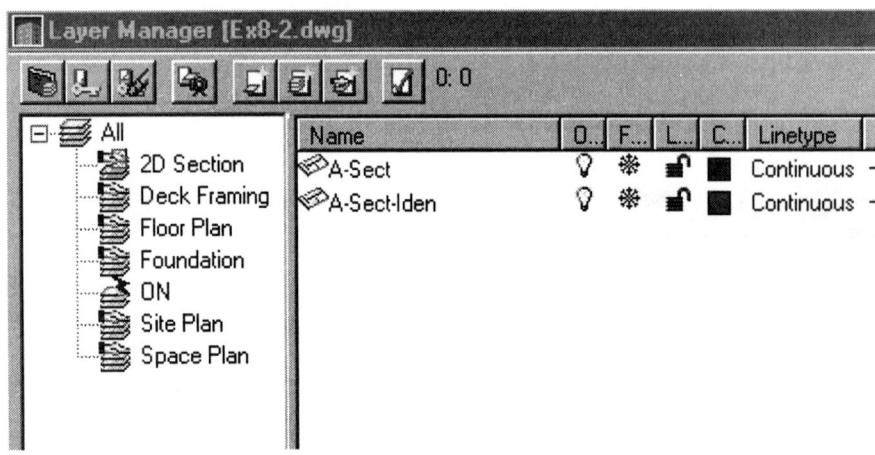

Freeze the two layers used for the 2D Section.

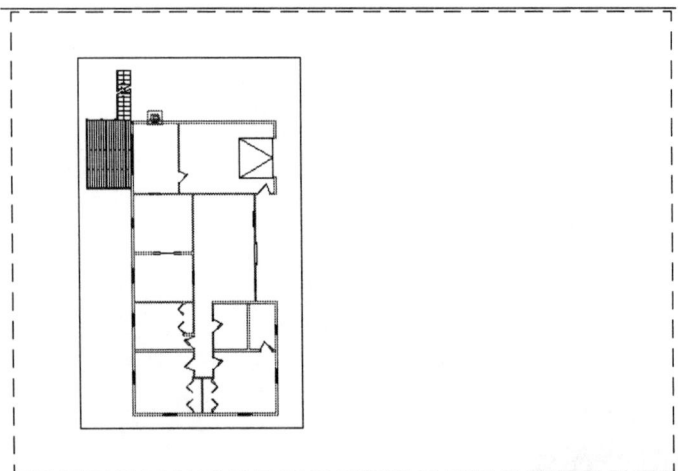

Use STRETCH to set the single viewport so it only occupies half the paper.
Set the viewport to TOP view.
Set the Scale to ¼" = 1'.
Lock the Display.

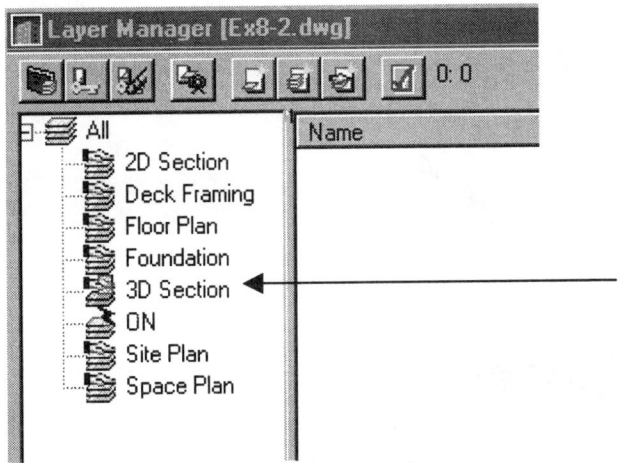

Create another Layer User Group called '3D Section'.

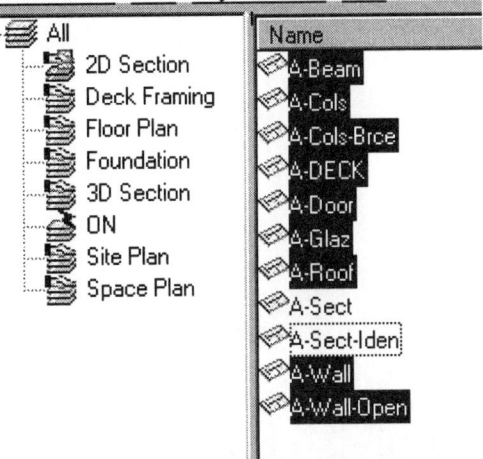

Go to the 2D Section Group and select all the Layers except A-Sect and A-Sect-Iden.

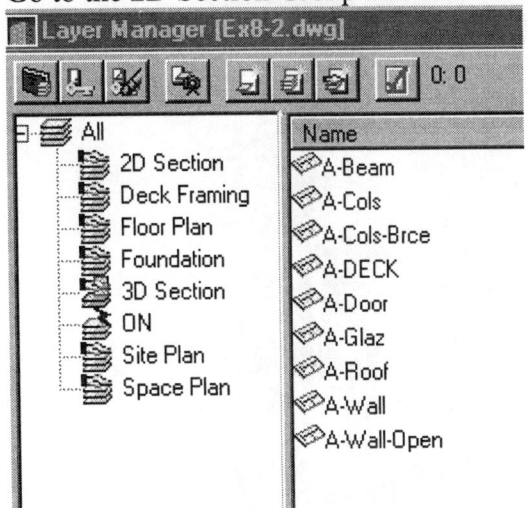

Drag and drop the selected layers into the 3D Section group.

Lesson 8
Layouts

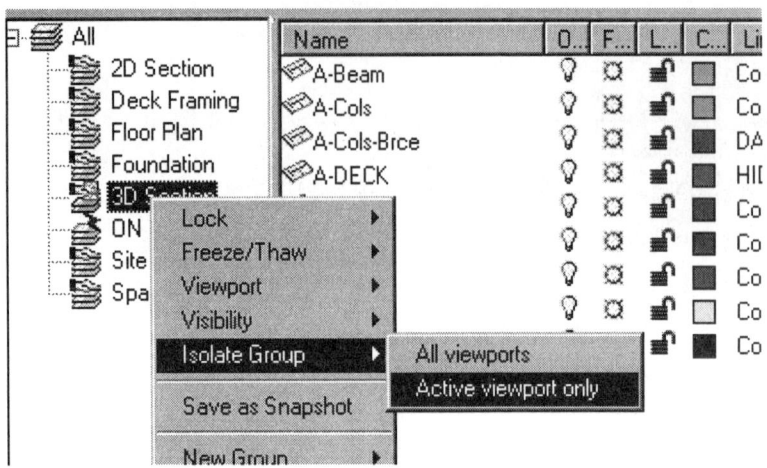

Highlight 3D Section.
Right click and select 'Isolate Group->Active Viewport only'.

Select the Viewport.

Right click and select 'Copy with Base Point'.

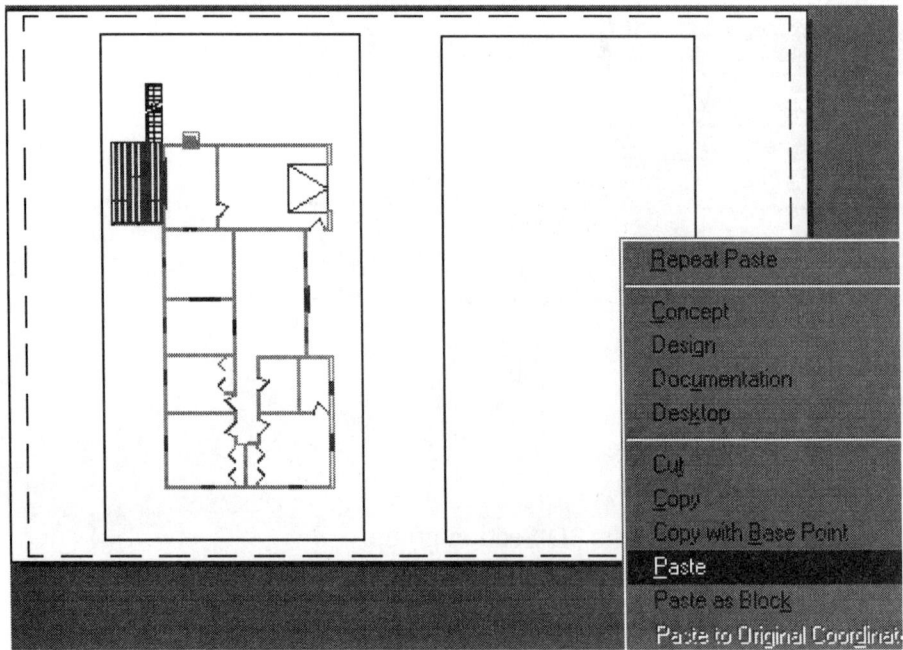

8-24

Right click and select 'Paste'.
Place the second viewport next to the first as shown.

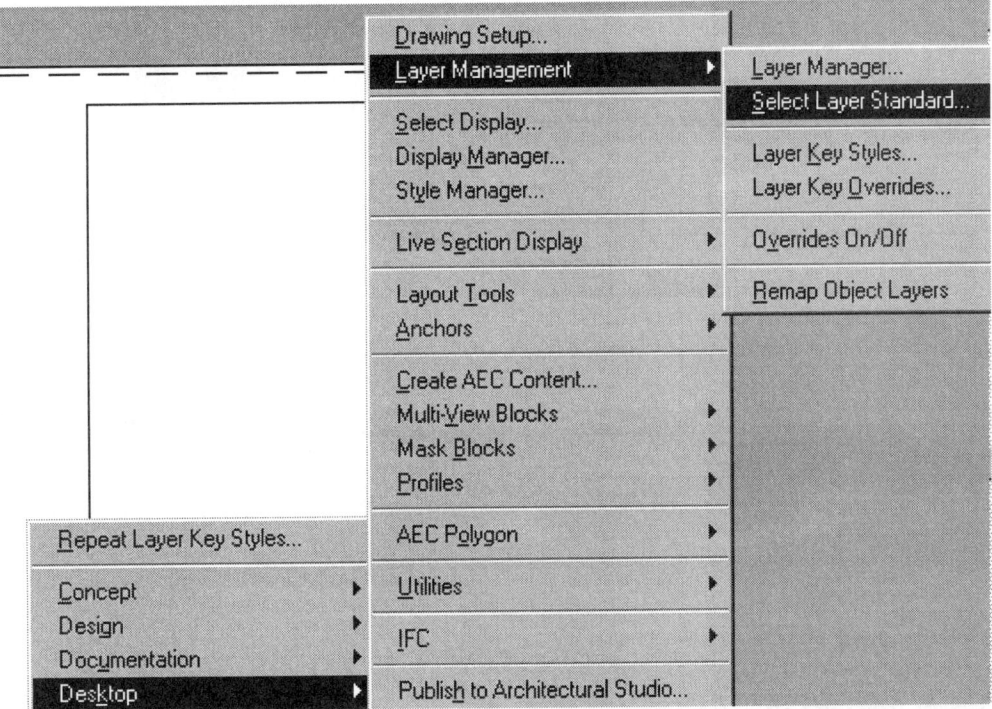

Right click anywhere in your graphics window.
Select Desktop->Layer Management->Select Layer Standard.

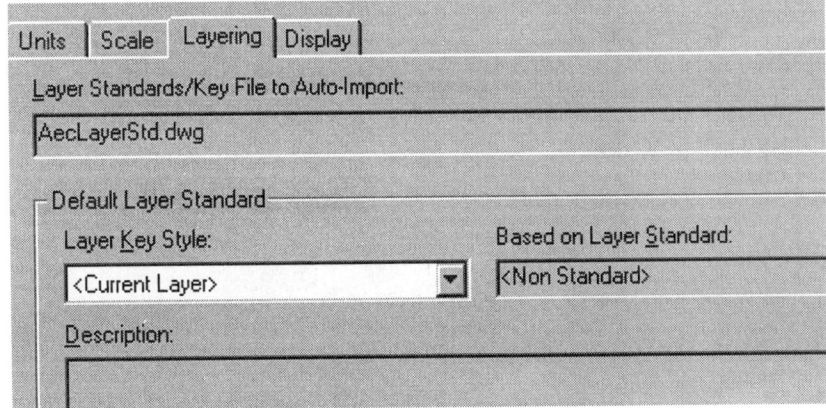

Select the Layering tab.
Set the Layer Key Style to Current Layer.

Press 'Apply' and 'OK'.

Lesson 8
Layouts

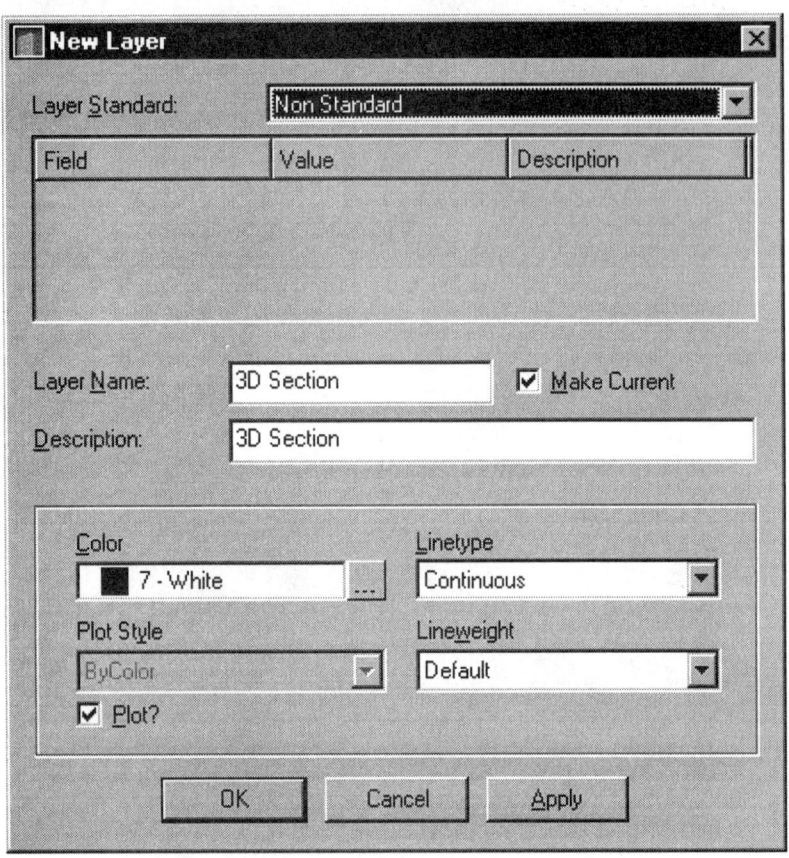

Create a New Layer called 3D Section in the 3D Section Layer User Group. Set the Color as White and the Linetype as Continuous.

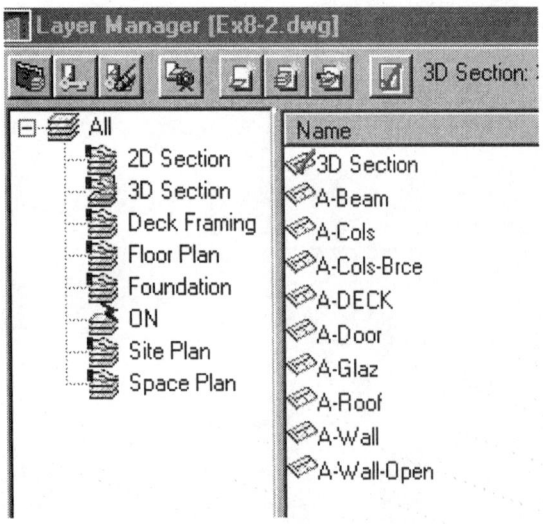

8-26

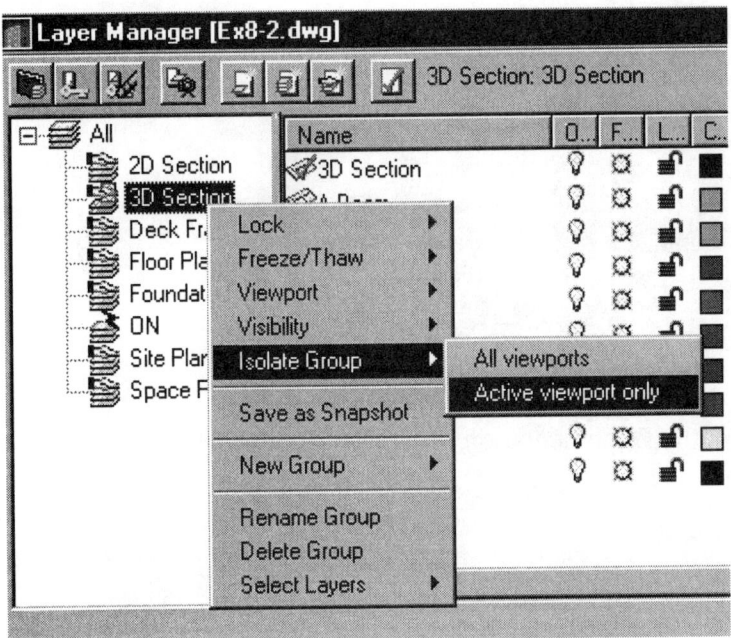

Activate the Right viewport.
Set the viewport for the 3D Section Layer Group.

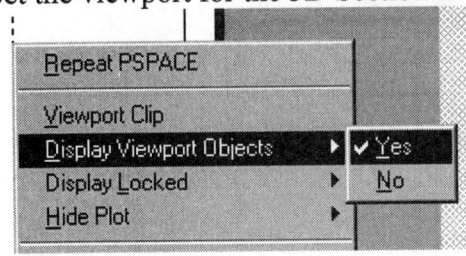

If you don't see anything in the Right Viewport, select and make sure the 'Display Viewport Objects' is set to 'Yes'.

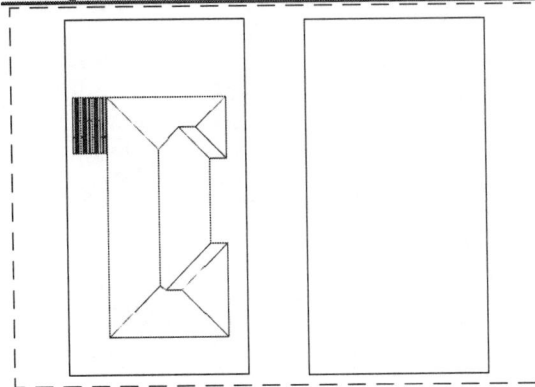

Pan the model space so that you have an empty area displayed in the right viewport.

Lesson 8
Layouts

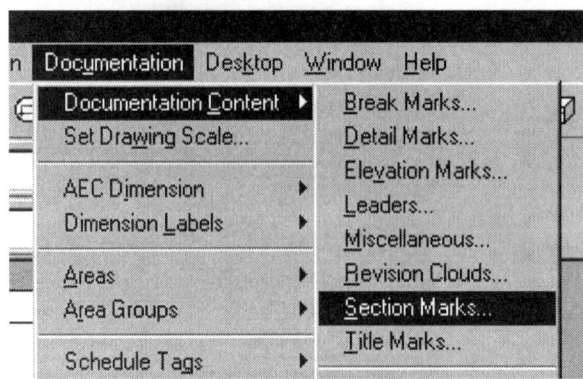

Select the Section Mark tool from the Documentation –Imperial Toolbar or select from the Documentation Content menu.

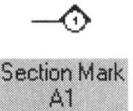

Activate the plan view viewport on the left.
Select Section Mark A1 to drag and drop into the left viewport.

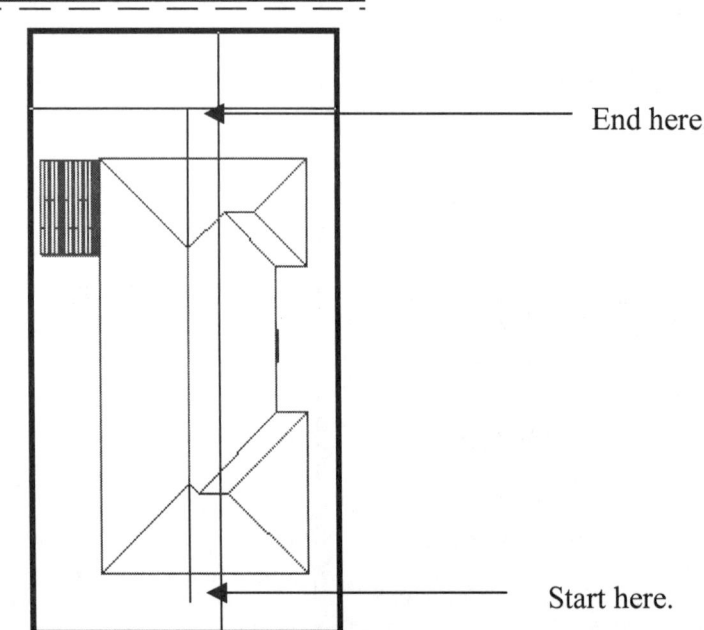

End here.

Start here.

Start the Section Mark line below the model and drag to a point above the model.

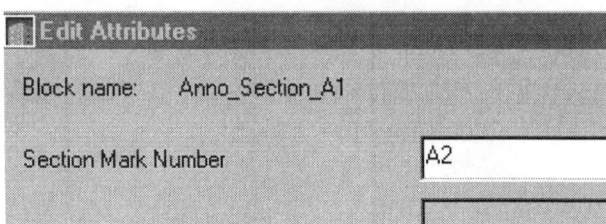

8-28

Type 'A2' for the Section Mark Number.

```
Select block reference:
Command:
Specify side for Arrow:
Add AEC section object? [Yes/No] <Y>: Y
```

Specify the right side of the model for the arrow side.
At the ADD AEC section object? Prompt, type 'Y' for YES.
(This creates the 2D/3D section object).

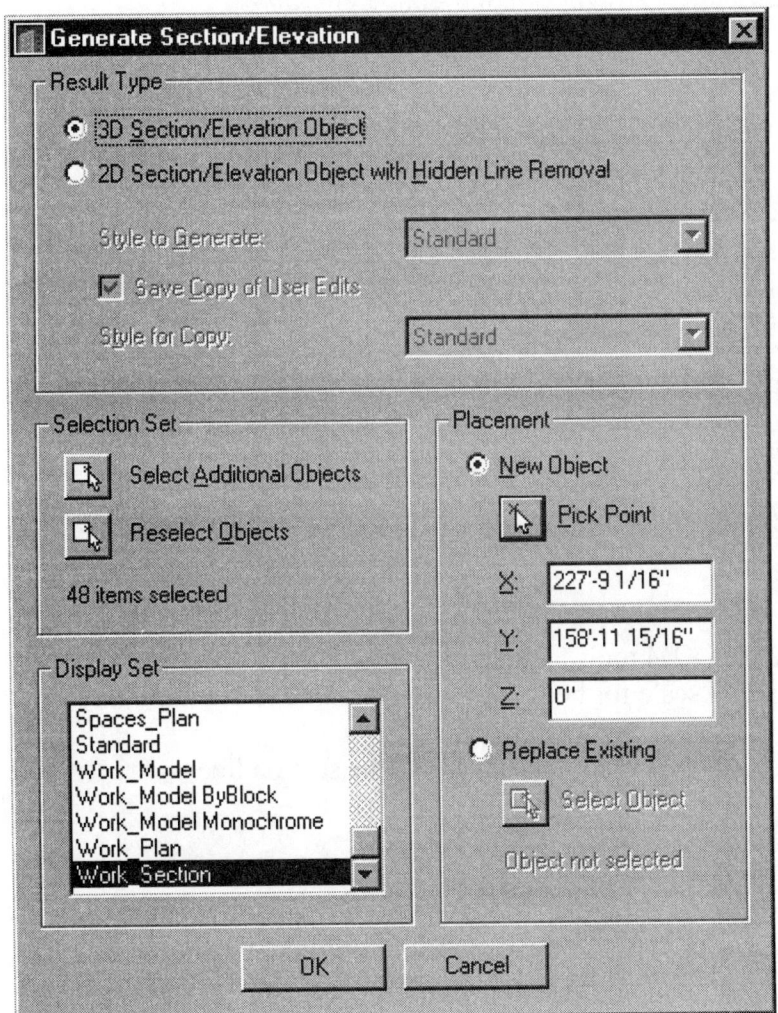

Select the 'Create Section' tool.
Select the section line you just created.
Enable the 3D Section/Elevation Object tool.
Select the house model.
Set the Display Set to Work Section.
For placement, select the viewport on the right.

With the right viewport activated, switch to an isometric view.

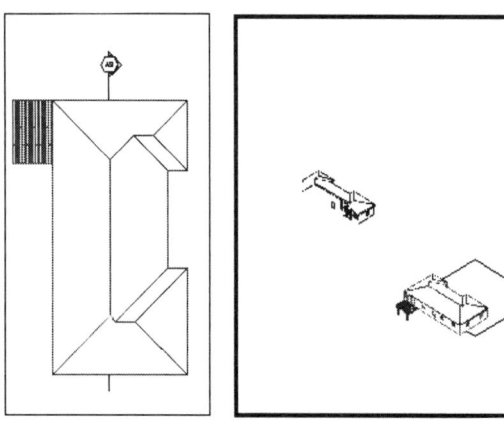

You can move the section view so it has a greater separation from the house view using the MOVE tool.

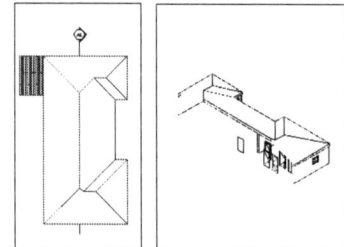

You can also use the PAN and ZOOM tools to position your 3D Section view.
Make sure that you set a standard scale for both your viewports and lock the display.

If you don't like the section view generated, you can move the section line and regenerate the section view.

Save as Ex8-3.dwg

Lesson 8
Layouts

Exercise 4:
Modifying Section Views

Drawing Name: Ex8-3.dwg saved from the previous lesson.
Estimated Time: 15 minutes

This exercise reinforces the following skills:

- Modifying Section Lines
- Creating Section Marks
- Modifying Section Views.

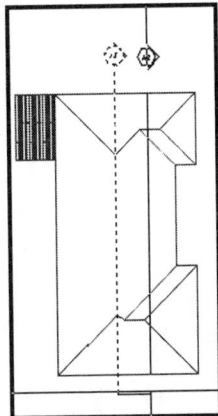

Activate the left viewport and shift the section line over to the right.

Enable the 3D Section/Elevation Object button.
Under Placement, select 'Replace Existing' and select the view shown in the right viewport.
Press 'OK'.

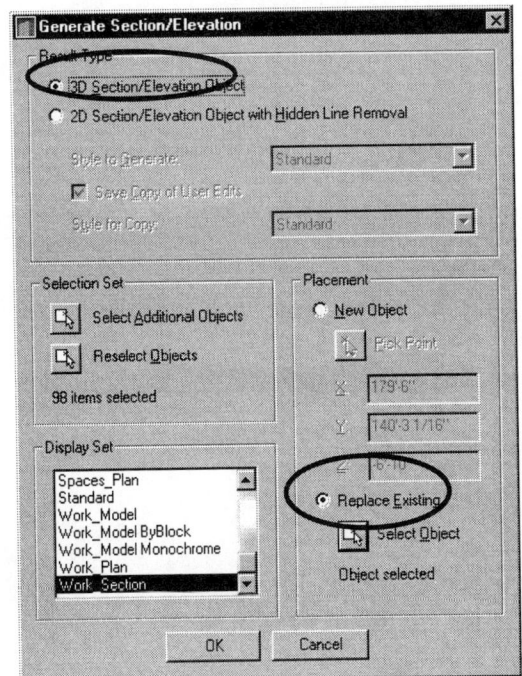

8-31

Lesson 8
Layouts

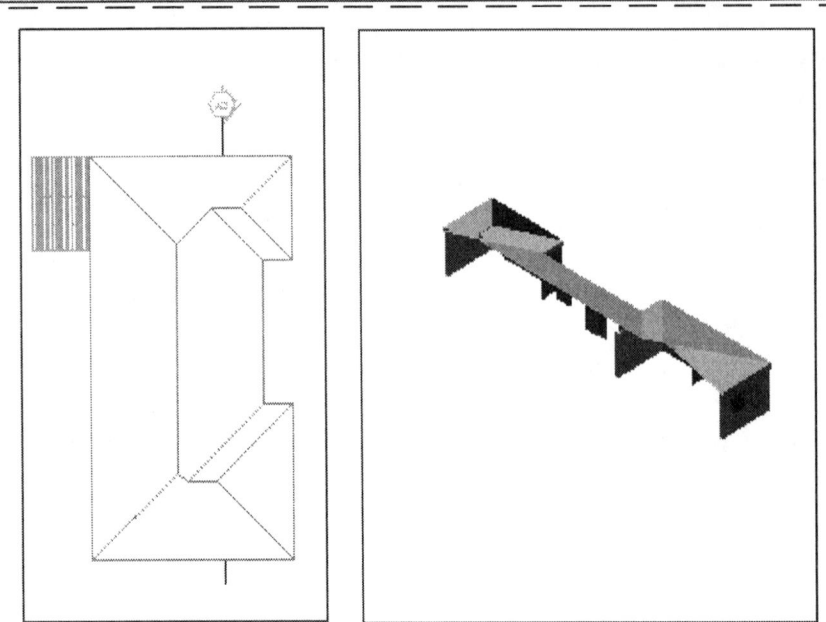

The right viewport section will automatically update.

Set your viewports so that they will not plot using your Layer Properties Manager.

Save the drawing as Lesson 8.dwg.

 TIP: Placing your viewports on the layer 'Defpoints' will ensure that they will never plot.

Quiz 4

True or False

1. Columns, braces, and beams are created using Structural Members.
2. When you isolate a Layer User Group, you are freezing all the layers in that group.
3. You can isolate a Layer User Group in ALL Viewports, a Single Viewport, or a Selection Set of Viewports.
4. When placing beams, you can switch the justification in the middle of the command.
5. Standard AutoCAD commands, like COPY, MOVE, and ARRAY can not be used in ADT.

Multiple Choice

6. Before you can place a beam or column, you must:

    A. Generate a Member Style
    B. Activate the Structural Member Catalog
    C. Select a structural member shape
    D. All of the above

7. Select the character that is OK to use when creating a Structural Member Style Name.

    A. -
    B. ?
    C. =
    D. /

8. Identify the icon shown.

    A. Add Brace
    B. Add Beam
    C. Add Column
    D. Structural Member Catalog

9. Setting a Layer Standard….

A. Controls which layer AEC objects will be placed on.
B. Determines the layer names created.
C. Sets layer properties.
D. All of the above.

10. There are two types of stairs used in residential buildings:

A. INSIDE and OUTSIDE
B. METAL and WOOD
C. MAIN and SERVICE
D. FLOATING and STATIONARY

11. Select the stair type that does not exist from the list below:

A. Straight-run
B. M Stairs
C. L Stairs
D. U Stairs

12. The first point selected when placing a set of stairs is:

A. The foot/bottom of the stairs.
B. The head/top of the stairs
C. The center point of the stairs
D. Depends on the user

ANSWERS:

1) T; 2) F; 3) F; 4) T; 5) F; 6) D; 7) A; 8) B; 9) D; 10) C; 11) B; 12) A

Lesson 9
Documentation

We can create the best 3D model in the world, but if we cannot produce working 2D drawings that can be read by a contractor or builder, then we have not done our job.

The AIA has established a set of rules for drawings, including how sheets are to be numbered, colors, and dimensioning standards.

Drawing Annotation includes

- Dimensions
- Notes
- Schedules
- Symbols

Prior to annotation being placed, you need to set the Drawing Scale. The drawing scale determines that size of symbols and text. The command is DWGSCALESETUP.

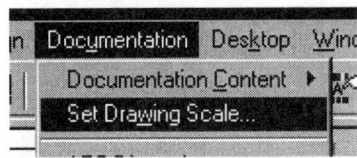

The drawing setup icon on the AEC Setup toolbar will activate the same dialog box, but the dialog box will open to whatever tab was last up.

Scale-dependent objects in an Autodesk Architectural Desktop drawing are scaled automatically to reflect the units set in a drawing. Any styles that you create in metric or imperial units are scaled appropriately. You can also set the scale for the plot size of annotation in your drawing. If you are using a drawing template, then you can customize the drawing scale in your template and save the changes.

TIP: The Annotation Plot Size value can be restricted by the Linear Precision setting on the Units tab. If the Annotation Plot Size value is more precise than the Linear Precision value, then the Annotation Plot Size value is not accepted.

Lesson 9
Documentation

The Documentation toolbars for Imperial and Metric look similar. The difference is basically the design content that is brought up and the scale of the symbols that are brought in. In Release 3.3, these commands seem to have different names beginning with AECDCSET . . . You do not have to type the AECDC prefix for the command to work.

	Break Marks	AnnoBreakMarkAdd SETIMPBREAKMARKS	Bar, Bar (Filled) Cut Line (1), Cut Line (2) Cut Line (Curved), Pipe Pipe (Filled)
	Detail Marks	AnnoDetailMarkAdd SETIMPDETAILMARKS	Detail Boundary A, Detail Boundary B Detail Boundary C, Detail Mark A1 Detail Mark A1T, Detail Mark A2 Detail Mark A2T

9-2

	Elevation Marks	AnnoElevationMarkAdd SETIMPELEVATIONMARKS	Elevation Mark A1 Elevation Mark A2 Elevation Mark B1 Elevation Mark B2 Elevation Mark C1 Elevation Mark C2
	Leaders	AnnoLeaderAdd SETIMPLEADERS	Spline (Circle) Spline (Diamond) Spline (Hexagon) Spline (Square) Spline (Text) Straight (Circle) Straight (Diamond) Straight (Hexagon) Straight (Square) Straight (Text)

	Miscellaneous	SetImpMiscellaneous	Dimensions	
			Aligned	Angular
			Baseline	Continue
			Linear	Radius
			Fire Rating Lines	
			1 Hr	2 Hr
			2 Hr - Smoke	4 Hr
			Smoke	
			Match Lines	
			Match Line	Match Line (Swiss)

	Miscellaneous	SetImpMiscellaneous	North Arrows (A–M)
	Revision Clouds	AnnoRevisionCloudAdd SETIMPREVISONCLOUDS	Large Arcs, Large Arcs & Tag, Medium Arcs, Medium Arcs & Tag, Small Arcs, Small Arcs & Tag
	Section Marks	AnnoSectionMarkAdd SETIMPSECTIONMARKS	Section Mark A1, Section Mark A1T, Section Mark A2, Section Mark A2T

◎	Title Marks	AnnoTitleMarkAdd SETIMPTITLEMARKS	Bar Scale (Inches) Title Mark A1 Title Mark A1 (Swiss)	
▽	Elevation Labels	AnnoElevationLabelAdd SETIMPELEVATIONLABELS	**2D Section** +10″ Elevation Label (1) +10″ Elevation Label (2) +10″ Elevation Label (3) +10″ Elevation Label (4) +10″ Elevation Label (5) +10″ Elevation Label (6) +10″ Elevation Label (7) +10″ Elevation Label (8) **Model** 3D Elevation Label (1) 3D Elevation Label (2) 3D Elevation Label (3) 3D Elevation Label (4) 3D Elevation Label (5) 3D Elevation Label (6) 3D Elevation Label (7) 3D Elevation Label (8)	

	Elevation Labels	AnnoElevationLabelAdd	Plan
	Chases	AECCannomVblockInterferenceAdd SETIMPCHASES	

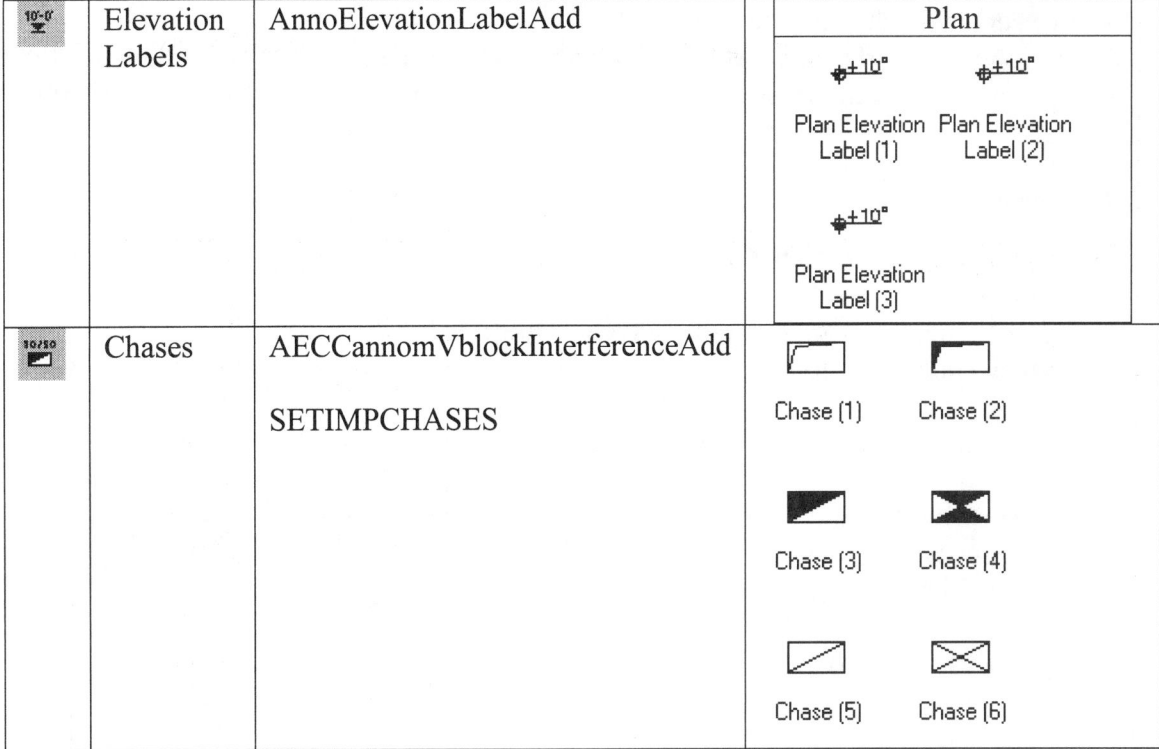

If you use any of the tools from the Documentation toolbars, it automatically opens the Design Center to the folder holding those symbols. You then must drag and drop the desired symbol into the drawing.

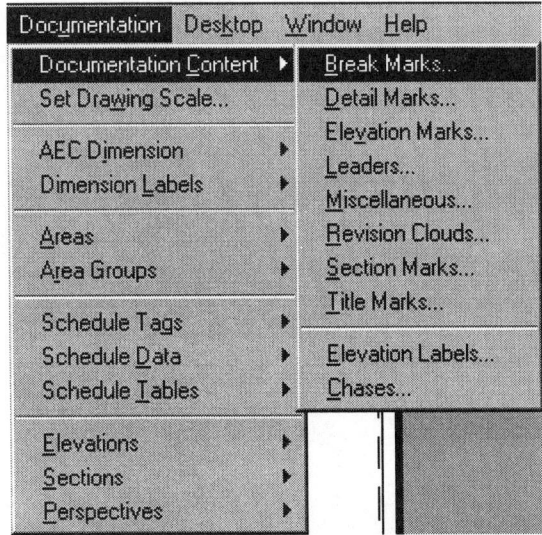

You can also access the symbols from the menu. Selecting the symbol class from the menu will also bring up the Design Center and require you to drag and drop.

Many users make the mistake of using the standard AutoCAD dimension tools to dimension their ADT plans. ADT comes with a set of tools specifically for dimensioning ADT objects. By using these tools, dimensions are automatically placed on the correct layers.

Exercise 1:
Dimensioning a Floor Plan

Drawing Name: Lesson 8.dwg saved from the previous lesson.
Estimated Time: 30 minutes

This exercise reinforces the following skills:

- Drawing Setup
- Dimension Styles
- AEC Dimensions

Open the Lesson 8.dwg.

Select the Plot-FLR tab. This is automatically set up to show only those layers pertaining to the floor plan of our residence.

Add AEC Dimension

You can apply AEC Dimensions to Walls with door and window openings and Grids. Before you can apply AEC Dimensions, you need to check several user system options.

✓ *Verify Dimension Settings*

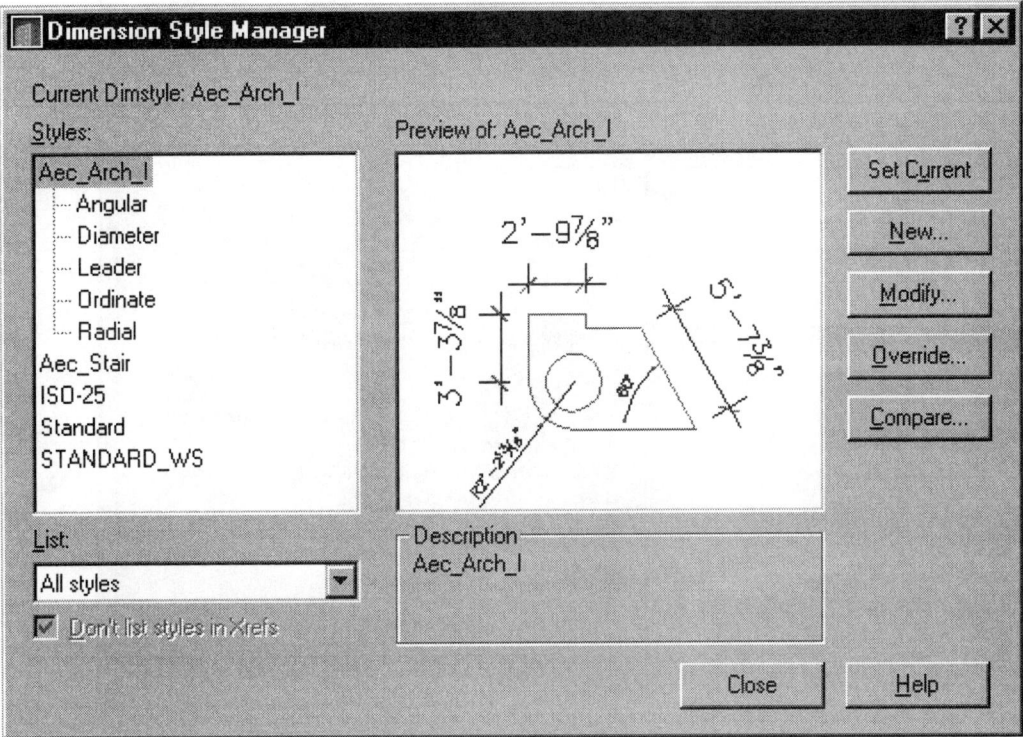

Verify that Aec_Arch_I is set as the Current Dimension Style.

✓ Verify Drawing Scale Settings

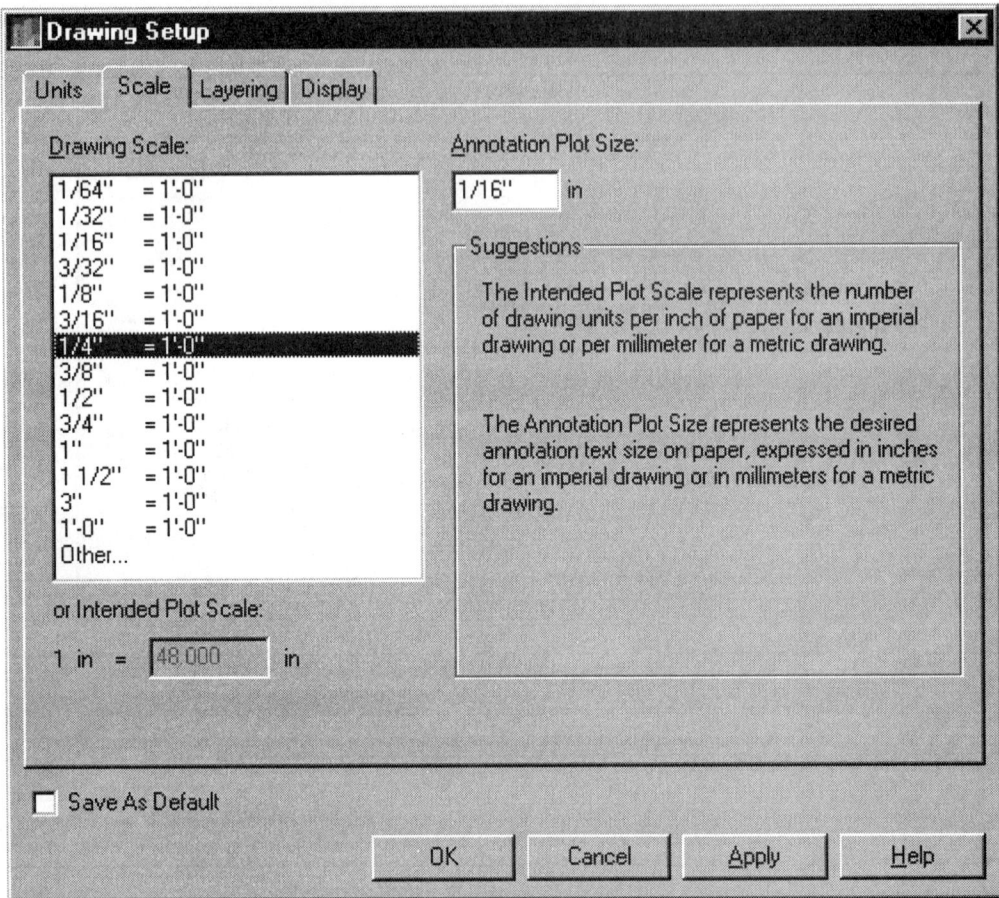

Make sure that the Drawing Scale is set appropriately for the paper size to be plotted. The default is ¼" = 1'-0". The Annotation Plot Size Default is 1/8".

The Save As Default option applies the selected Drawing Scale to any new drawings. The selected scale will not be applied to the active drawing unless the 'Apply' button is pressed. If you have several files open, only the active one will be affected by a drawing scale change. (You can have different drawing scales for different drawing files open simultaneously.)

The Annotation Plot Size Value determines the text size and symbol size.

✓ Verify AEC DwgDefaults Settings

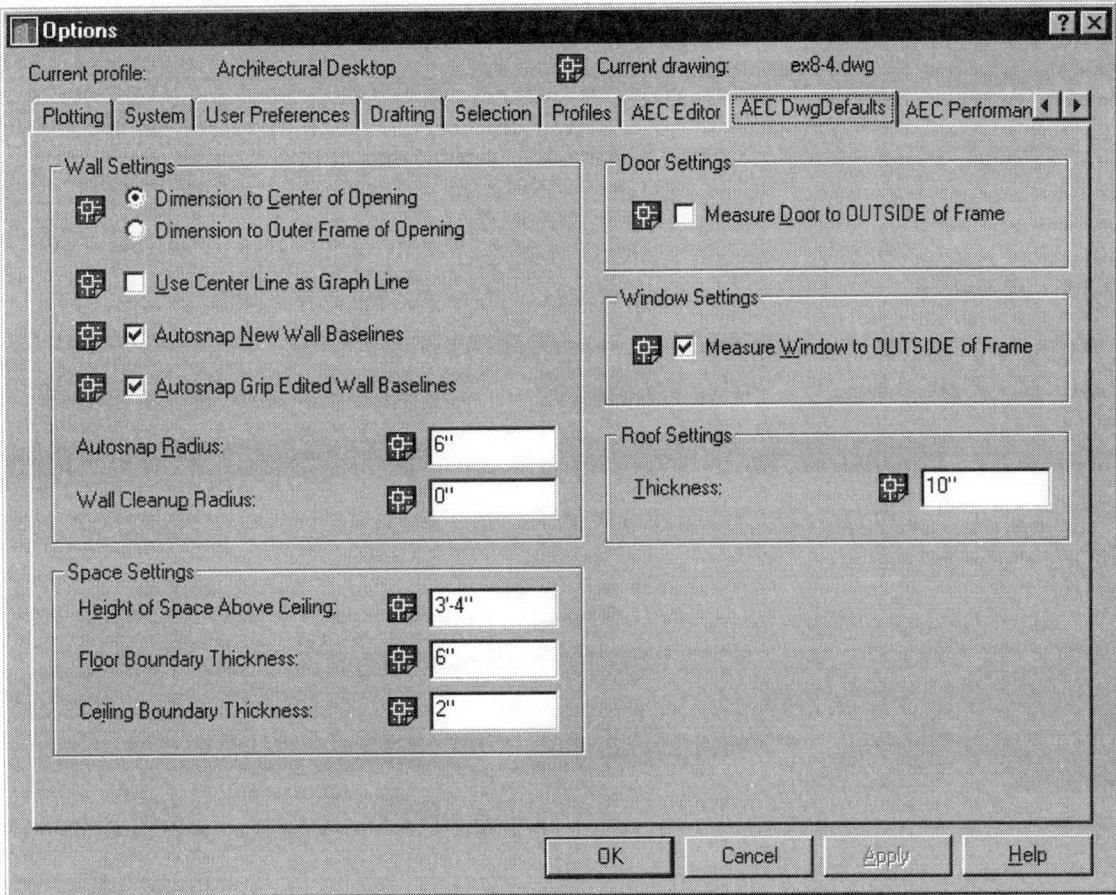

Access this dialog using Tools->Options. Then select the AEC DwgDefaults. Make sure that the Wall Settings is set to Dimension to Center of Opening.

✓ Verify Display Manager Settings

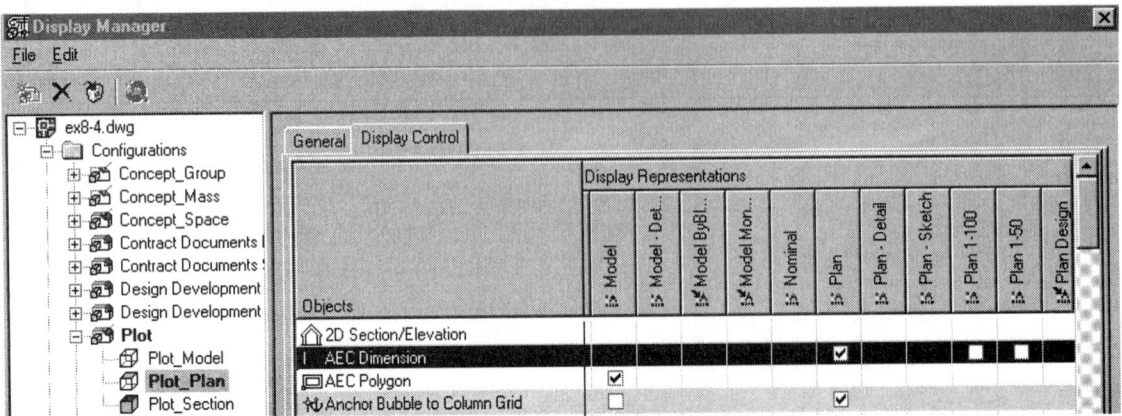

To make AEC dimensions visible, you must have the correct display configuration applied to the active viewport.

1. From the Desktop menu, choose Display Manager.
2. In the Display Manager, expand the Configurations option.

   The current configuration is **bold**.

3. Expand the current configuration option.

   The current display is **bold**.

4. Select the current display to open the Display Control page in the right pane of the Display Manager.
5. In the Objects column, find and select AEC Dimensions.
6. Find the Plan representation toggle box in the corresponding line.
7. Check the box to include the AEC Dimensions in the Plan representation.

Choose Apply and OK to accept changes and end the command.

Add Wall Dimensions

Menu	Design->Wall Tools->Dimension Walls
Wall Tools Toolbar	
Command Line	WallDim

Add Wall Dimensions is only available in Model Space. Dimensions are automatically placed on A-Anno-Dims layer when the Layering Standard is set to AIA.

Select the wall(s) you wish to dimension.
Select the side of the wall where you wish the dimension to be placed.
All dimensions relevant to the wall will be placed.
You may need to edit some of the dimensions to make it easier for them to be read. You can use GRIPS to shift the dimension lines or text around.

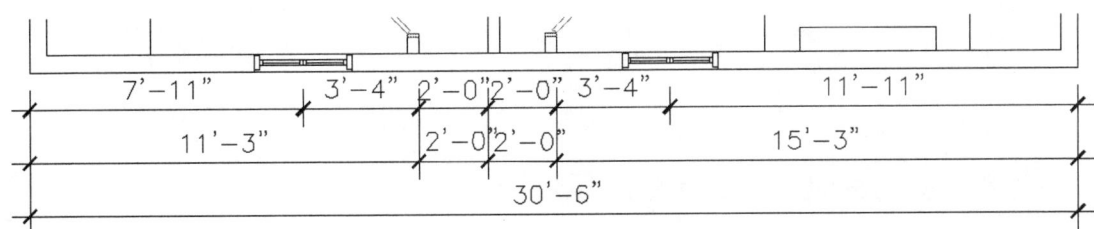

Here's what the dimensions look like at ¼" scale, with the text set to 1/8" tall. 1/8" is actually a bit big—a lot of firms use 3/32" text height as standard.

AEC Dimension

Add AEC Dimension is similar to the Add Wall Dimensions tool.

Menu	Documentation->AEC Dimension ->Add AEC Dimension
AEC Dimension Toolbar	
Command Line	AecDimAdd

By default AEC Dimensions are configured to use a fixed length extension line. Use the following procedure to turn this option off:

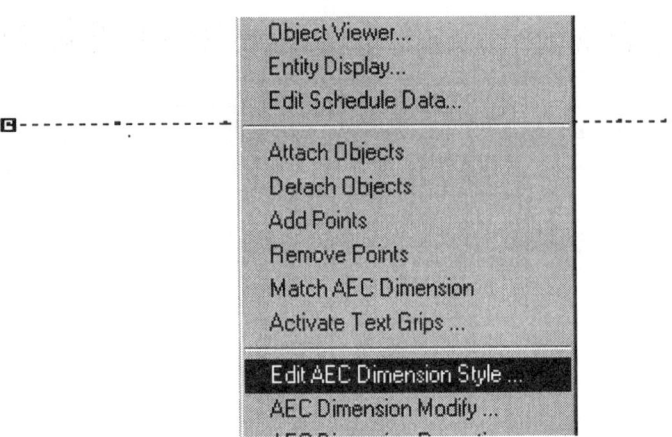

1. Select the AEC Dimension, right-click and choose Edit AEC Dimension Style.

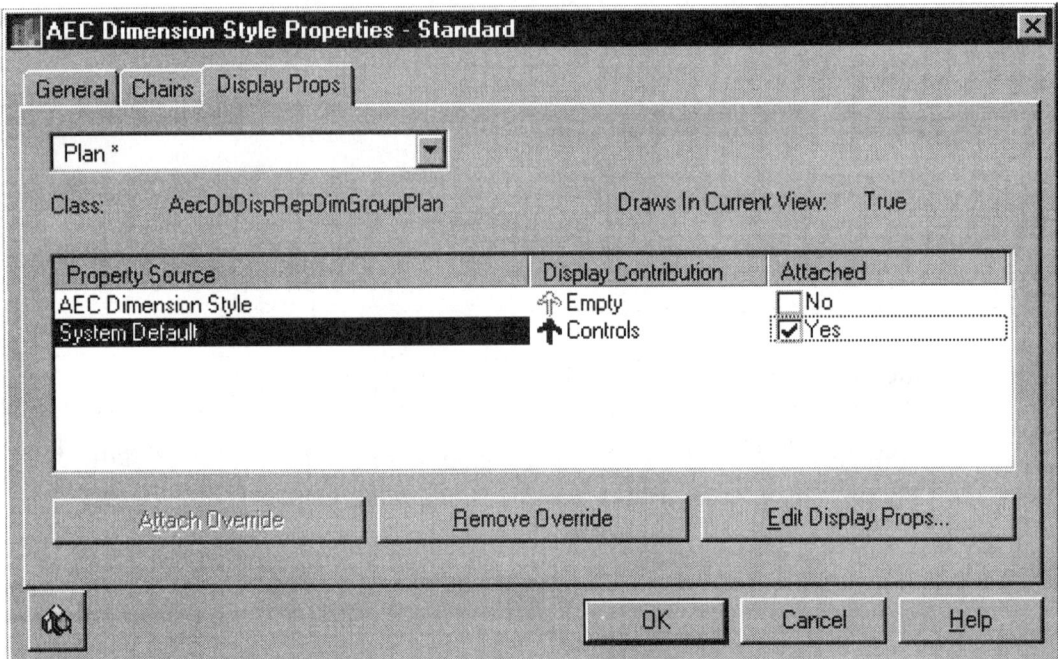

2. In the AEC Dimension Style Properties dialog box, set the Property Source to System Default and then choose Edit Display Props.

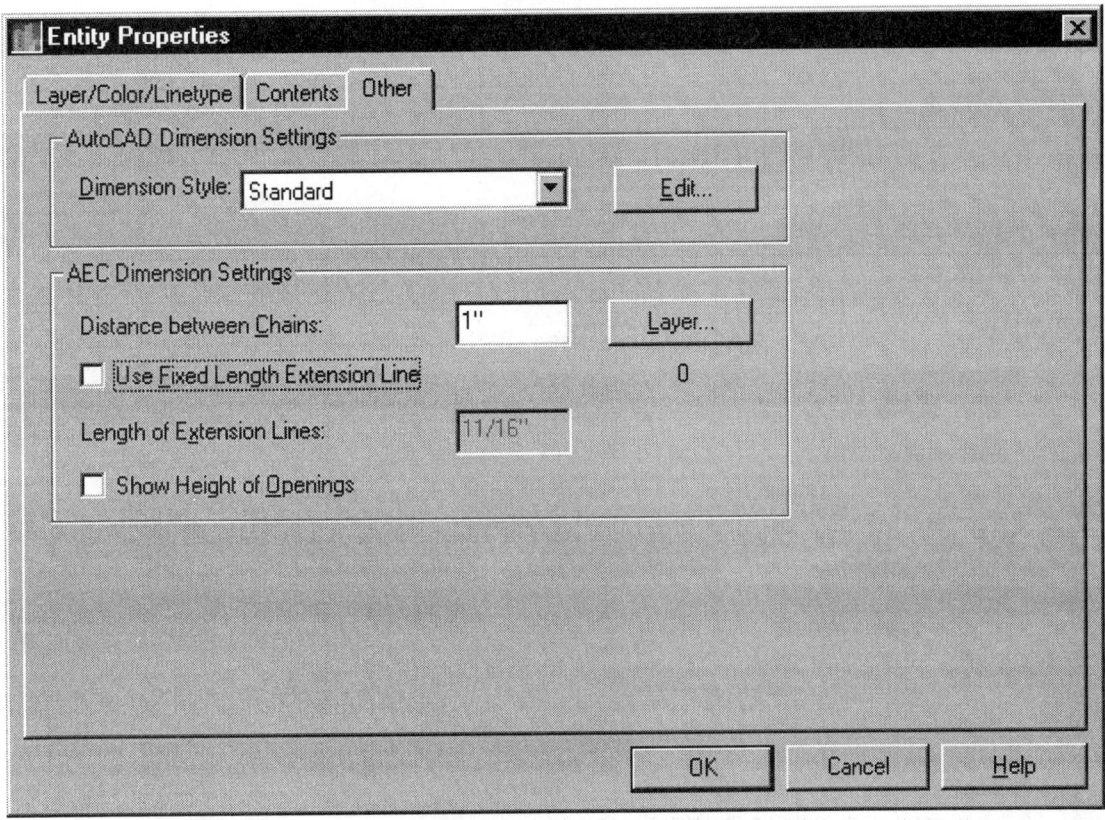

3. In the Entity Properties dialog box, select the Other tab.
4. Under AEC Dimension Settings area, clear the Use Fixed Length Extension Line option and choose OK twice.
5. Type **re** on the command line to regenerate the drawing.

The extension lines will become visible.

Save as Ex9-1.dwg

Lesson 9
Documentation

Exercise 2:
Adding Wall Dimensions

Drawing Name: Ex9-1.dwg saved from the previous lesson.
Estimated Time: 30 minutes

This exercise reinforces the following skills:

- Drawing Setup
- Dimension Style
- Add Wall Dimensions

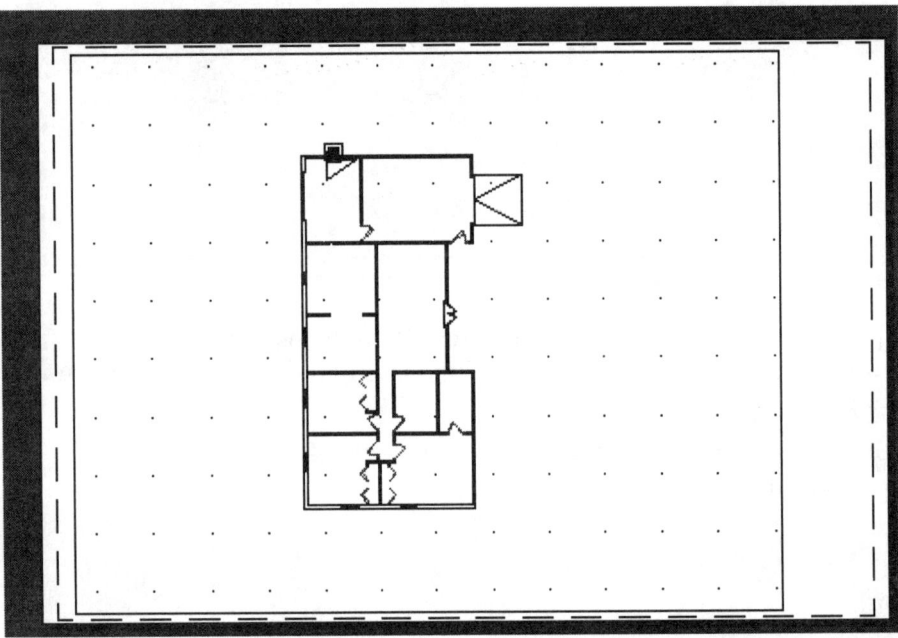

Select the Work-FLR tab.

In the viewport, setup your model so you can see the floor plan as shown.

Imperial (foot-inch) drawings will nearly always use ¼" = 1'-0" (1:48).

Set the Display to 'Locked'.

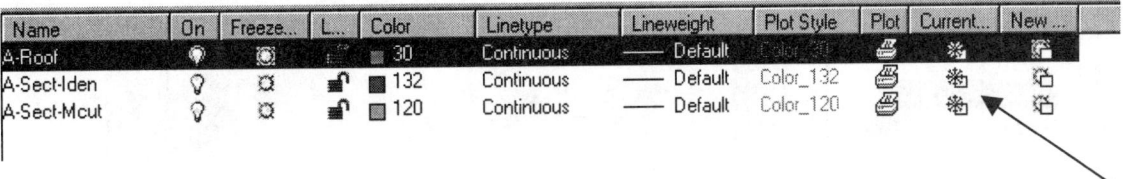

One way to control visibility is to Freeze Layers in the Current Viewport.
That way they will be seen in other viewports on other sheets, but not on the viewport on this sheet.

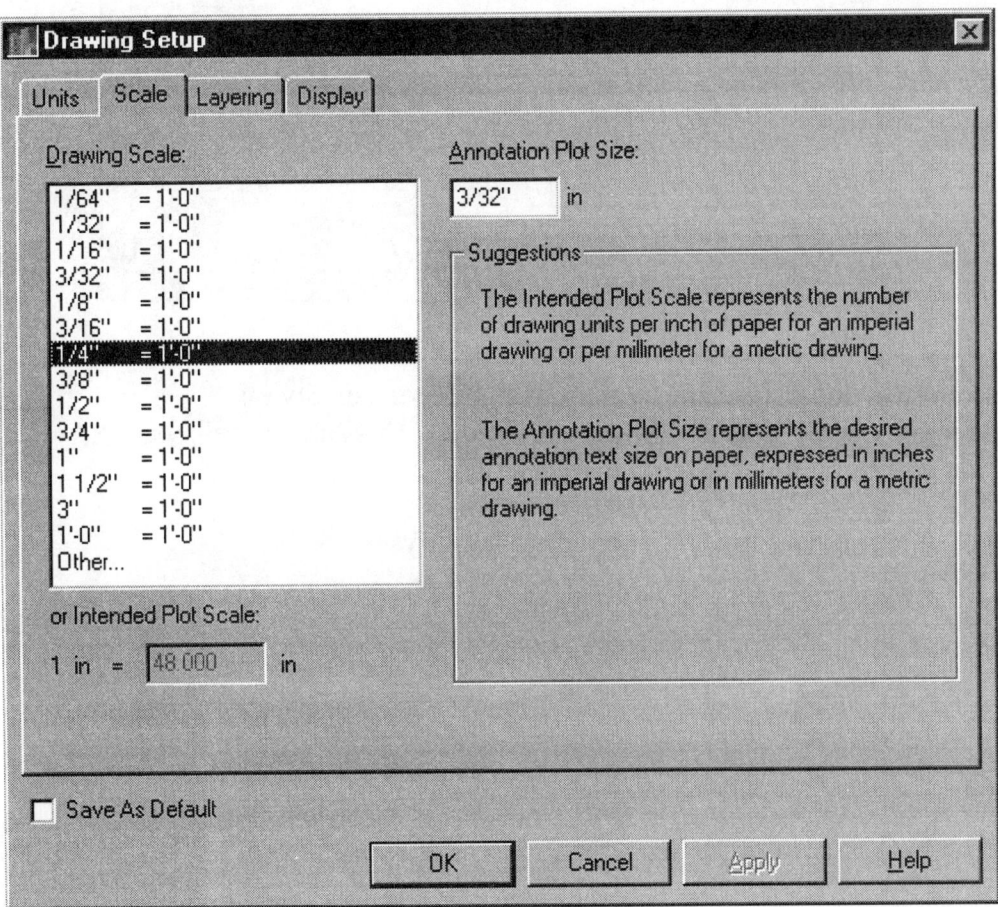

Go to Drawing Setup.
Set the Drawing Scale to ¼" = 1'-0".
Set the Annotation Plot Size to 3/32".

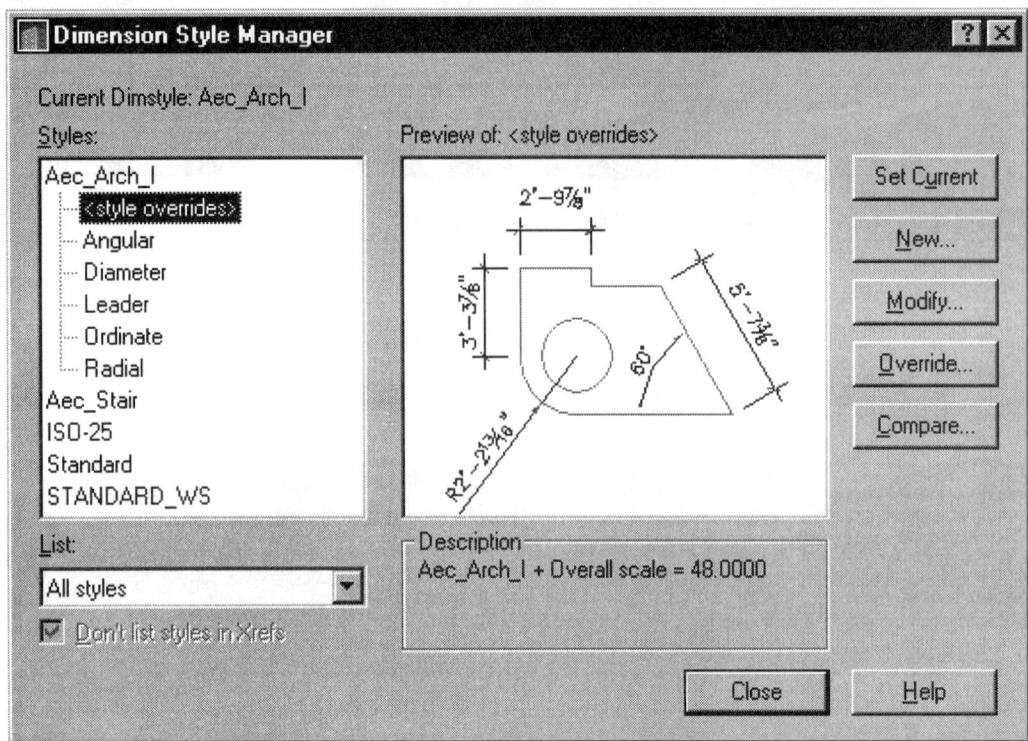

Go to the Dimension Style Manager.
(You can access this under the menu using Format->Dimension Style.)
Notice that the dimension style now has style overrides. This automatically occurs whenever you change the scale in the Drawing Setup area.

Highlight the style overrides as shown and select 'Modify'.

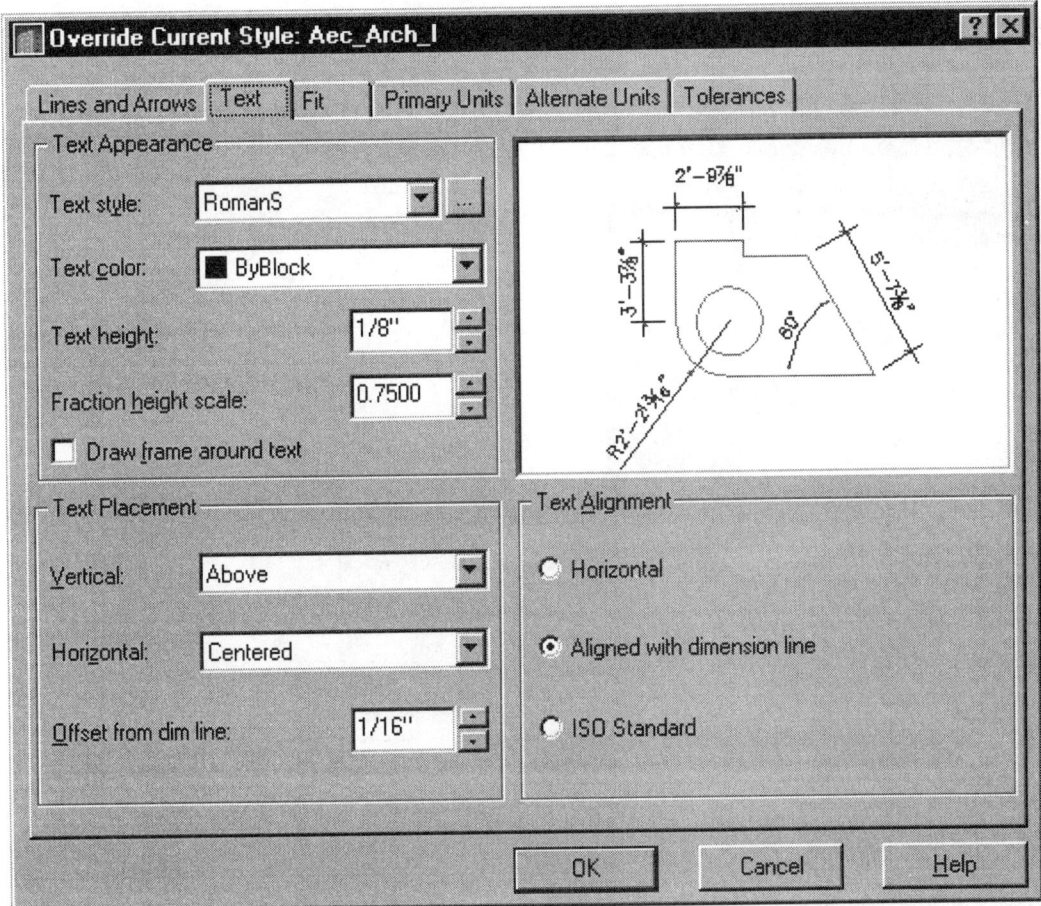

Select the Text tab. Notice that the Text Height has not been changed. This must be manually changed to 3/32".

Lesson 9
Documentation

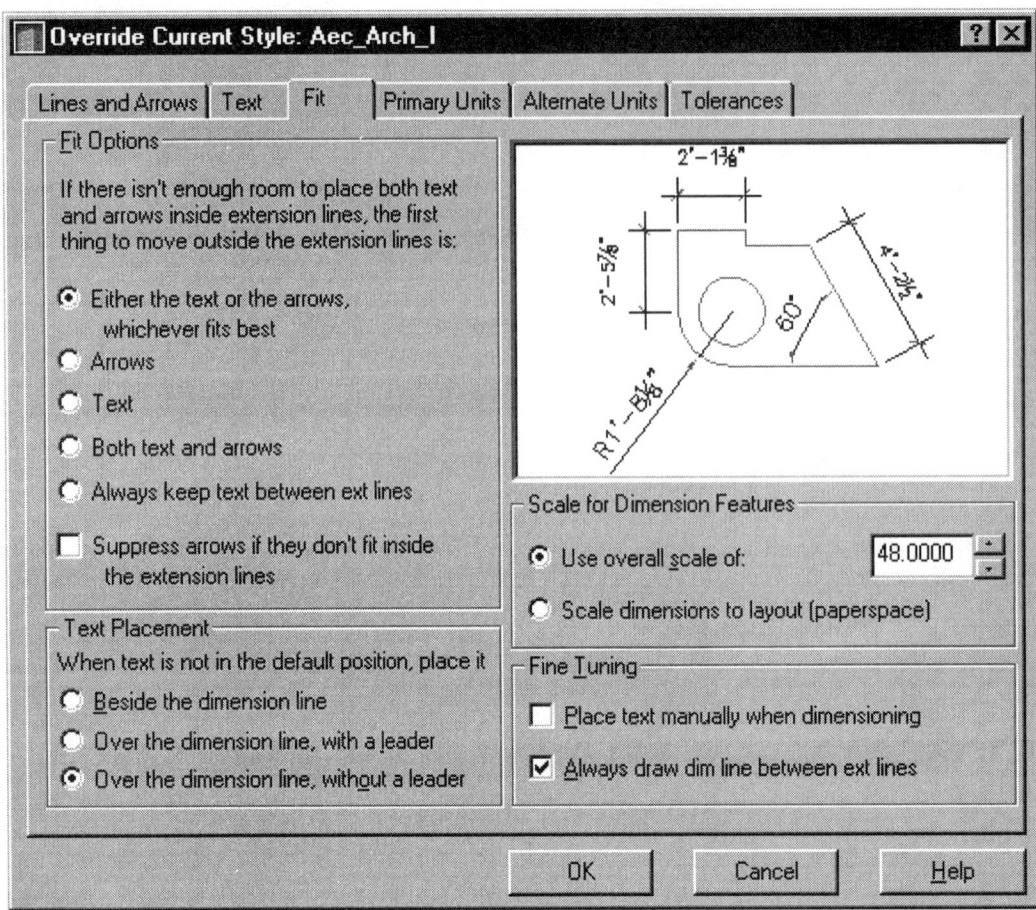

If you select the Fit tab, you will notice that the Scale has been set correctly.

Display the Wall Tools toolbar.

Activate Model Space for the viewport.

Select the Create Wall Dimensions tool and pick the wall at the top of the screen.

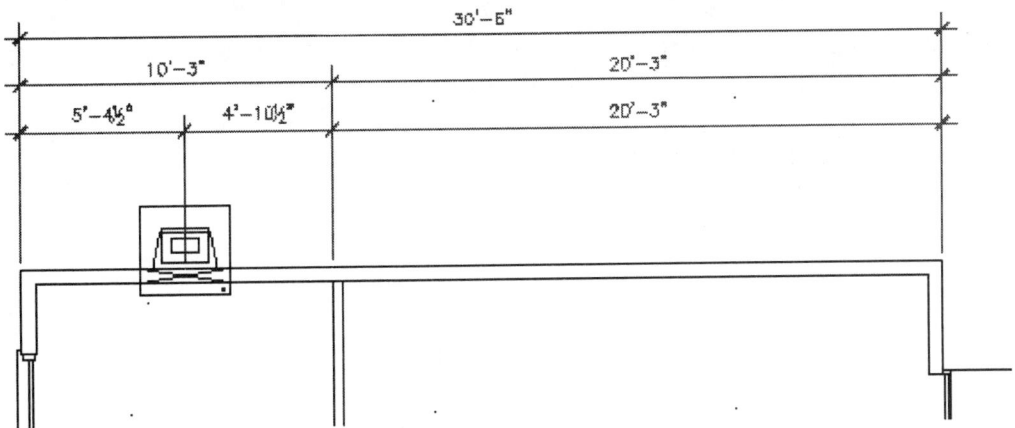

When asked to pick a side to dimension, pick the location of the dimension line that will be closest to the wall. When asked to indicate the second point, enter 'P' for Parallel.

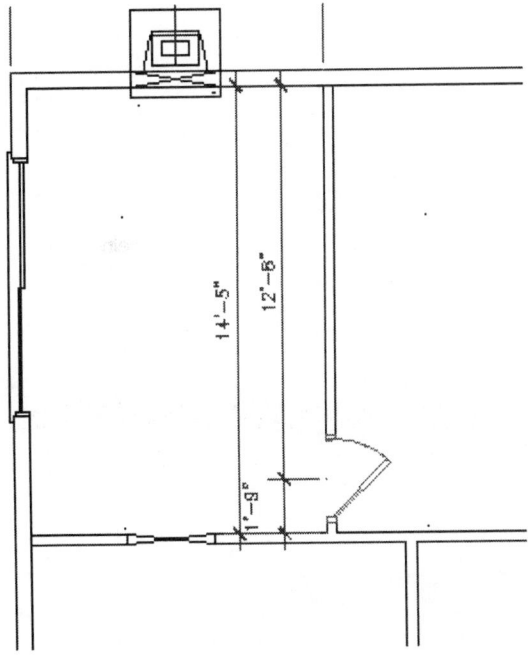

Dimension the inside wall using the same method.
Note that the inside wall dimensions do not go to the center of the wall as they should.

AEC Wall Dimensions are not very useful for interior dimensions yet--they do not pick up the baseline of the wall, only the end points or openings. Interior dimensions only pick up necessary dimensions that are not defined by the exterior dimensions, which in the case of this building are mostly just door openings, so they do not need to be complete strings from wall to wall, and in fact will be more practical if they are not.

Interior dimensions that are complete strings usually run from inside face of outside wall either to centers of interior walls or to one side of each wall, keeping to the same side all the way down the string to the opposite exterior wall. Single strings most often suffice. I would recommend that you use regular AutoCAD dimensions for the interior strings of dimensions, and only dimension walls (such as hallways) and openings that are not picked up by the exterior dimensions.

Use AEC dimensions for exterior features and whenever possible.

Save the file as Ex9-2.dwg.

Lesson 9
Documentation

Exercise 3:
Add Drawing Scale

Drawing Name: Ex9-1.dwg saved from the previous lesson.
Estimated Time: 30 minutes

This exercise reinforces the following skills:

- Drawing Setup
- Dimension Style
- Add TitleMarks

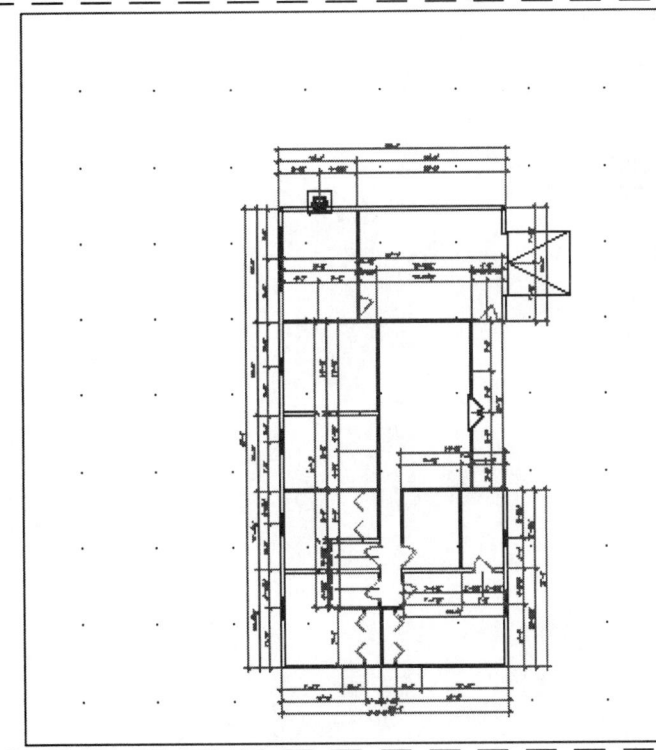

Once you have fully dimensioned your floor plan, it is time to add a Drawing Scale.

Lesson 9
Documentation

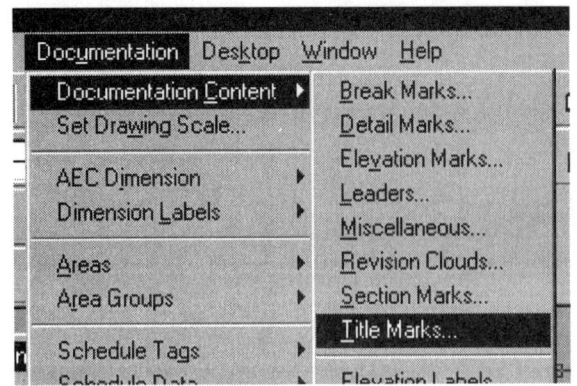

Select the Title Mark tool from the Documentation-Imperial toolbar or from the menu as shown.

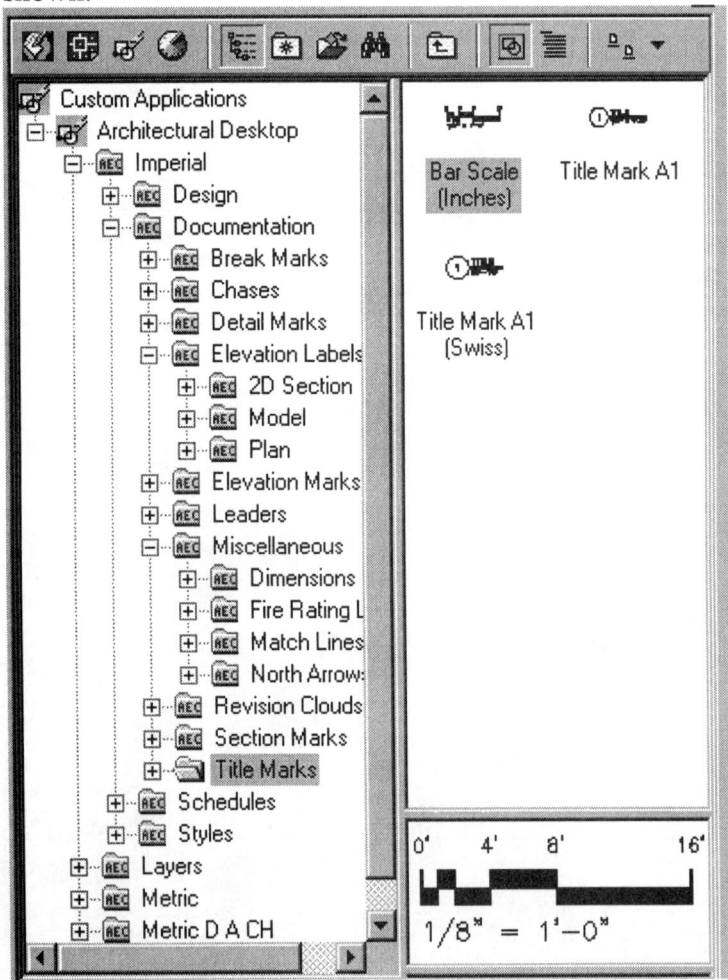

The Design Center will open.

Highlight the Bar Scale (Inches) block.
Drag and drop into your drawing.

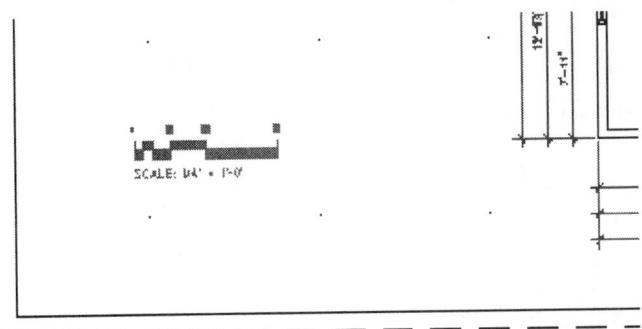

Place the scale in the lower left hand corner of your drawing. Set the rotation to 0.

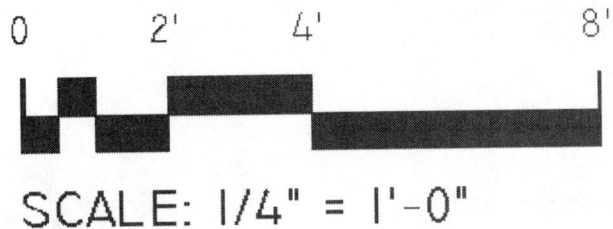

The scale automatically uses the scale set using DrawingSetUp.

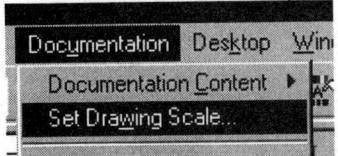

Go to Documentation->Set Drawing Scale.

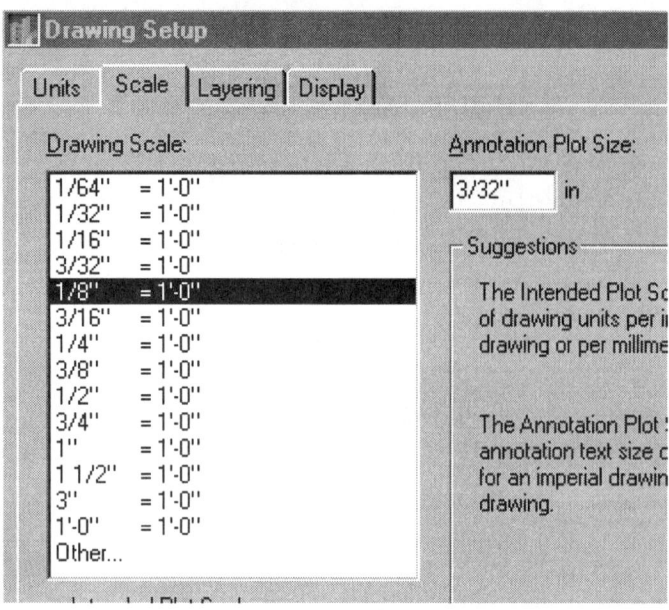

Change the scale to 1/8" = 1'-0".
Press 'Apply' and 'OK'.

Note that the scale and your dimensions are unchanged.

To modify the scale, you must edit the attributes:

Menu	Modify->Object->Attribute->Single
Modify II Toolbar	
Command Line	EATTEDIT

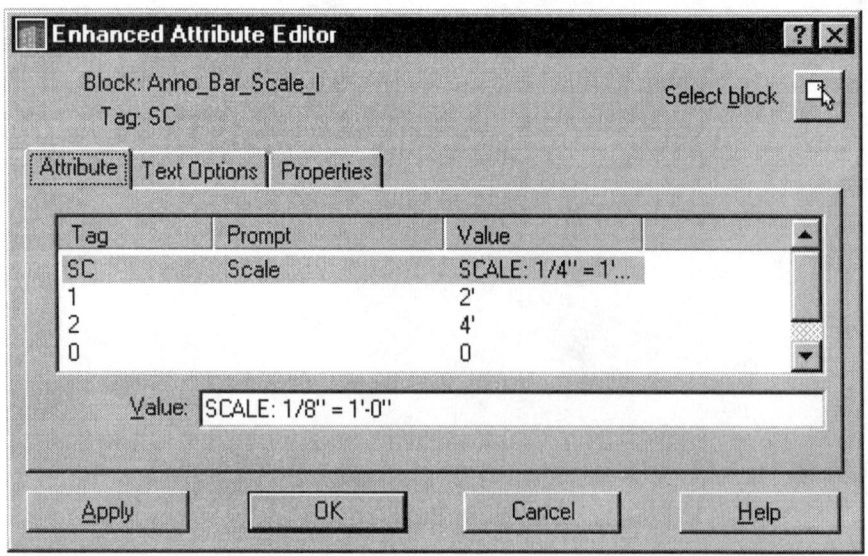

The Enhanced attribute editor appears.
Change the Value in the Edit Box to SCALE: 1/8" = 1'-0".

Press 'Apply' and 'OK'.

>
> **TIP:** You can insert blocks into model or paper space. Note that you did not need to activate the viewport in order to place the Scale block.

Save the drawing as Lesson 9.dwg.

Notes:

Lesson 10
Schedules

Schedules can be created to list doors, windows, walls, or any other object in your drawing. The schedule provides detail information regarding the size, construction, and vendor for each object type. Each object in a schedule is usually assigned a mark, which is placed as a tag, next to the object. Schedules form the basis of a Bill of Materials or shopping list for the contractor/builder.

The information included in the schedule is extracted from the property data attached to the drawing object (window/door/wall/etc.). Some properties are automatically generated when the object is inserted and placed in the drawing. However, other properties may be edited manually. You can also export schedule data to a .TXT file, a .XLS file or a .CSV file for use in a database. This would allow you to perform a cost analysis on required building materials.

Schedule Table Styles

The format of schedule tables can be controlled using three methods:

- Use the default schedule table
- Import a style from another drawing
- Create a Custom Style

Lesson 10
Schedules

Using the Style Manager

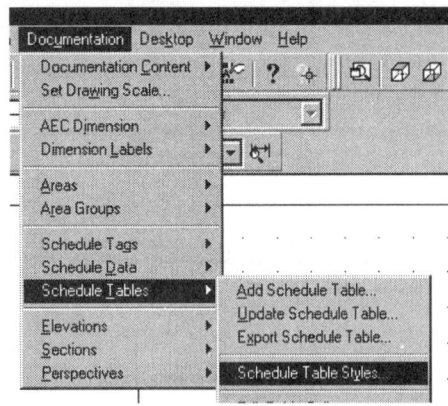

Menu	Documentation->Schedule Tables->Schedule Table Styles
	Schedule Table Styles
Command Line	TableStyle

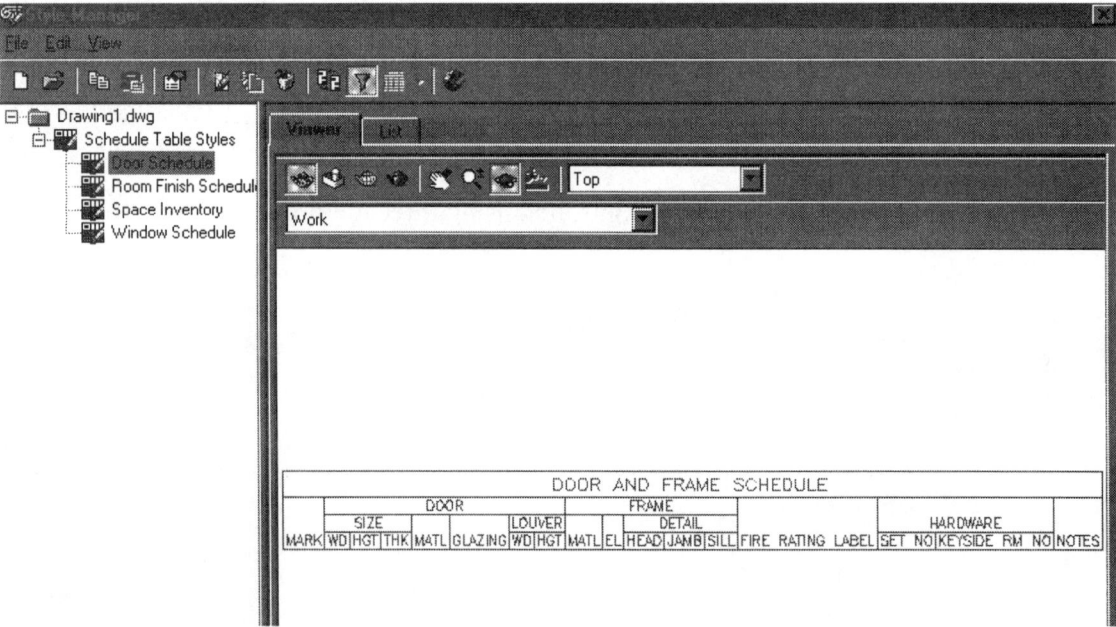

The TableStyle tool brings up the Style Manager dialog box. This dialog has its own toolbar and menu.

Lesson 10
Schedules

The first four icons in the dialog are standard WINDOW icons: NEW, OPEN, COPY, and PASTE. These allow users to create new styles, open an existing style, copy a style from another source, or paste from another source.

Edit Style

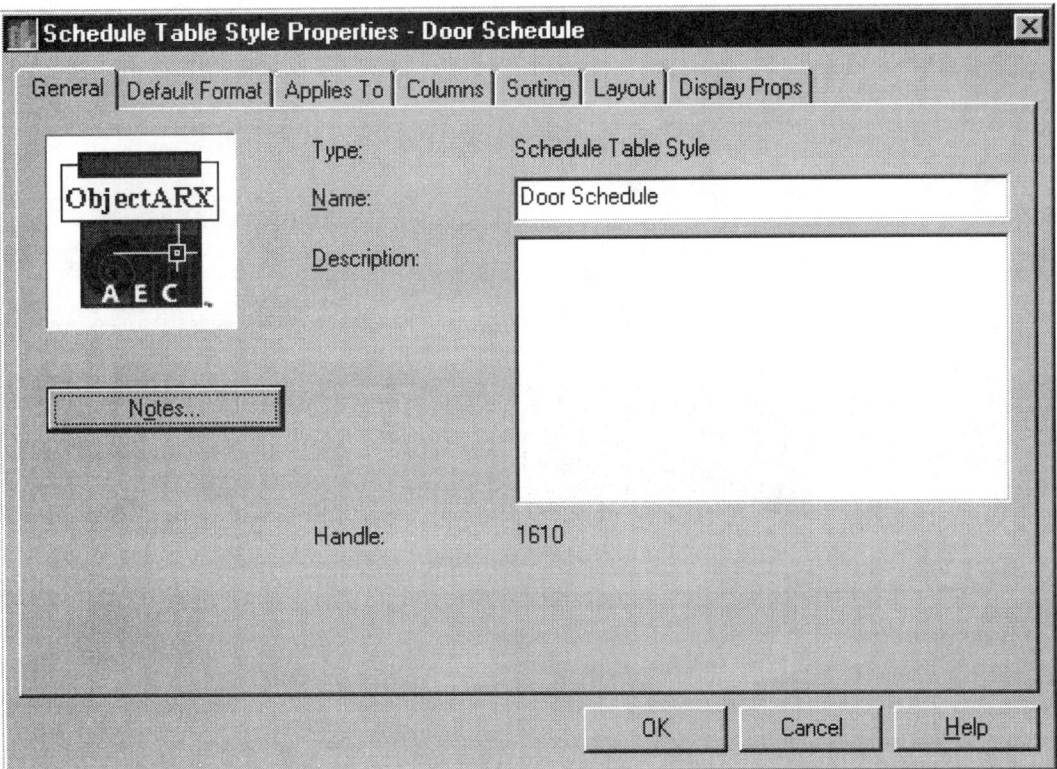

The Edit Style button allows you to modify an existing schedule style. There are six tabs, plus a Notes button.

10-3

Lesson 10
Schedules

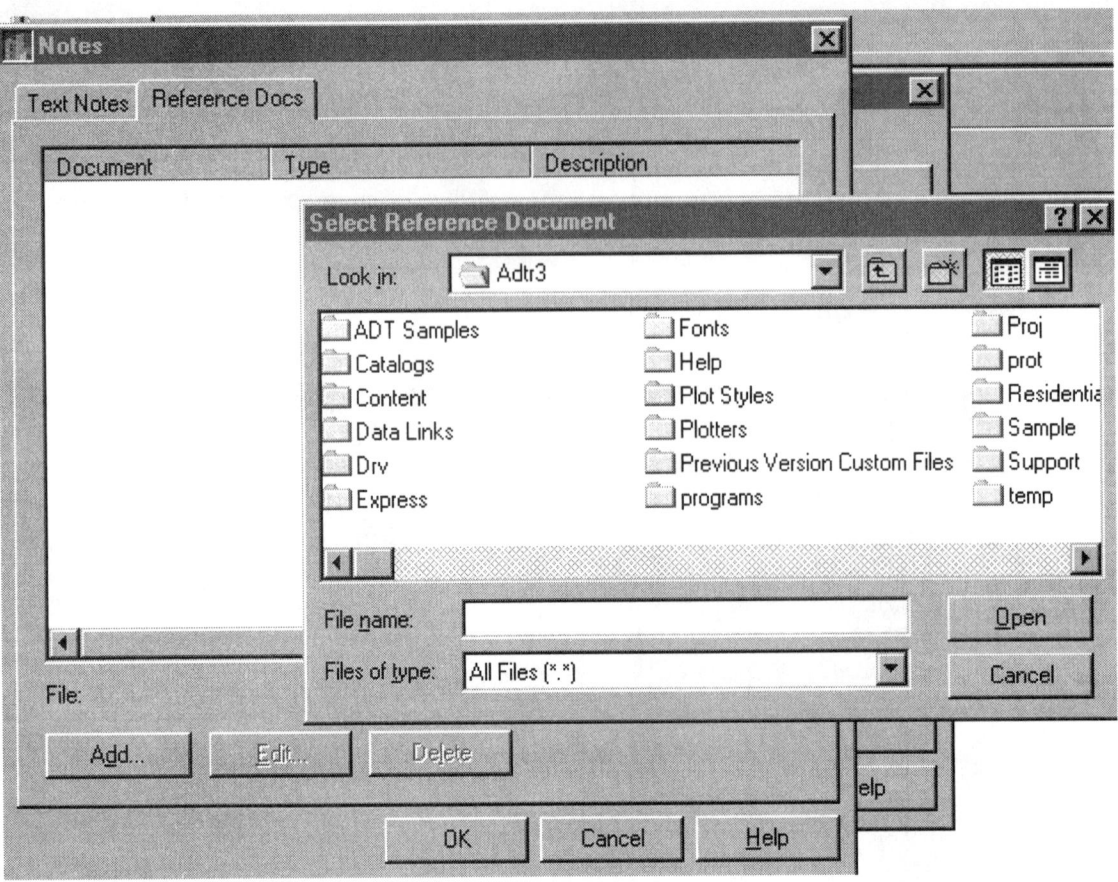

Pressing the Notes button brings up a Notes dialog box. The user can reference specifications to support the schedule definition style or create a note to explain the schedule style.

Lesson 10
Schedules

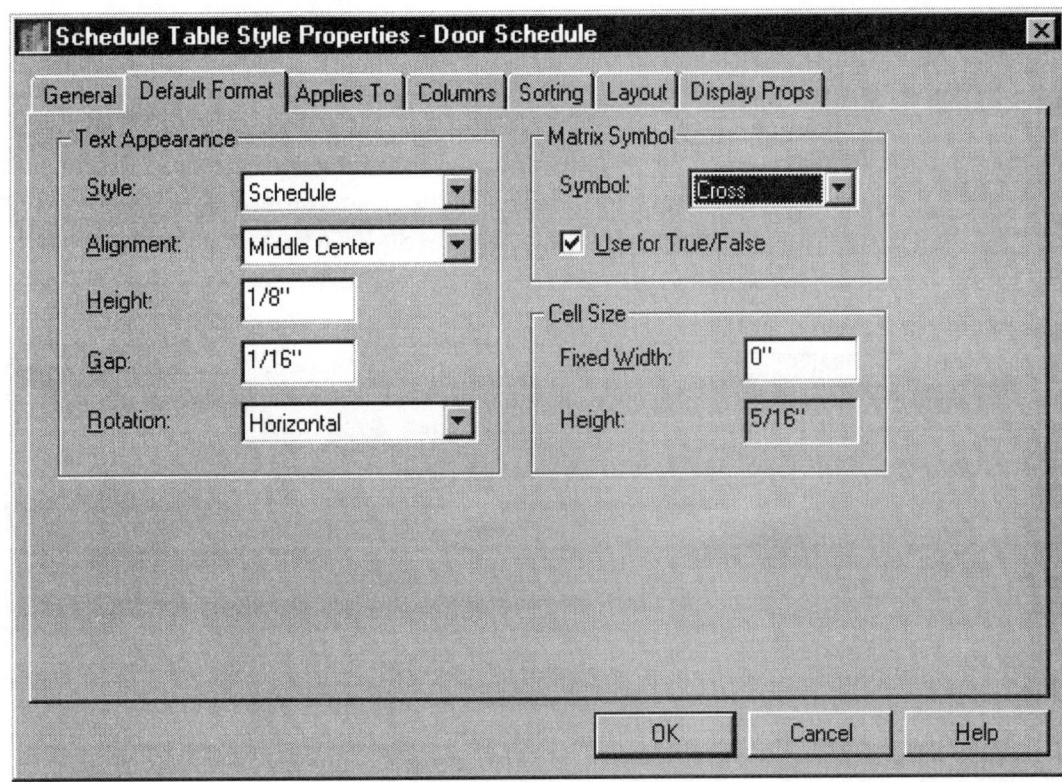

Style	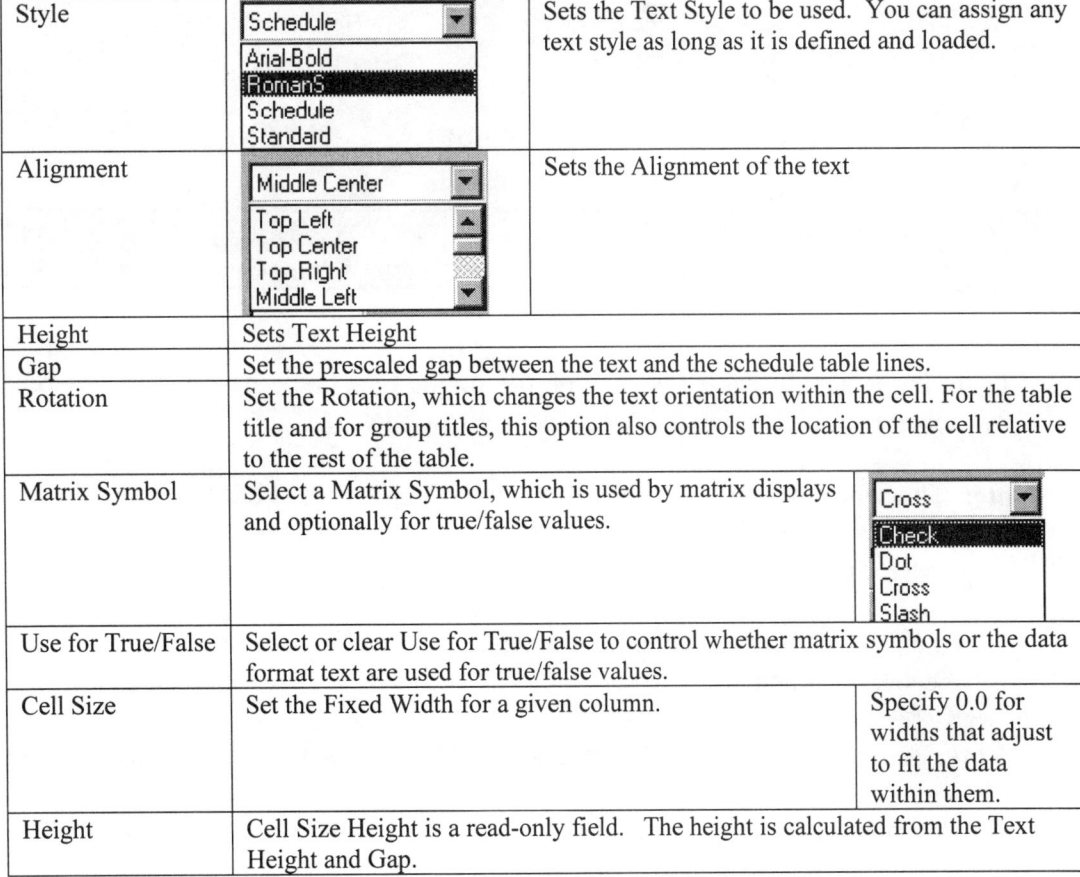	Sets the Text Style to be used. You can assign any text style as long as it is defined and loaded.	
Alignment		Sets the Alignment of the text	
Height		Sets Text Height	
Gap		Set the prescaled gap between the text and the schedule table lines.	
Rotation		Set the Rotation, which changes the text orientation within the cell. For the table title and for group titles, this option also controls the location of the cell relative to the rest of the table.	
Matrix Symbol		Select a Matrix Symbol, which is used by matrix displays and optionally for true/false values.	
Use for True/False		Select or clear Use for True/False to control whether matrix symbols or the data format text are used for true/false values.	
Cell Size		Set the Fixed Width for a given column.	Specify 0.0 for widths that adjust to fit the data within them.
Height		Cell Size Height is a read-only field. The height is calculated from the Text Height and Gap.	

10-5

TIP: The text style must be defined as an AutoCAD text style before you can apply it to the schedule table. If you want to use a different font for part of your schedule, you must first define an AutoCAD text style that uses the desired font. The text style must be loaded before it can be accessed by the Schedule Styles Manager.

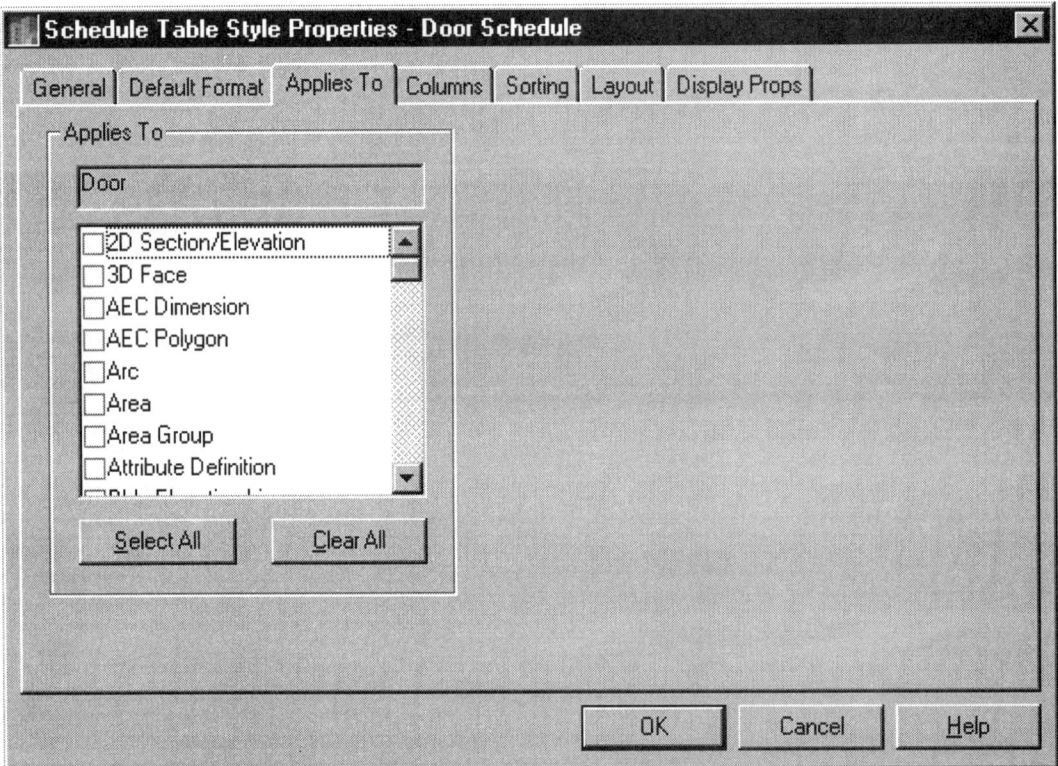

1. In the Schedule Table Style Properties dialog box, click the Applies To tab.
2. Select the objects the schedule table style applies to.

Note: The schedule table can apply to all objects or to any of the objects in the list. However, you should limit your selection as much as possible to speed performance.

3. Click Select All to select all of the objects on the list, or Clear All to clear all of the selection boxes.

Note: The area at the top of the tab displays the objects that are currently selected.

4. When you finish making changes, click OK to return to the Style Manager.

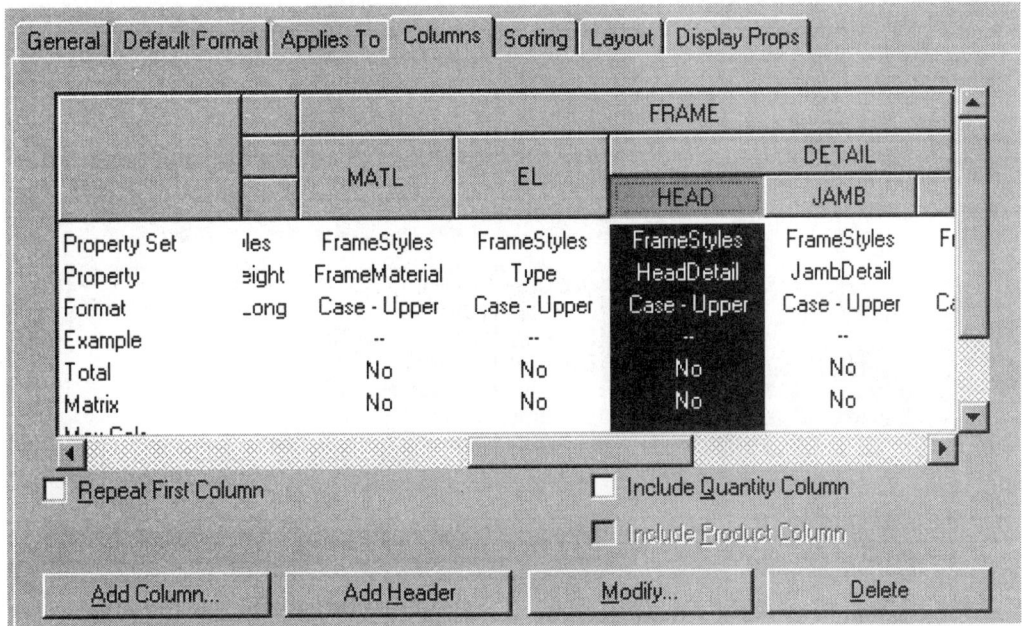

Select the columns you want to include under a heading by clicking the leftmost column heading and then pressing CTRL while you select the other columns.

Note: A heading can only be added above consecutive columns.

Click Add Header.

Type a title for the heading, and press ENTER.

To remove the heading, select it, and click Delete. In the Remove Columns/Headings dialog box, click OK.

When you finish making changes, click OK to return to the Style Manager.

Lesson 10
Schedules

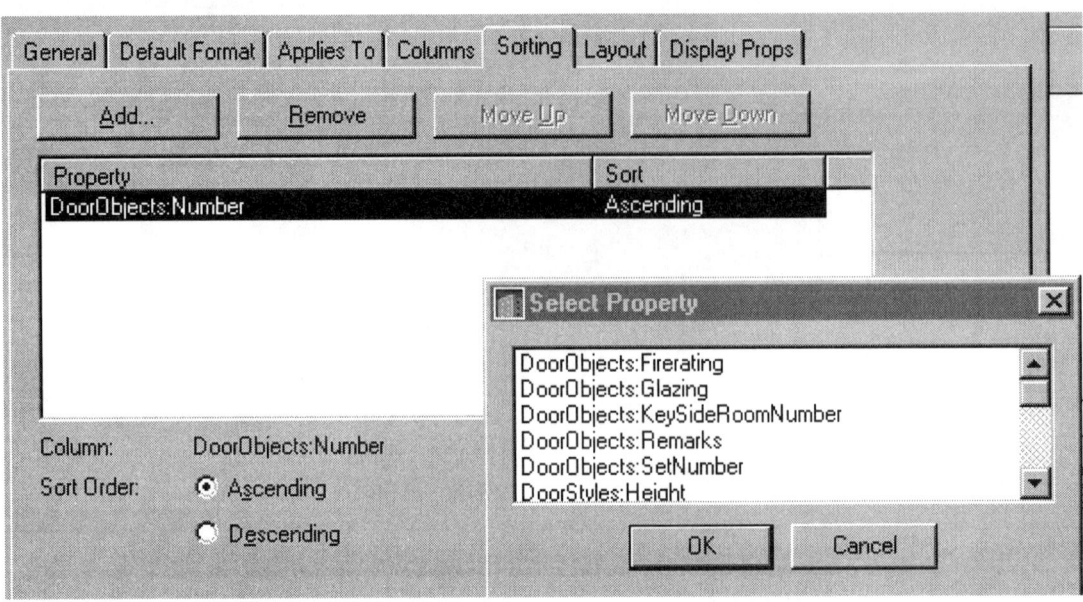

If assigned, the current property for sorting the schedule displays.

Click Add to add additional properties to sort by. The "Select Property' dialog will appear.

Select a property to sort by, and click OK. The selections presented are the columns defined on the Columns tab that are not already selected for sorting.

Select the property from the list, and then select either Ascending or Descending as the sorting method for that property.

To remove a property from the sorting list, select the property and then click Remove.

To move a property higher or lower in the sorting order, select the property and then either Move Up or Move Down.

When you finish making changes, click OK to return to the Style Manager.

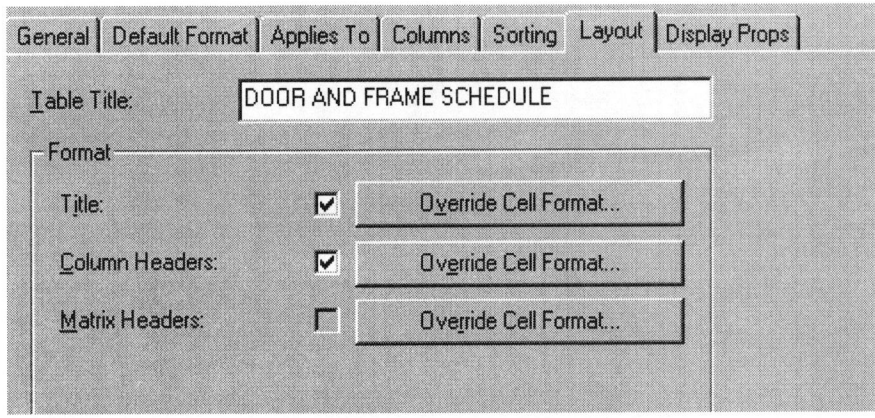

To modify the title for your schedule, click the Layout tab and edit the text in the Table Title edit box.

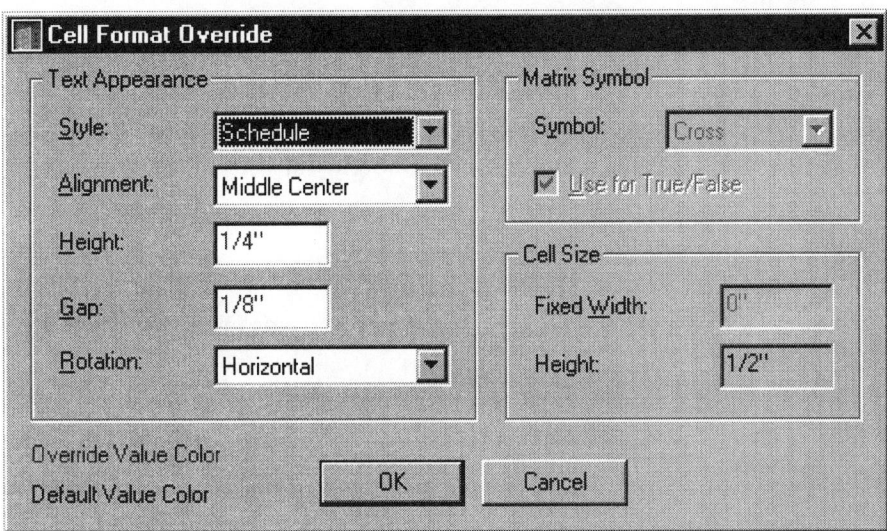

Pressing the Override Cell Format will allow you to modify the properties for each cell.

Lesson 10
Schedules

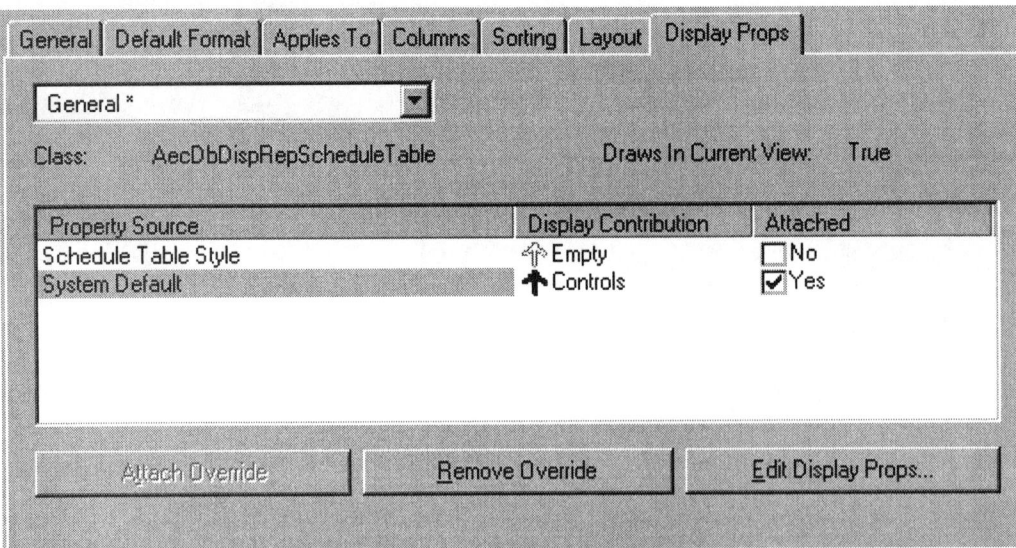

The display properties of an object affect the way the object appears in the drawing. You can override the default display for the current display configuration by setting a different visibility, layer, color, linetype, hatching, and cut plane height for the selected object.

New Style

This icon allows the user to create a new Schedule Style. Styles are listed in alphabetical order in the style tree.

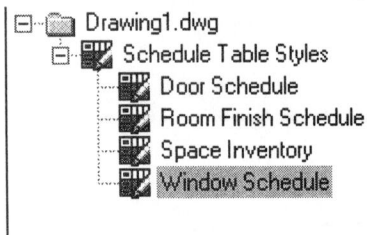

Purge

Purges unused schedule tables from the drawing.

Lesson 10
Schedules

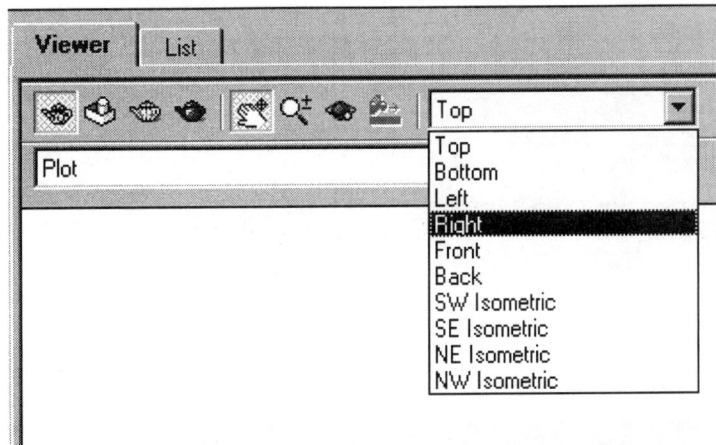

The Style Manager includes a Viewer that allows the user to change the view of the schedule that will be inserted into the drawing. You can zoom in, window, change shading, orbit, etc.

Exercise 1:
Adding a Schedule Table

Drawing Name: Ex10-1.dwg
Estimated Time: 15 minutes

This exercise reinforces the following skills:

- Schedule Table Styles
- Add Schedule Table

Open Ex10-1.dwg. (This can be downloaded from www.schroff.com) or use Lesson 9 where you have added the dimensions to your floor plan.

Activate the Plot-FLR tab.

Lesson 10
Schedules

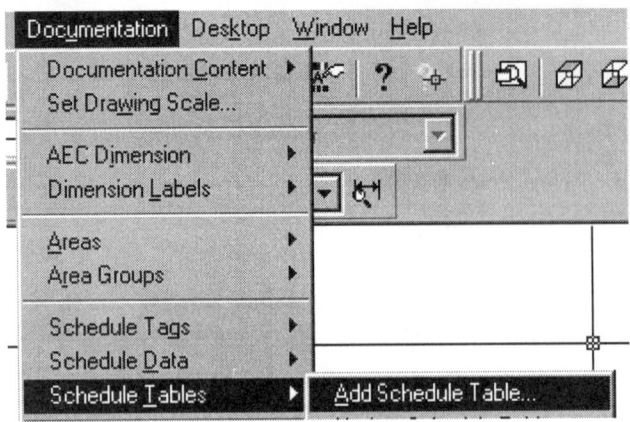

Menu	Documentation->Schedule Tables->Add Schedule Table
Schedules Toolbar	
Command Line	TableAdd

Select the Add Schedule Table tool from the Schedules toolbar.

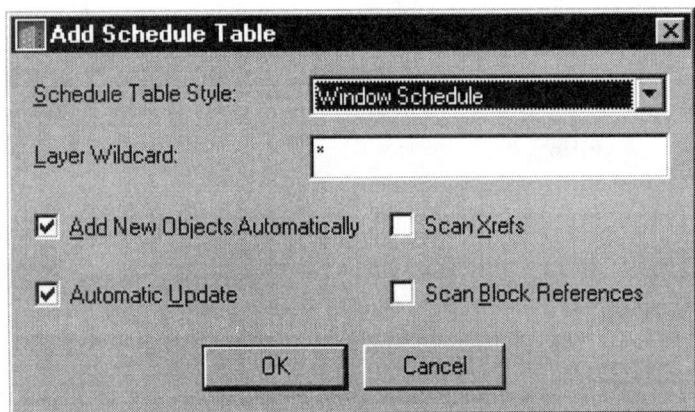

Select the 'Window Schedule' from the drop down.
Enable the Add New Objects Automatically.
Enable Automatic Update.
Press 'OK'.

Lesson 10
Schedules

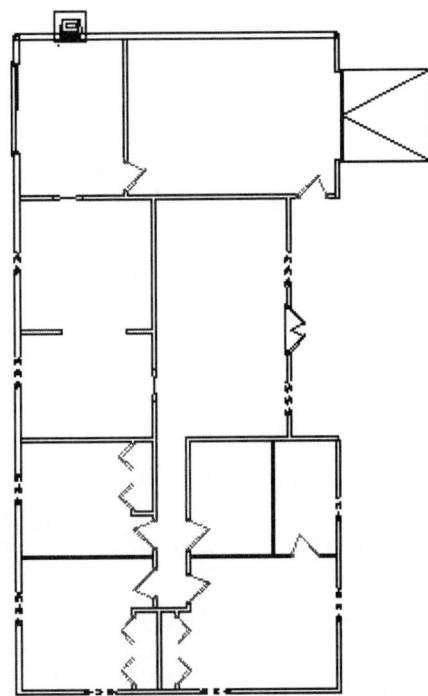

Window around your floorplan.
Notice that all the windows are automatically highlighted and all the other objects are ignored.
Press ENTER to complete the selection.

WINDOW SCHEDULE					
MARK	SIZE		TYPE	MATERIAL	NOTES
	WIDTH	HEIGHT			
?	2'-10"	3'-0"	--	--	--
?	2'-10"	3'-0"	--	--	--
?	2'-10"	3'-0"	--	--	--
?	3'-0"	4'-4"	--	--	--
?	2'-0"	3'-0"	--	--	--
?	2'-10"	3'-0"	--	--	--
?	2'-10"	3'-0"	--	--	--
?	3'-0"	4'-5"	--	--	--
?	2'-0"	3'-0"	--	--	--
?	3'-0"	4'-5"	--	--	--

```
Upper left corner of table:
Lower right corner (or RETURN):
```

When prompted for the upper left corner of table, pick to the right of your floor plan in a clear area. Press ENTER when prompted for the lower right corner.

Notice that the schedule is missing some information and there are no tags placed on any of the windows.

Save as Ex10-1.dwg.

Lesson 10
Schedules

Exercise 2:
Adding Window Tags

Drawing Name: Ex10-1.dwg
Estimated Time: 15 minutes

This exercise reinforces the following skills:

- Schedule Table Styles
- Add Schedule Table

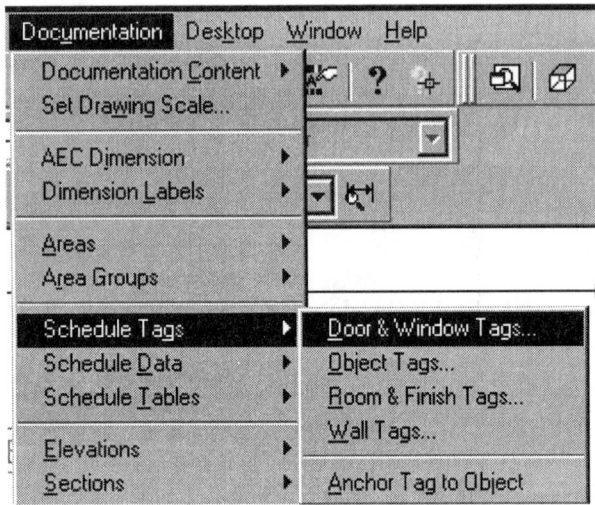

Menu	Documenation->Schedule Tags->Door & Window Tags
Schedules Toolbar	
Command Line	AnnoScheduleTagAdd

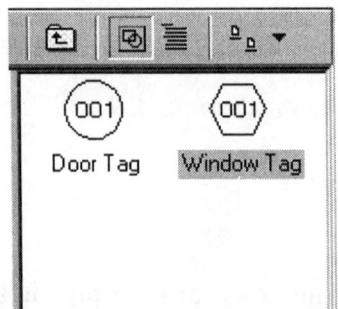

The Door and Window Tag section of the Design Center opens.

10-14

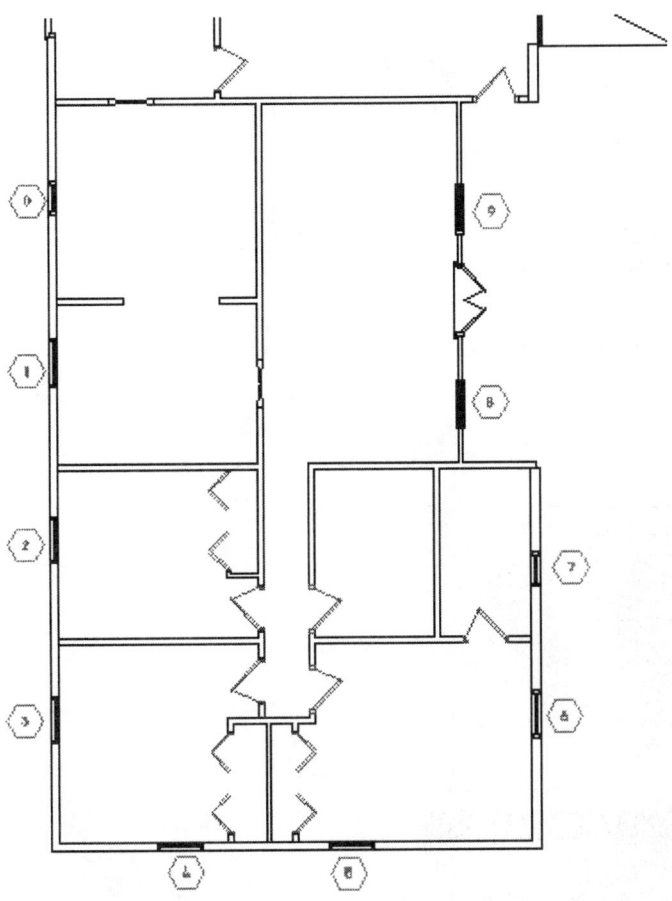

Drag and drop a Window Tag next to each window. Press ENTER to place. The Window Tag number will automatically increment.

TIP: The Window Tags should automatically be placed on the A-Glaz-Iden Layer which is displayed as GREEN in Model Space, black on the Plot-* tabs.

Notice that the Window Schedule you placed in Lesson 1 automatically updated with the Mark Numbers.

Save as Ex10-2.dwg.

Lesson 10
Schedules

Exercise 3:
Editing a Schedule

Drawing Name: Ex10-1.dwg
Estimated Time: 15 minutes

This exercise reinforces the following skills:

❑ Edit Table Style

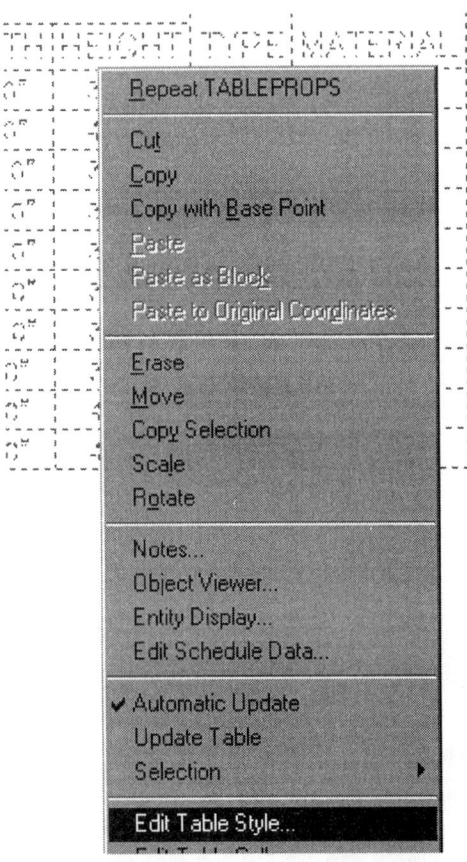

You can edit an existing schedule by picking, right click and select 'Edit Table Style'.

Lesson 10
Schedules

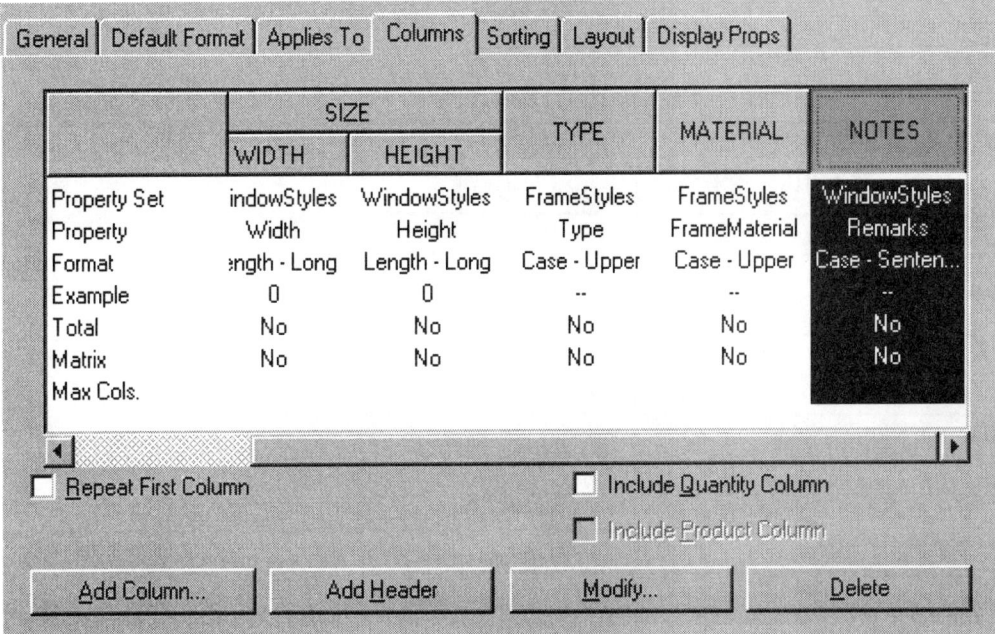

Highlight the Notes Column. Press 'Delete'.

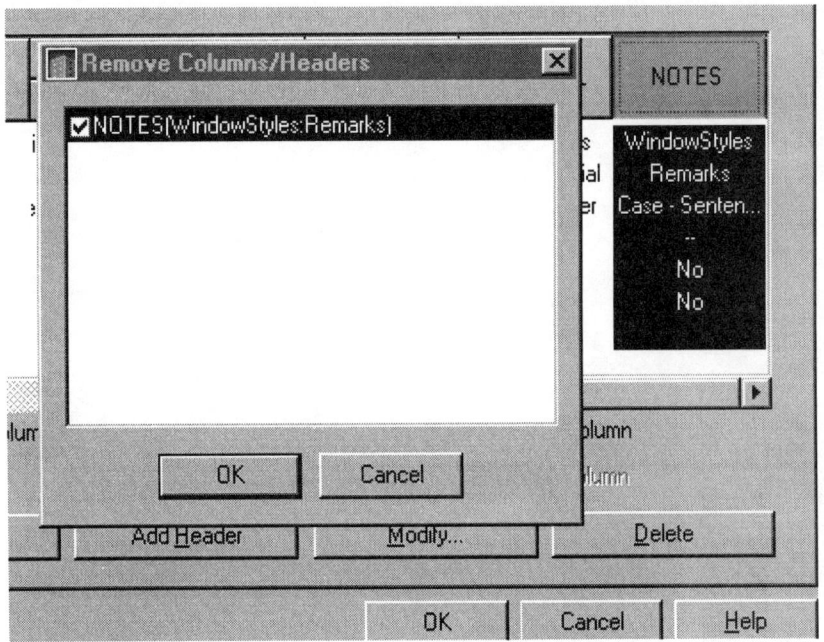

The Remove Columns/Headers dialog appears.
Press 'OK'.

WINDOW SCHEDULE				
MARK	SIZE WIDTH	SIZE HEIGHT	TYPE	MATERIAL
0	2'-0"	3'-0"	--	--
1	3'-0"	4'-4"	--	--
2	2'-10"	3'-0"	--	--
3	2'-10"	3'-0"	--	--
4	2'-10"	3'-0"	--	--
5	2'-10"	3'-0"	--	--
6	2'-10"	3'-0"	--	--
7	2'-0"	3'-0"	--	--
8	3'-0"	4'-5"	--	--
9	3'-0"	4'-5"	--	--

The schedule automatically updates.

Save as Ex10-3.dwg

Lesson 10
Schedules

Exercise 4:
Creating a New Table Style

Drawing Name: Ex10-1.dwg
Estimated Time: 15 minutes

This exercise reinforces the following skills:

- Table Styles
- Creating a New Table Style
- Applying a New Table Style
- Modifying an Exiting Schedule Table

Menu	Documentation->Schedule Tables->Schedule Table Styles
	Schedule Table Styles
Command Line	TableStyle

Initiate the TableStyle command.

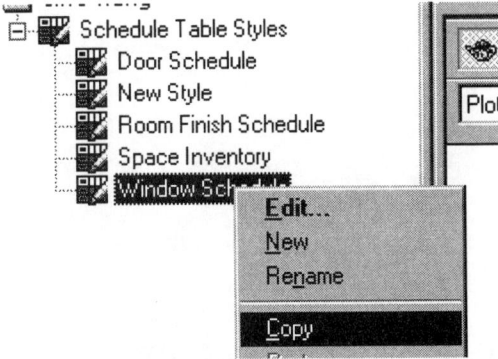

Highlight the Existing default Window Schedule.
Right click and select 'Copy'.

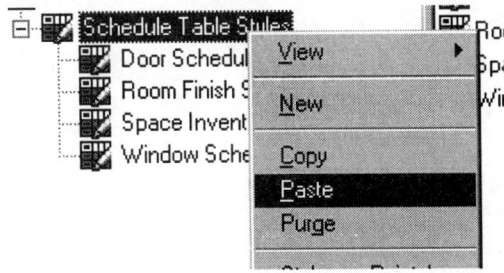

Highlight the Schedule Table Styles.
Right click and select 'Paste'.

Lesson 10
Schedules

A second Window Schedule appears in the tree.

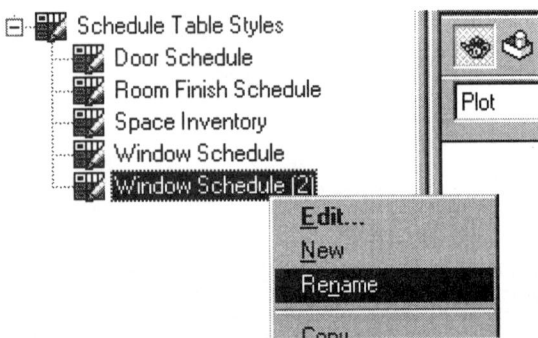

Highlight the second Window Schedule.
Right click and select 'Rename'.

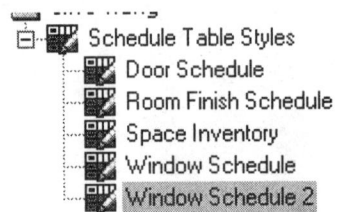

Edit the schedule to read 'Window Schedule 2'.

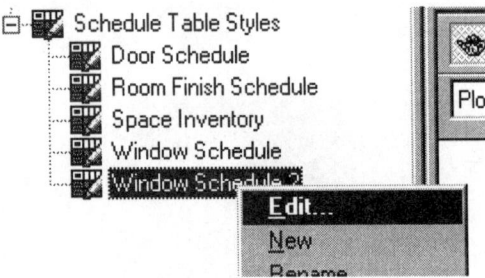

Highlight the Window Schedule 2.
Right click and select 'Edit'.

Lesson 10
Schedules

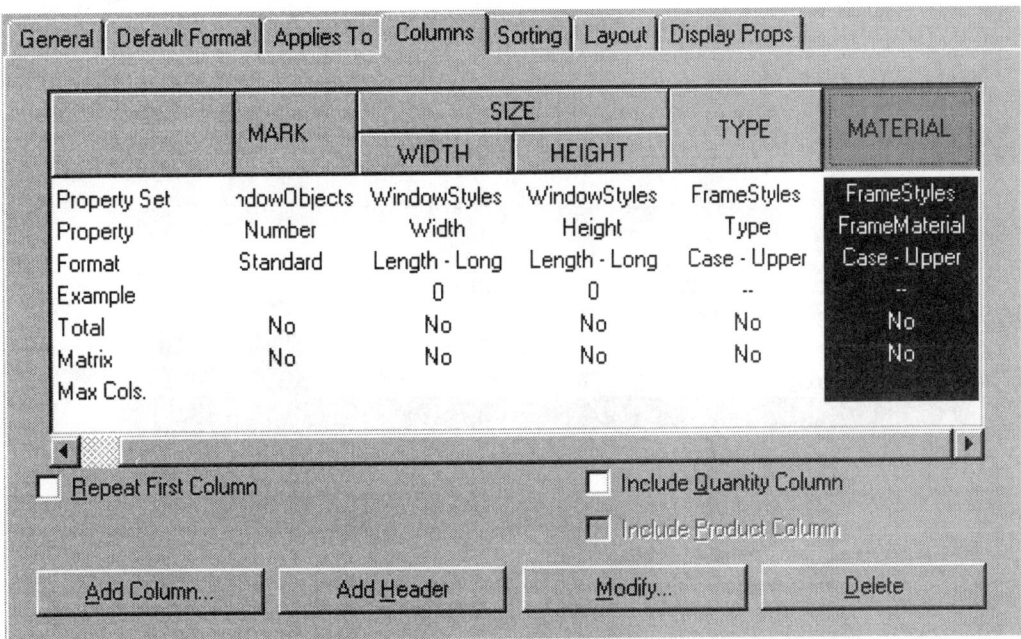

Highlight the Material Column.
Press the 'Add Column' button.

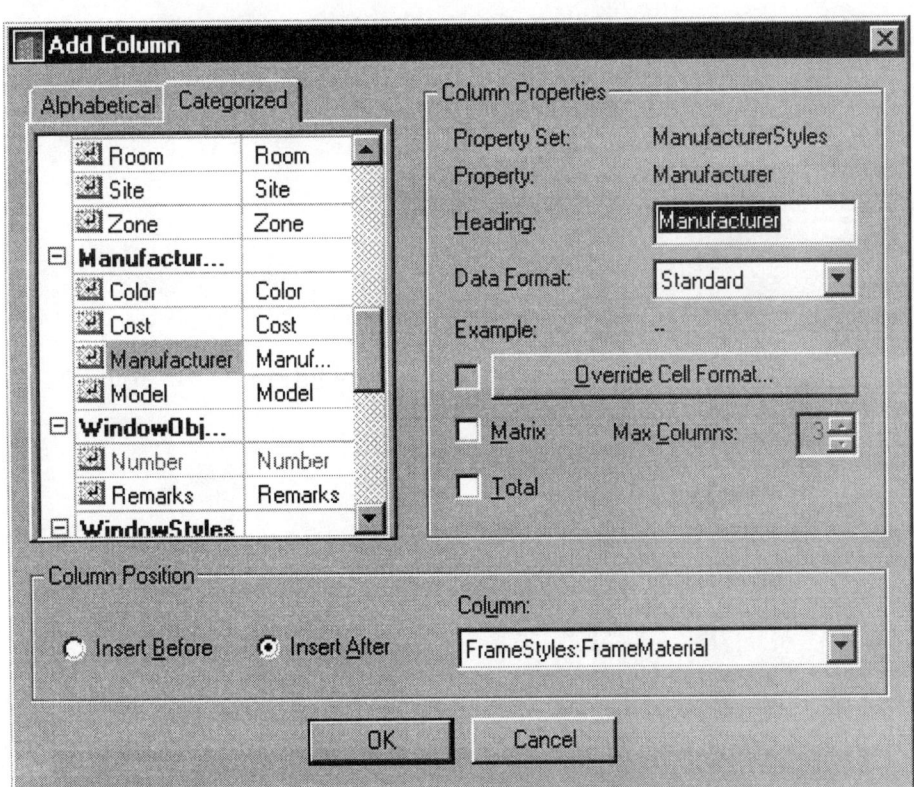

Locate the 'Manufacturer' column in the slider window.
Enable 'Insert After' and verify that the column FrameStyles:FrameMaterial is shown in the drop-down.
Press 'OK'.

Lesson 10
Schedules

Click the General tab, and click Property Sets.

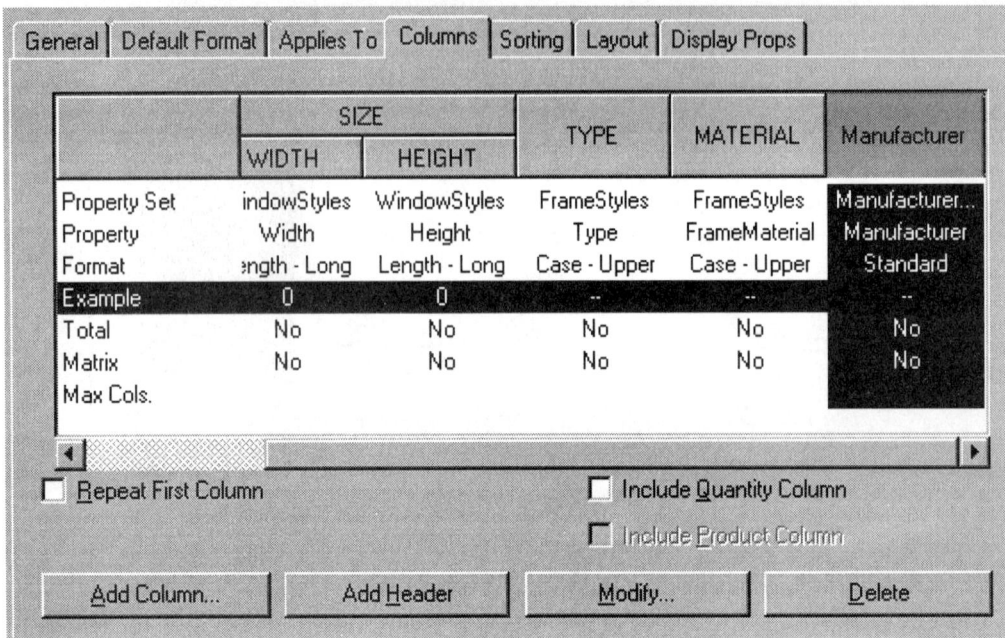

The Manufacturer Column is now added.

Highlight the Manufacturer column.
Press 'Modify'.

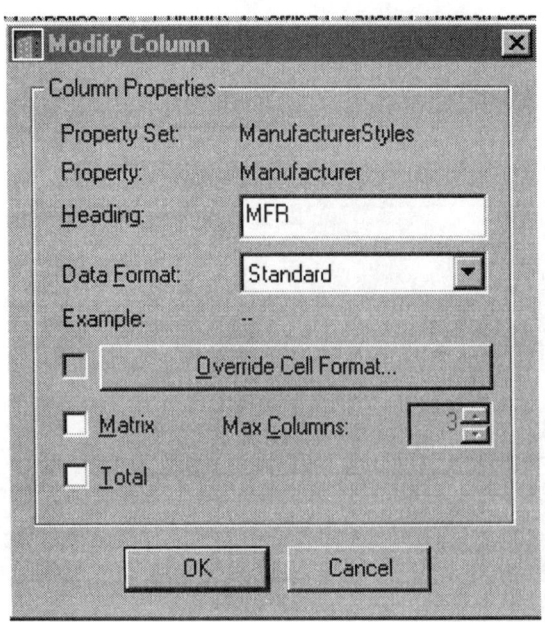

In the Heading text box, edit it to read 'MFR'.
Press 'OK'

10-22

Lesson 10
Schedules

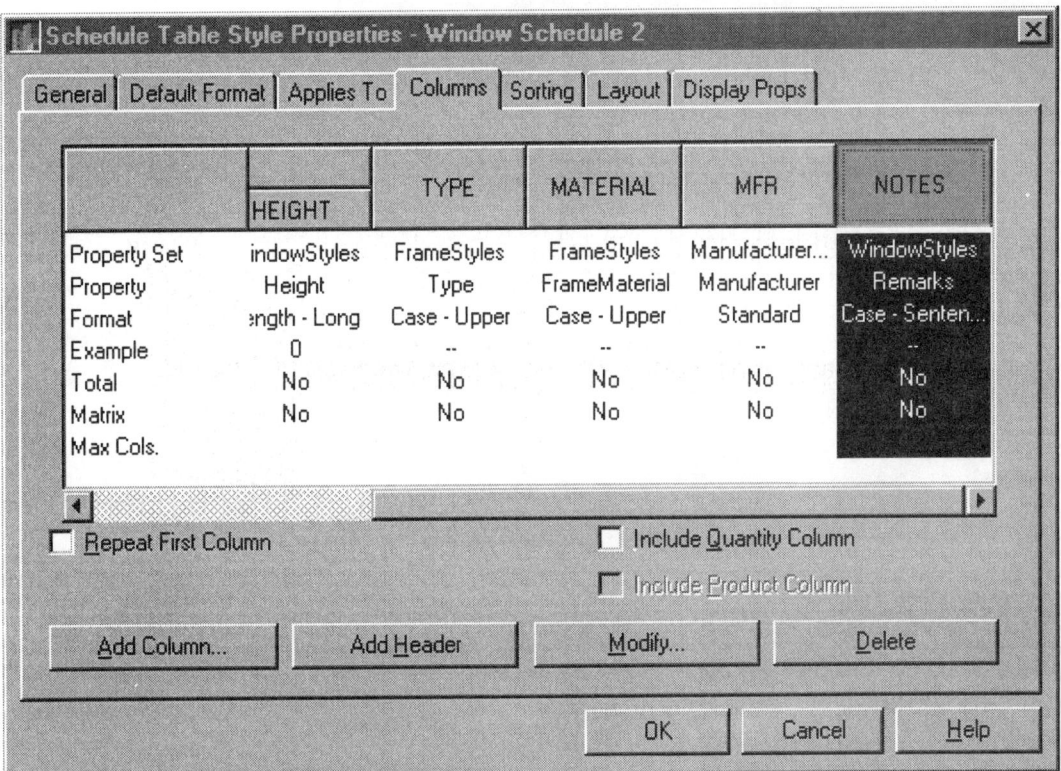

Highlight the NOTES column.
Press the 'Delete' button.

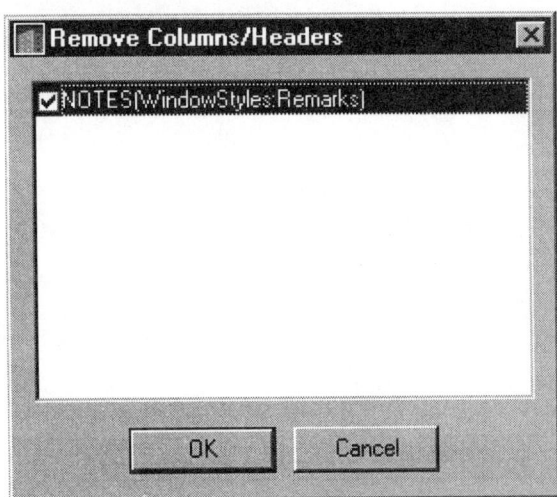

The Remove Columns/Headers dialog will appear.
Press 'OK'.

10-23

Lesson 10
Schedules

WINDOW SCHEDULE					
	SIZE				
MARK	WIDTH	HEIGHT	TYPE	MATERIAL	MFR

You should see the modification in the right pane of the Table Styles window.

Press 'Apply'
Press 'OK.'

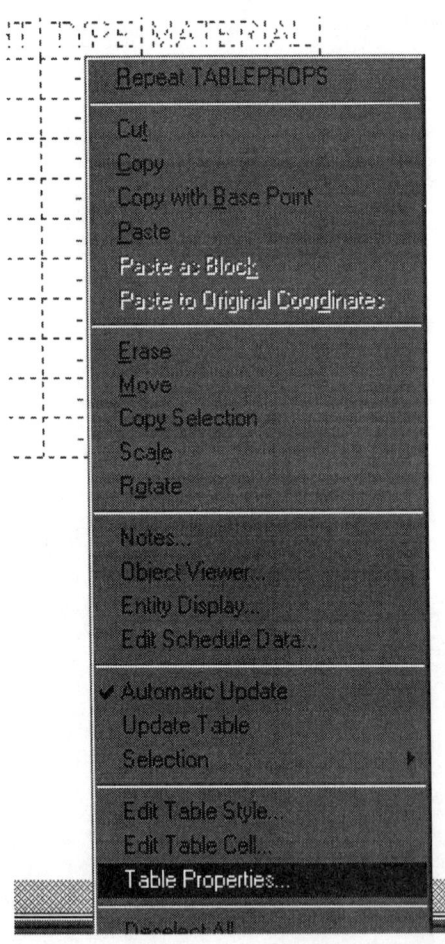

Highlight the existing schedule table in the drawing.
Right click and Select 'Table Properties'.

10-24

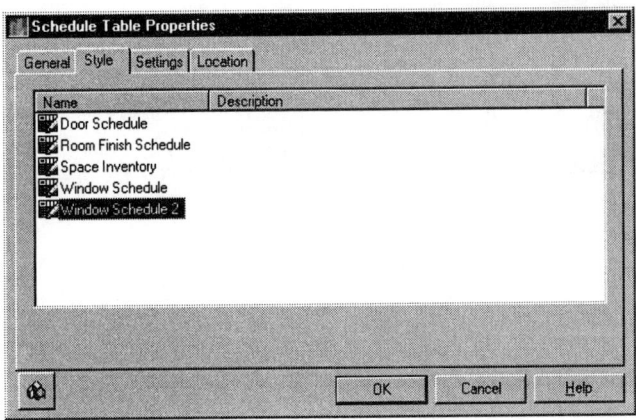

Select the Style tab.
Highlight Window Schedule 2.
Press 'OK'.

WINDOW SCHEDULE					
MARK	SIZE		TYPE	MATERIAL	MFR
	WIDTH	HEIGHT			
0	2'-0"	3'-0"	--	--	?
1	3'-0"	4'-4"	--	--	?
2	2'-10"	3'-0"	--	--	?
3	2'-10"	3'-0"	--	--	?
4	2'-10"	3'-0"	--	--	?
5	2'-10"	3'-0"	--	--	?
6	2'-10"	3'-0"	--	--	?
7	2'-0"	3'-0"	--	--	?
8	3'-0"	4'-5"	--	--	?
9	3'-0"	4'-5"	--	--	?

The table updates to the selected Window Table Style.

In order to have the Window Tags hold the information that we want to appear in the schedule, we need to edit the Window Styles for the various windows. This is because the schedule we created is based on the Window Styles Property Set for the Width and Height values, the Frame Style Property Set for the Type and Material values and the Manufacturer Style Property Set for the MFR column we added.

Pick the window next to Tag Number 1. Right click and pick Edit Window Style. . . .

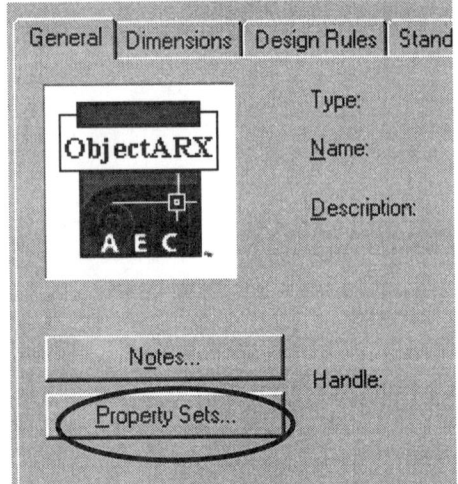

On the General tab pick the Property Sets button.

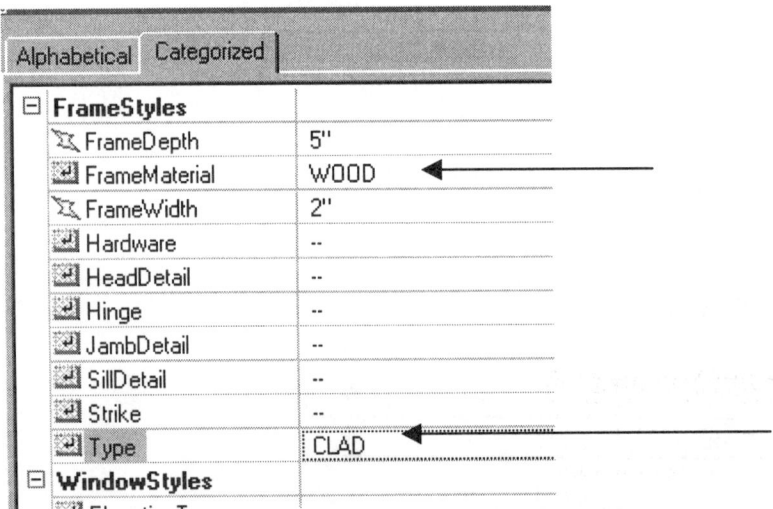

In the Edit Schedule Data dialogue, in the FrameStyles property set type in Wood for the FrameMaterial value and Clad for the Type. These values will automatically become upper case after you type them in.

Lesson 10
Schedules

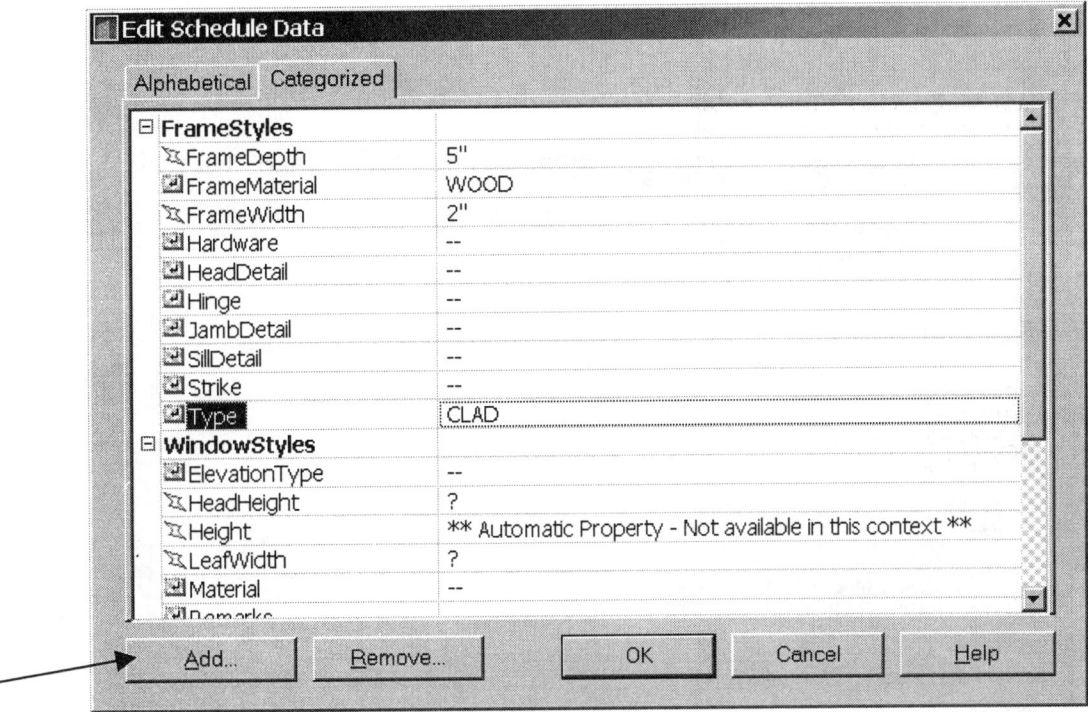

Pick the Add Button.

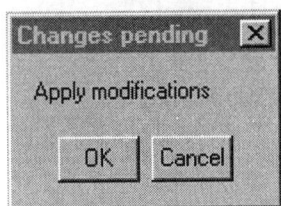

Pick OK to the Changes Pending box.

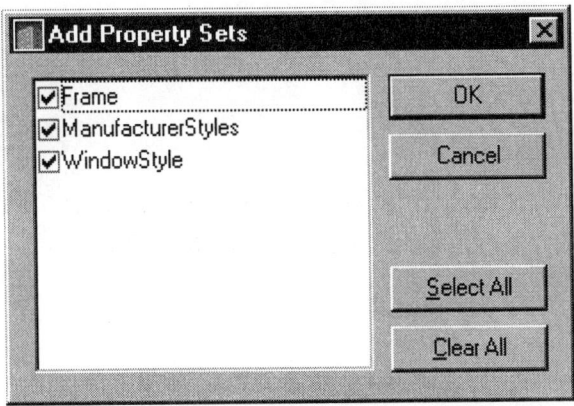

Pick OK to the Add Property Set box.
Verify the ManufacturerStyles check box is checked.

In the Edit Schedule Data dialogue, scroll down. Under the ManufacturerStyles set which now appears, type in 'Smith' for the Manufacturer.

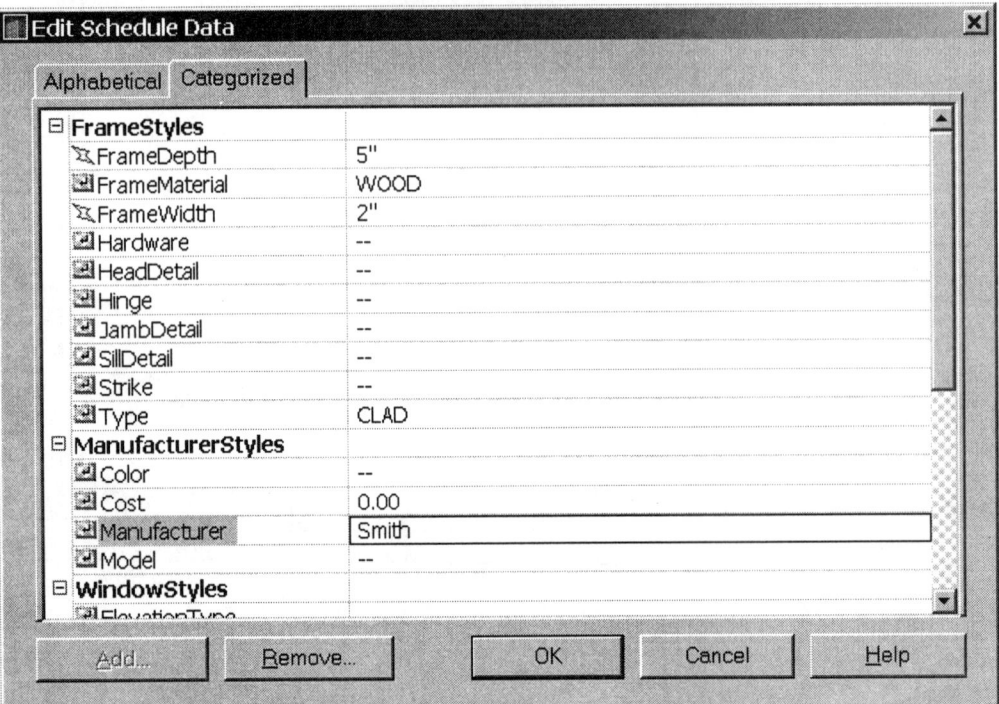

Pick OK twice. The schedule will update for all the windows of that type.

Select each window type, right click and select 'Edit Window Style'.
Repeat the steps for each window type to fill in the schedule completely.

WINDOW SCHEDULE					
MARK	SIZE		TYPE	MATERIAL	MFR
	WIDTH	HEIGHT			
1	2'-0"	3'-0"	CLAD	WOOD	Smith
2	2'-10"	3'-0"	CLAD	WOOD	SMITH
3	3'-0"	4'-5"	CLAD	WOOD	SMITH
4	2'-10"	3'-0"	CLAD	WOOD	SMITH
5	2'-10"	3'-0"	CLAD	WOOD	SMITH
6	2'-10"	3'-0"	CLAD	WOOD	SMITH
7	2'-10"	3'-0"	CLAD	WOOD	SMITH
8	2'-0"	3'-0"	CLAD	WOOD	SMITH
9	3'-0"	4'-5"	CLAD	WOOD	SMITH
10	3'-0"	4'-5"	CLAD	WOOD	SMITH

Save the file as Ex10-4.dwg

Lesson 10
Schedules

Exercise 5:
Exporting a Schedule

Drawing Name: Ex10-4.dwg
Estimated Time: 15 minutes

This exercise reinforces the following skills:

- Table Styles
- Creating a New Table Style
- Exporting an Exiting Schedule Table

Often, you want to export a schedule to a spreadsheet so you can calculate costs.

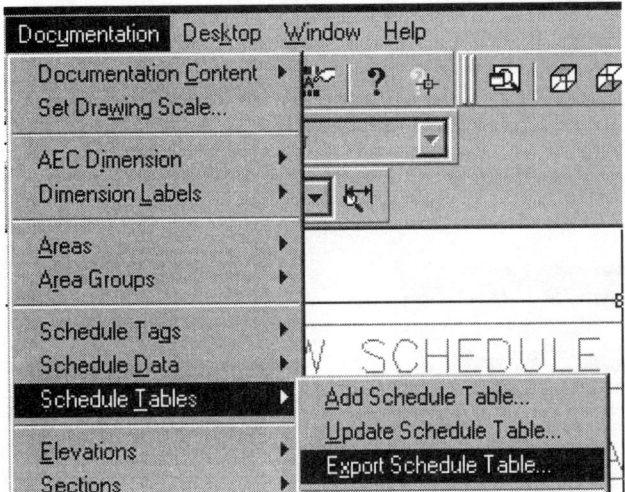

Menu	Documentation->Schedule Tables->Export Schedule Table
Schedules Toolbar	
Command Line	TableExport

10-29

Lesson 10
Schedules

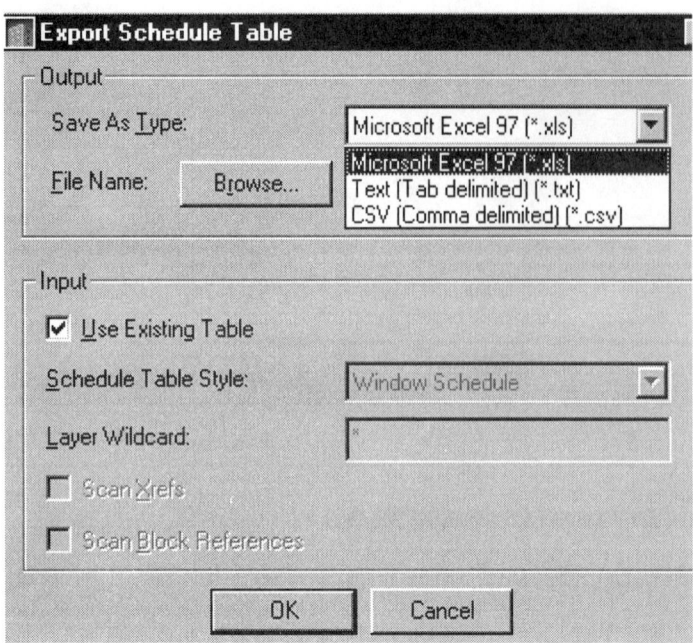

You have three options of file type you can export: Excel, text, or csv.

Select .txt.

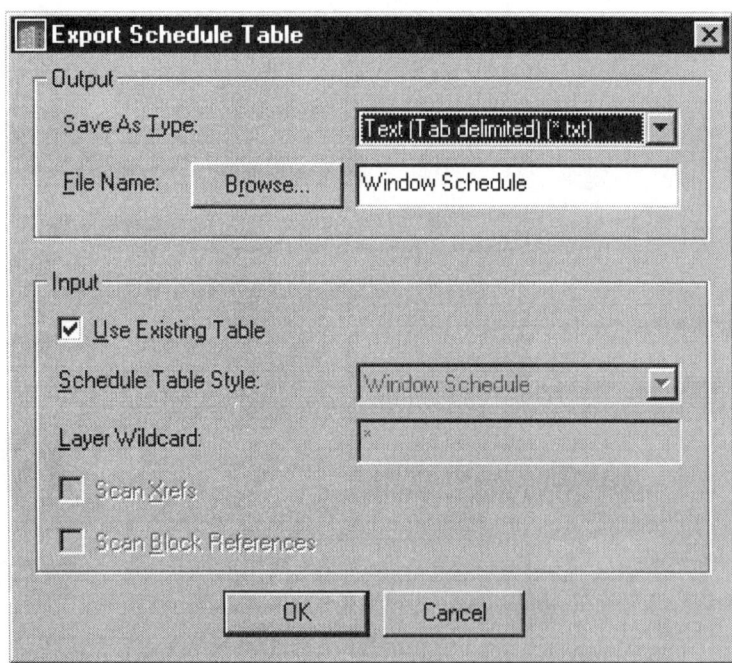

Pressing the Browse button will allow you to set the path where the export file will be saved.

Change the file name to Window Schedule.

Press 'OK'.

10-30

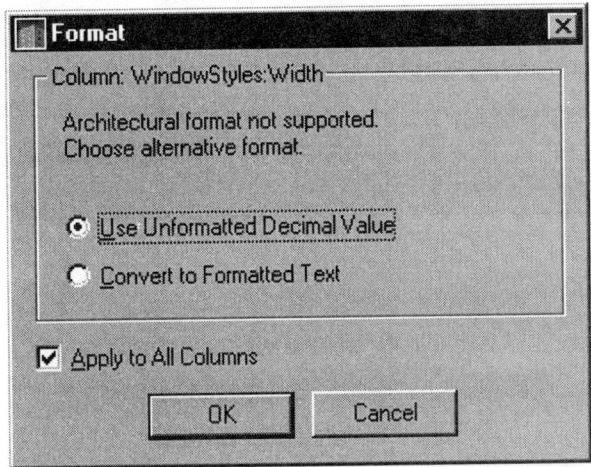

When prompted to 'Select existing table to export:', pick the Window Table.
Enable the check box to apply formatting to all columns.
(If you don't, you will be prompted on format type for each column.)

Press 'OK'.

WALFCTS3.dwg	45KB	AutoCAD Drawing
Window Schedule	1KB	File
WS4848_3.dwg	26KB	AutoCAD Drawing

Locate the file you just created and open using 'Notepad.'

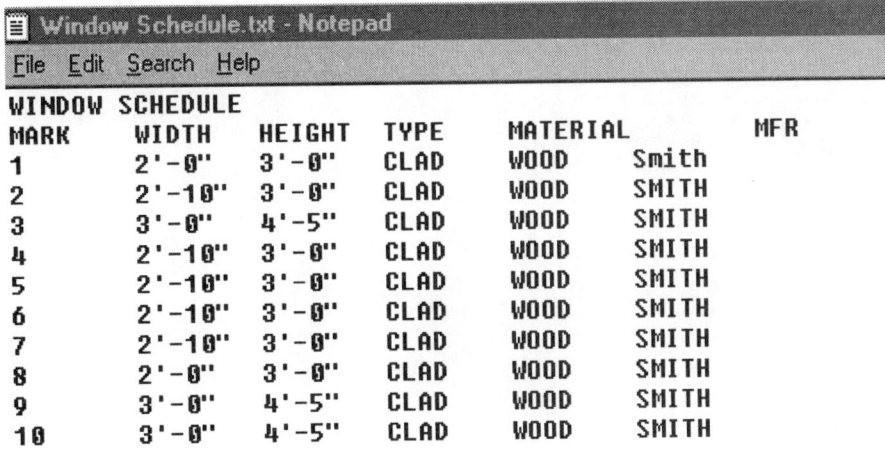

The file shows the data.

Save your drawing as Lesson 10.dwg.

Notes:

Lesson 11
Creating a Video

One of the more fun (and the more frustrating) tools within Architectural Desktop is the ability to create a video.

A video requires two basic things:

A path – this is the path, which the camera will travel along. The path must be an open or closed polyline, circle, or rectangle.

A camera – the camera can be set to point to different angles and elevations along the path.

Exercise 1:
Adding a Path

Drawing Name: Ex11-1.dwg
Estimated Time: 30 minutes

This exercise reinforces the following skills:

- Creating a path

Open the Ex11-1 file available on www.schroff.com or open your existing drawing.

Model

Activate the Model tab and set it to a top view of the house model only.

Lesson 11
Creating a Video

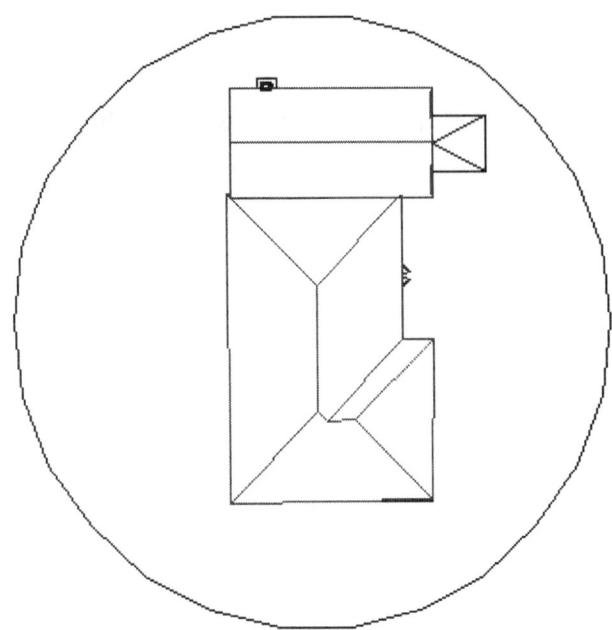

Draw a circle around the house.

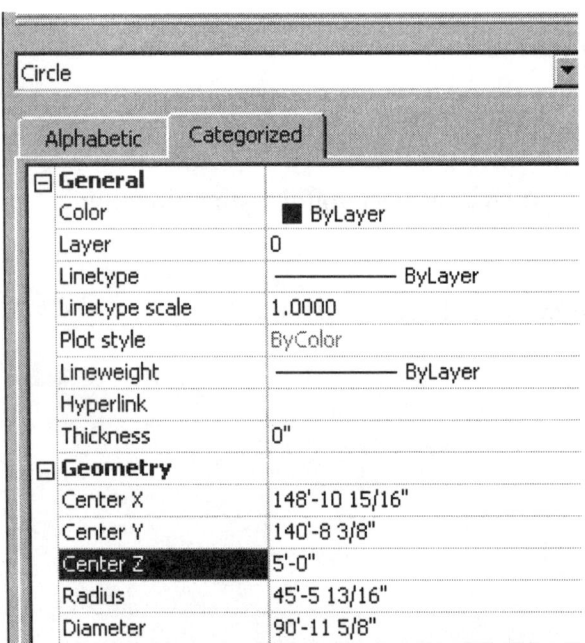

Select the circle and Change the Properties.
Set the Center Z to 5'-0". This changes the elevation of the circle.

Lesson 11
Creating a Video

Exercise 2:
Adding a Camera

Drawing Name: Ex11-1.dwg
Estimated Time: 30 minutes

This exercise reinforces the following skills:

- Adding a Camera
- Modify Camera Properties
- Adding a Camera View
- Adjust Camera View

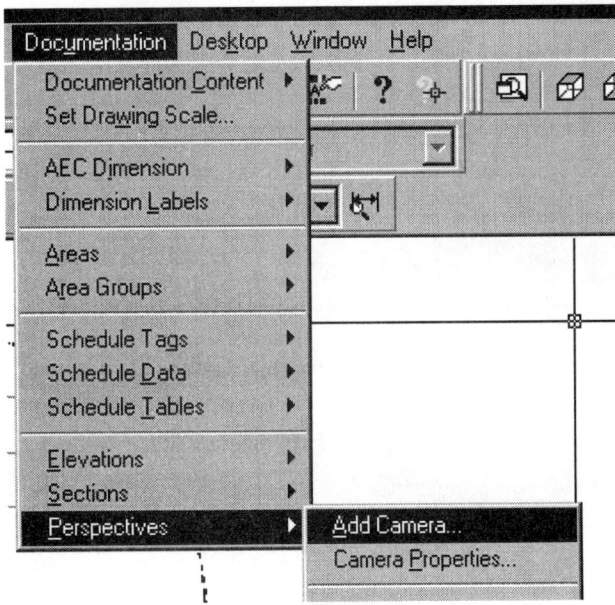

Menu	Documentation->Perpectives->Add Camera
Perspectives Toolbar	
Command Line	

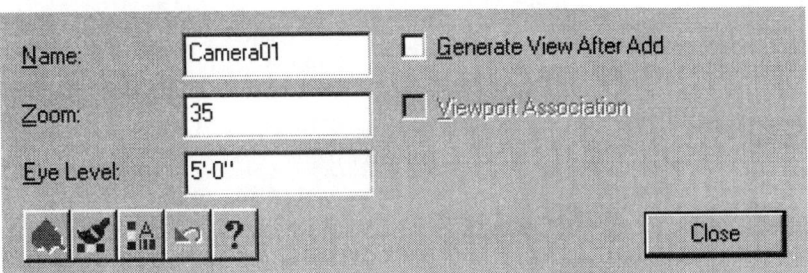

11-3

Lesson 11
Creating a Video

It is important to give the camera a name so you can keep track when you want to switch from one camera to another when creating your video.

The Zoom indicates a lens size. A value of 50 represents a 50 mm lens. This is similar to what the average human eye sees. Above 50 is considered a telephoto lens. Below 50 is considered a wide-angle (panoramic) lens, allowing you to see more of the scene.

The Eye level sets the initial height of the camera.

The Generate View After Add box allows you to apply the camera view to a selected viewport after the camera is placed.

The Viewport Association box associates the camera with a particular viewport.

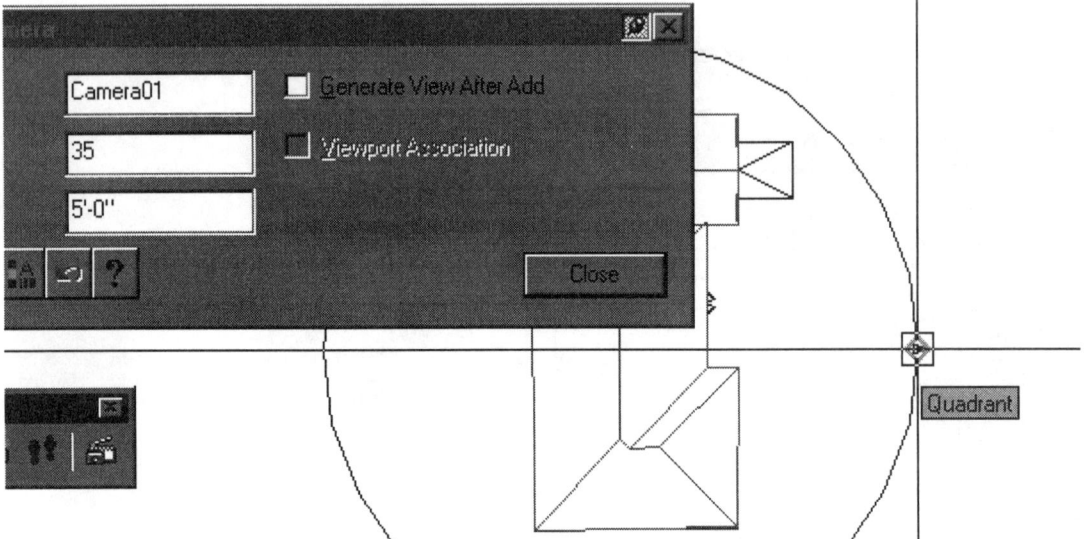

Select the right quadrant as the insertion point.

Lesson 11
Creating a Video

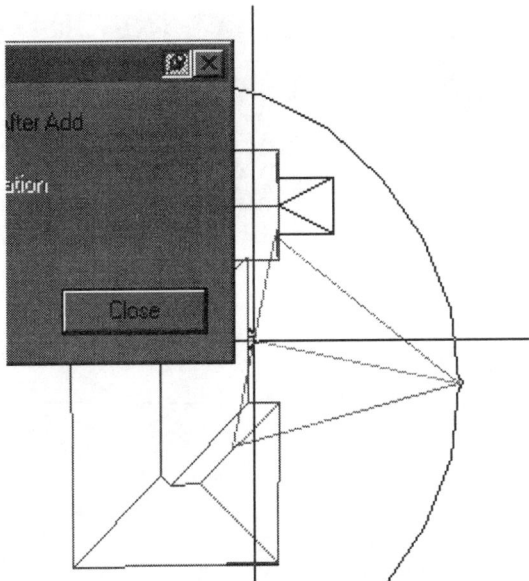

Select the front door as the target point.

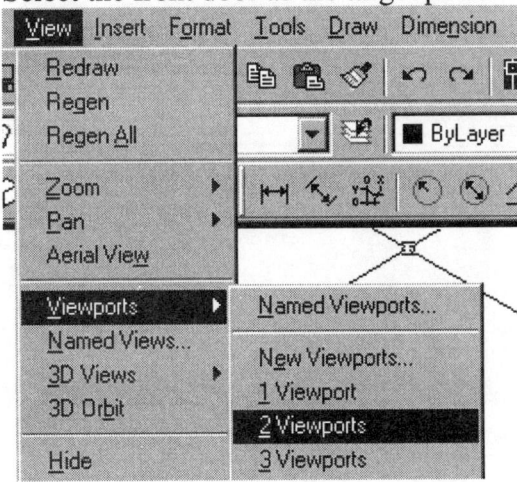

Set up two viewports. You can use the viewport on the right to check your camera view.

Lesson 11
Creating a Video

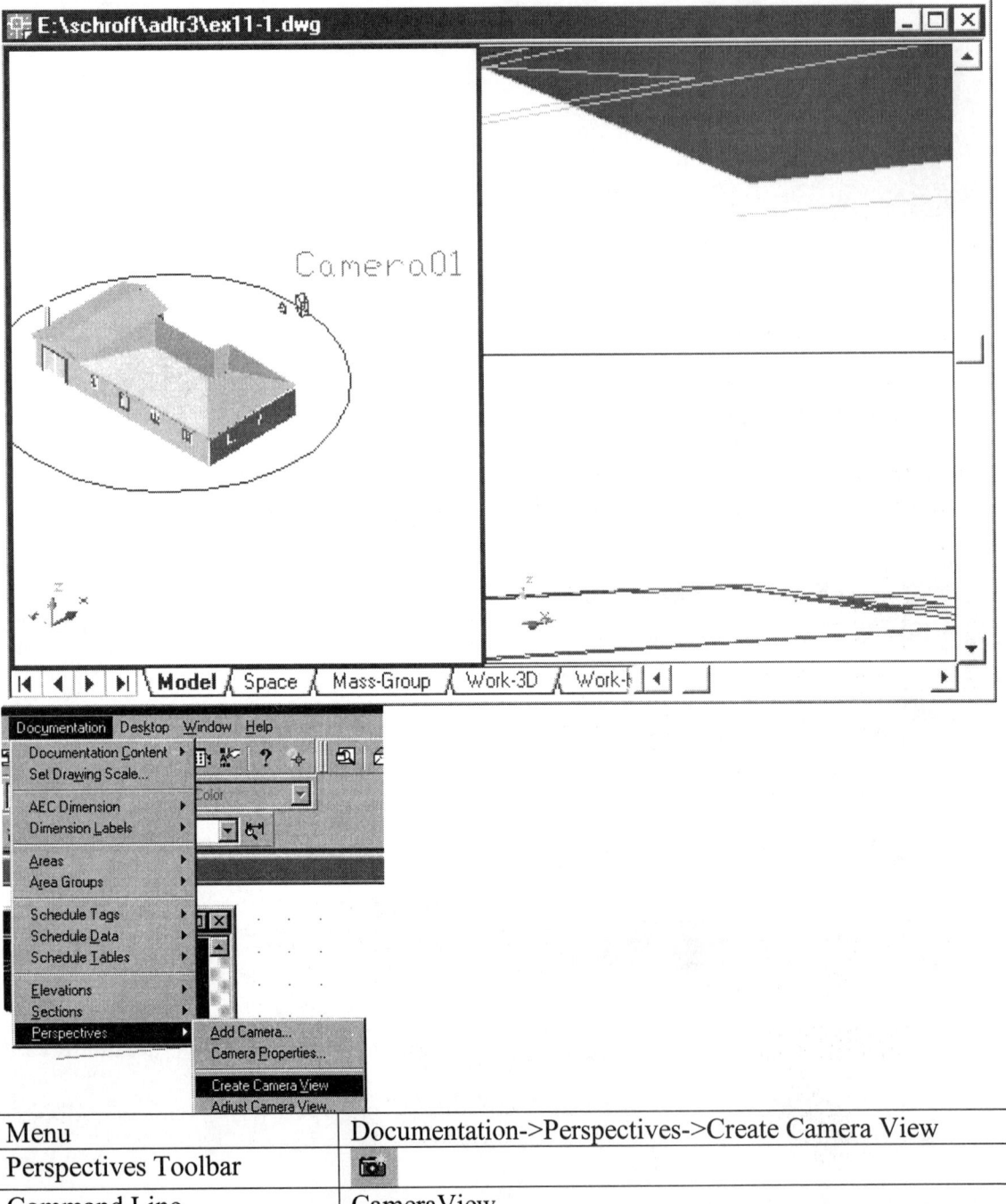

Menu	Documentation->Perspectives->Create Camera View
Perspectives Toolbar	
Command Line	CameraView

Select the CameraView tool. When prompted to select a viewport, select the viewport on the right.

Menu	Documentation->Perspectives->Camera Properties
Perspectives Toolbar	
Command Line	CameraProps

Select the CameraProps tool.

Select the existing Camera.

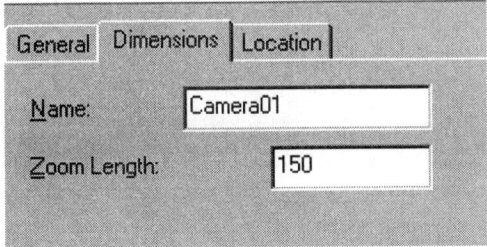

Change the Zoom Length to 150.

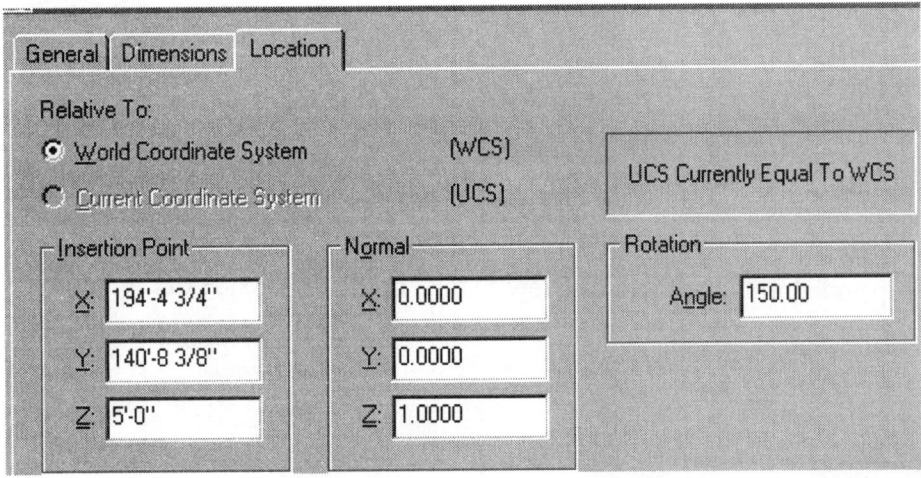

Change the Rotation Angle to 150.

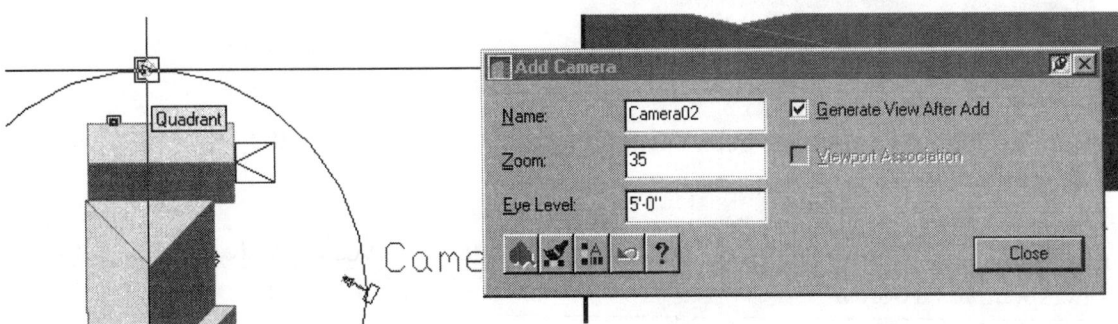

Add a second camera at the top quadrant.
Set the Zoom to 35.
Enable Generate View After Add.

Lesson 11
Creating a Video

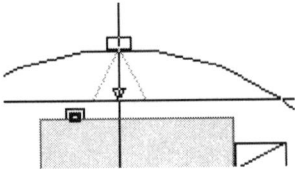

Set the target so it is toward the side of the house as shown.

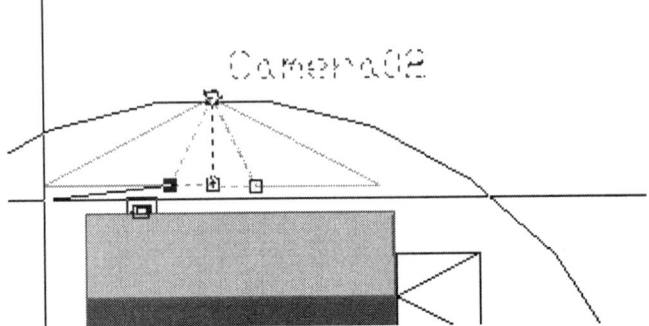

Highlight Camera02.
Select the left grip and stretch the Field of View to take in the entire side of the building.

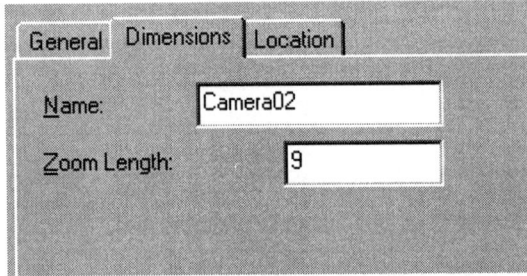

Note the Zoom Length has been changed to a value less than 35, the original value you set.

Menu	Documentation->Perspectives->Create Camera View
Perspectives Toolbar	📷
Command Line	CameraView

Select the CameraView tool. When prompted to select a Camera, pick Camera02. When prompted to select a viewport, select the viewport on the right.

Note that the camera view is updated.

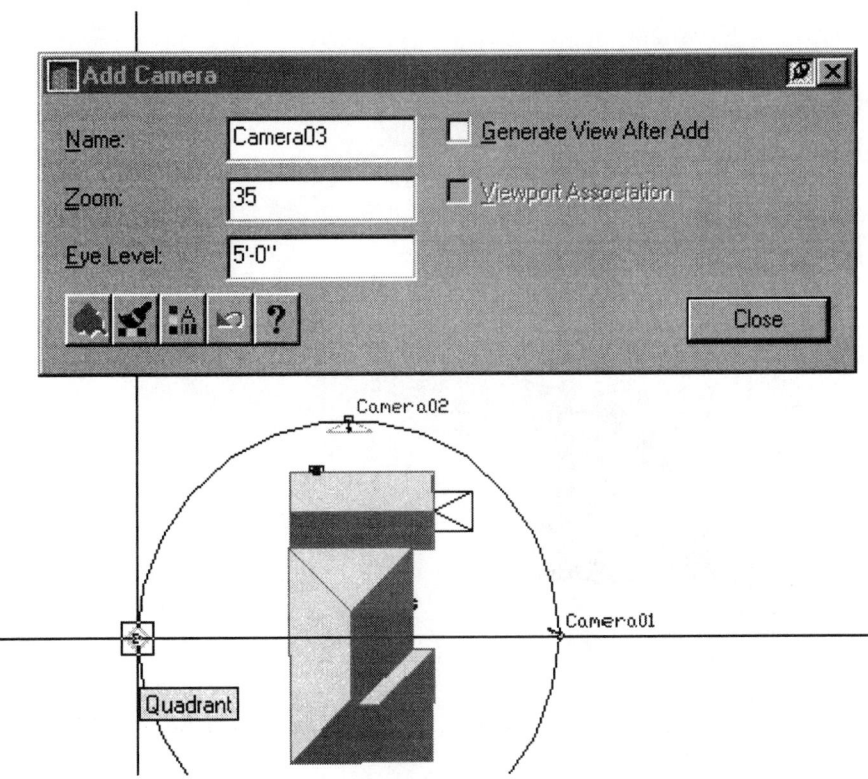

Add a third camera at the left quadrant as shown.

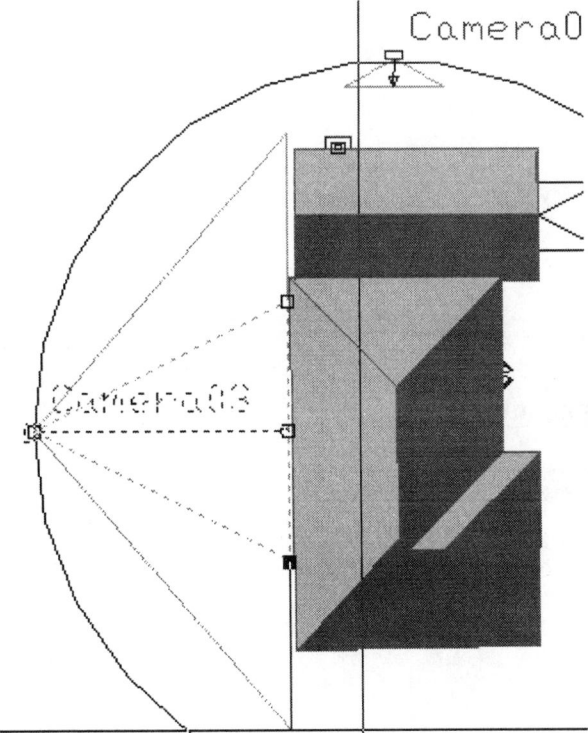

Use the grip tools to extend the field of view. Confirm that the Zoom Length is around 16.

Lesson 11
Creating a Video

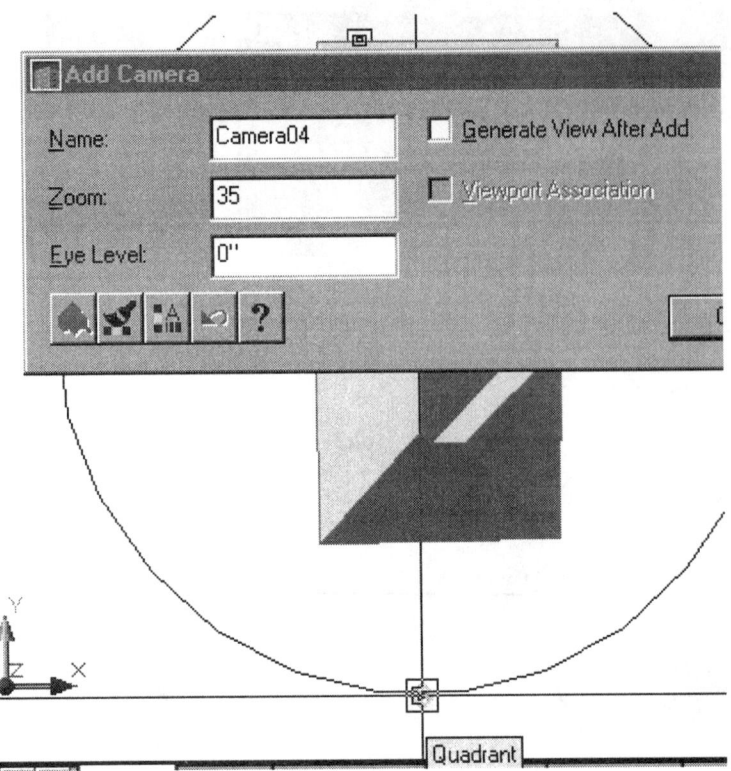

Add a fourth camera at the bottom quadrant.
Set the Eye Level to 0".

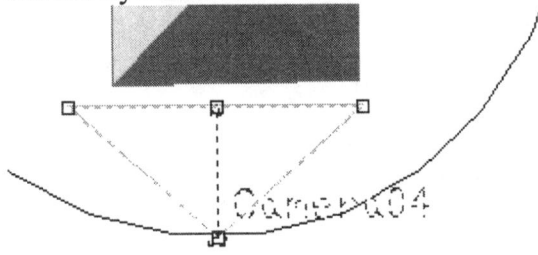

Use your grips to extend the field of view to a Zoom of 15.

Menu	Documentation->Perspectives->Create Camera View
Perspectives Toolbar	
Command Line	CameraView

Select the CameraView tool. When prompted to select a Camera, right click the mouse.

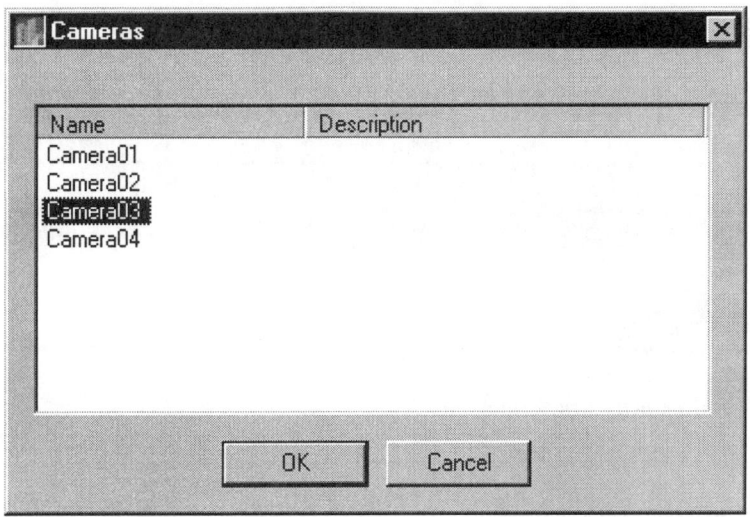

Select Camera03.
When prompted to select a viewport, select the viewport on the right.
Note that the camera view is updated.

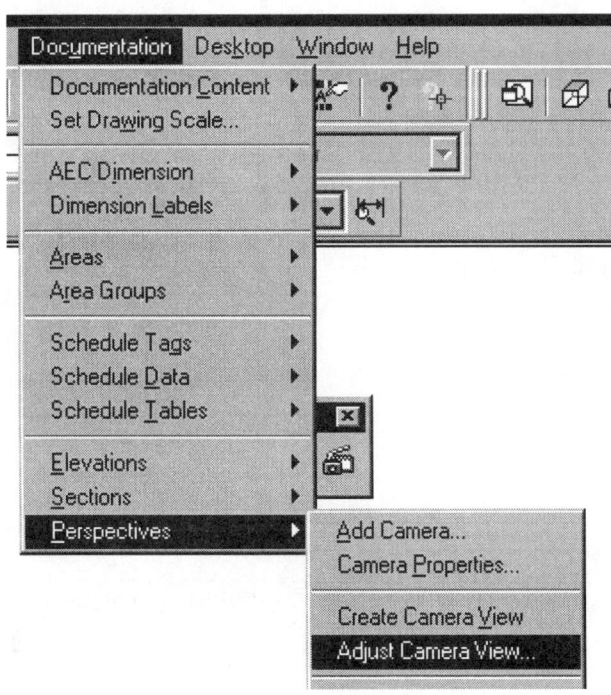

Menu	Documentation->Perspectives->Adjust Camera View
Perspectives Toolbar	
Command Line	CameraAdjust

Select the Adjust Camera View tool.
When prompted for camera, select Camera03.

Lesson 11
Creating a Video

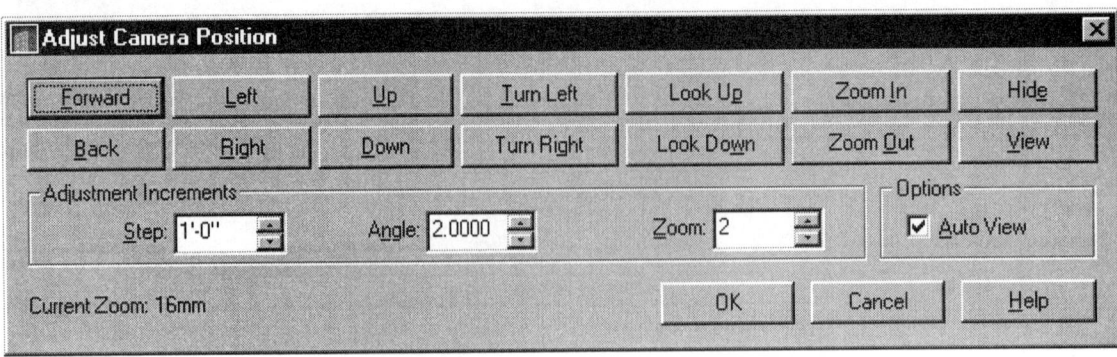

The Adjust Camera Position Dialog appears.

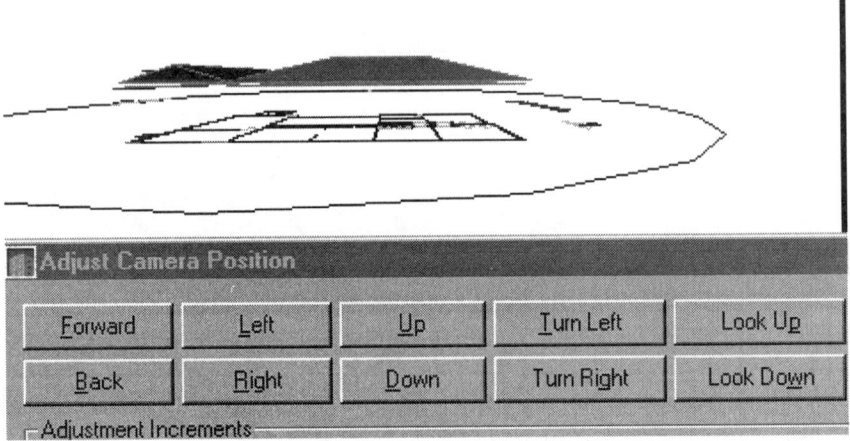

Notice how the view changes depending on the buttons pushed.
Experiment with different views until you find a camera view you like.

Exercise 3:
Creating a Video

Drawing Name: Ex11-3.dwg
Estimated Time: 30 minutes

This exercise reinforces the following skills:

Creating a Video

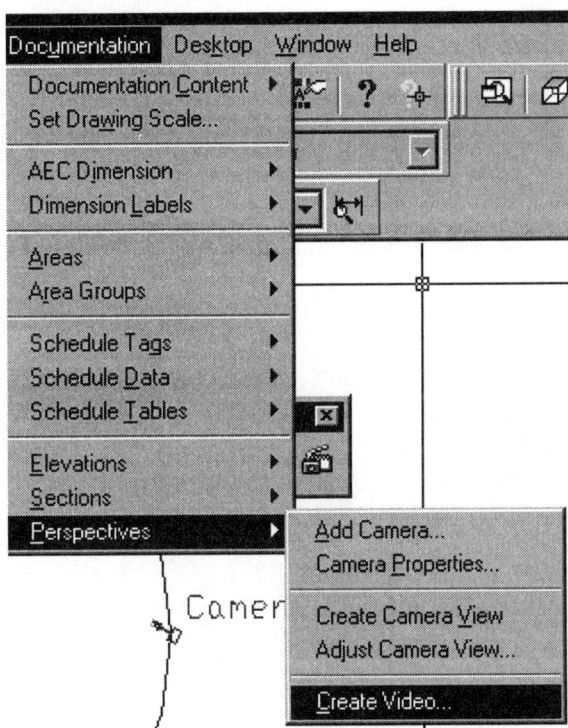

Menu	Documentation-> Perspectives->Create Video
Perspectives Toolbar	
Command Line	CameraVideo

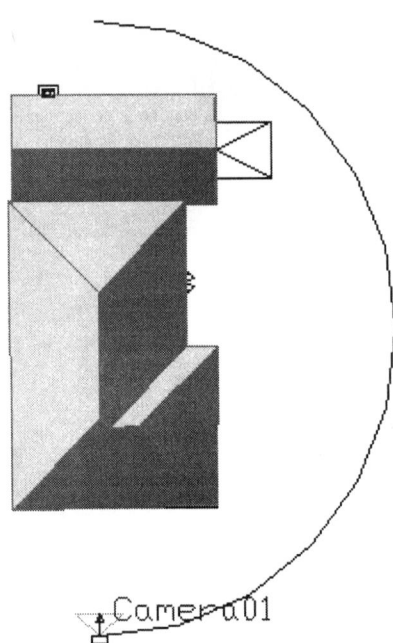

Open Ex11-3 from www.schroff.com.
Or set up the drawing as shown.
Draw an arc on the front side of the home.

Name	On	Freeze...	L..	Color
2D_ELEV				White
2D_PLAN	♀	❄	🔓	White
A-Anno-Dims	♀	❄	🔓	221
A-Anno-Note	♀	❄	🔓	211
A-Contour-Line	♀	❄	🔓	42
A-Elev-Point	♀	❄	🔓	42
A-Equip	♀	❄	🔓	Blue
A-Foundation	♀	❄	🔓	170
A-Furnishings	♀	❄	🔓	White
A-Gas-Line	♀	❄	🔓	White
A-Glaz	♀	❄	🔓	92
A-House-Outline	♀	❄	🔓	White
A-Sect	♀	❄	🔓	240
A-Sect-Iden	♀	❄	🔓	132
A-Sect-Mcut	♀	❄	🔓	120
A-Sewer-Line	♀	❄	🔓	White
A-Site-Dims	♀	❄	🔓	Red
A-Site-Property-Lines	♀	❄	🔓	Green
A-Site-Text	♀	❄	🔓	White
A-Street-Line	♀	❄	🔓	White
A-Topo	♀	❄	🔓	92
A-Water-Line	♀	❄	🔓	White

Freeze layers as shown.

Lesson 11
Creating a Video

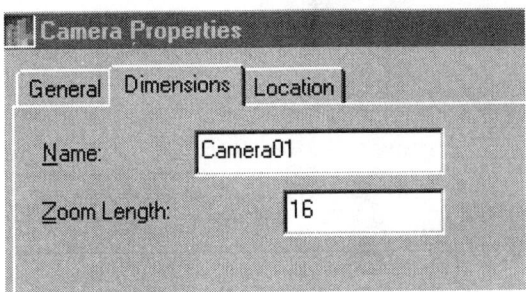

Place a camera at the lower endpoint of the arc.
Set the Zoom Length to 16.

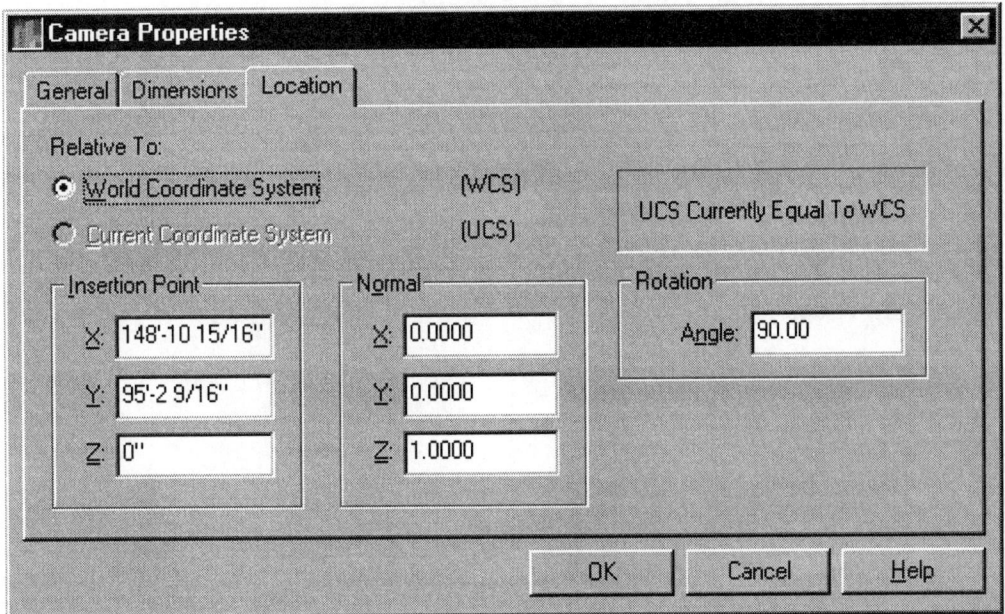

Set the Z to 0".
Set the angle to 90.00.

Select CameraVideo.

11-15

Lesson 11
Creating a Video

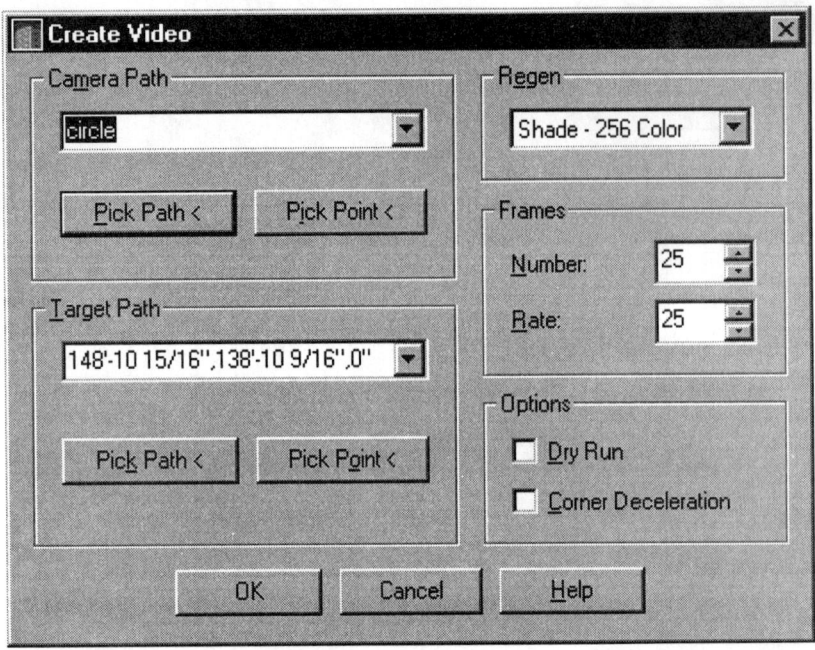

Select the 'Pick Path' button under Camera Path and select the arc.
Select 'Pick Point' under Target Path and select a point approximately in the center of the house.
Set the Regen to Shade – 256 Color.
Set the Frames to 25 and Rate to 25.
Pick 'OK'.

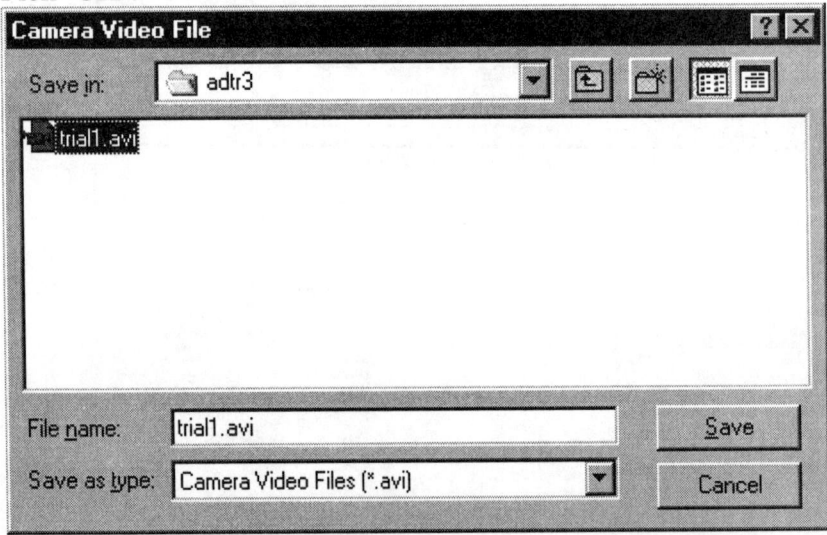

Locate a folder to hold your avi file and name your file 'trial1.avi.'

Lesson 11
Creating a Video

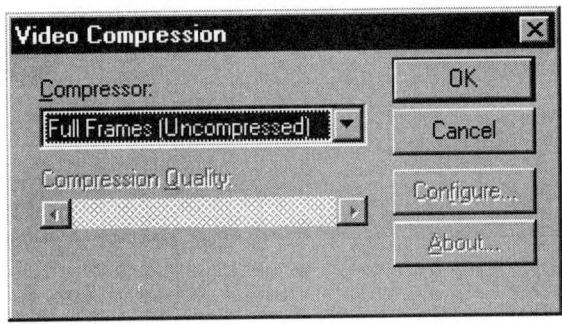

Select Full Frames (Uncompressed) and Press 'OK'.

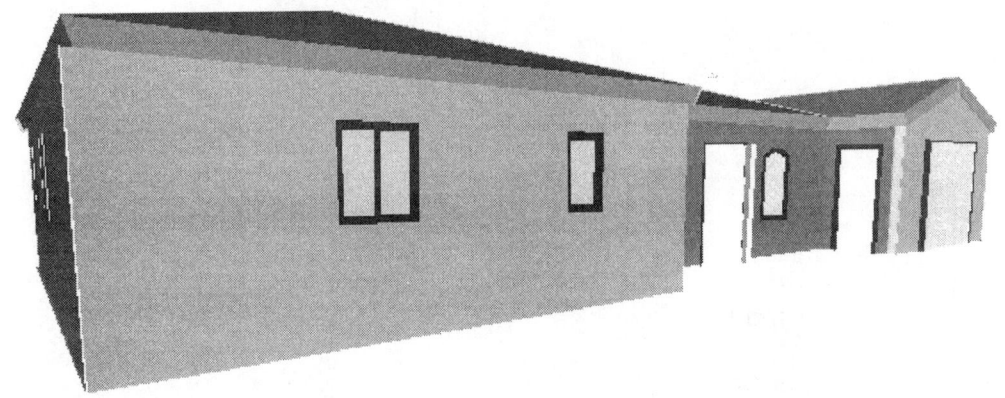

Playing your video shows an animated movie of your home. It looks a bit cartoonish, but that's OK.

ADT only allows you to animate shaded, hidden, or wire frame. It will not allow you to create a rendered animation of your model. To do that, you need software like 3D Studio Viz. Likewise, you can only select one camera at a time. You can modify the path to change elevations (using a 3D polyline), but you cannot change camera angles or perspectives.

Notes:

COMPREHENSIVE EXAM

True or False

1. Doors, windows, and stairs are inserted into a drawing as objects.
2. The Object Viewer is a dialog box that allows you to view an object prior to placement.
3. The handle for placing a wall is located at the bottom of the Wall Justification line.
4. The Offset option of the Add Wall command shifts the justification handle off the justification line the Offset Value distance.
5. Wall Grips are located at the beginning, middle, and end on the bottom of the justification line of a wall.
6. AutoCAD commands such as OFFSET, TRIM, and EXTEND can be used to create and modify walls.
7. You can use wall grips to copy, stretch, or move walls.
8. Polylines can be converted to walls.
9. Door styles can not be copied.
10. Flip Hinge can be used to change the location and swing of a door or window.
11. You can change the location of a door or window using GRIPS.
12. If you remove a door or window from a wall, it will leave a gap. You will then need to "heal" the wall to close the opening.
13. Openings can be created with a specified width and height.
14. You can import door and window styles from external sources.
15. Roof properties define the thickness of the roof plane and the angle of the fascia.
16. The angle of a roof can be defined by the ratio of the rise to run or by a slope angle.
17. The CONVERT TO ROOF command is used to create a roof from a roof slab.
18. The ADDRAILING command can place a railing attached to stairs ONLY.
19. The scale for symbols and annotation is set with the OPTIONS dialog.
20. The Create AEC Content Wizard is used to add blocks, drawings, and custom commands to the Design Center.
21. The Model Explorer allows you to view mass groups and elements.
22. The Model Explorer allows you to name mass groups and elements.
23. Grouping Mass Elements is the same as Joining them together.
24. A camera can be used to obtain views of a building.
25. In order to create a video, you must create a path.

FINAL EXAM

Multiple Choice

1. The Menu Group that includes Architectural Desktop toolbars is:

A. ACAD
B. ACCOV
C. AECARCHX
D. EXPRESS

2. The layout tab selected for developing elevations is:

A. MODEL
B. WORK-3D
C. WORK-SEC
D. PLOT-SEC

3. The layout tab selected for developing a finished floor plan is:

A. MODEL
B. WORK-3D
C. WORK-FLR
D. PLOT-FLR

4. An object's display is controlled by:

A. Display Representation
B. Entity Properties
C. Layer
D. Color

5. The default layer standard for the Imperial template is:

A. None
B. AIA (256 Colors)
C. Generic Architectural Desktop
D. Standard

6. The limits of the AEC_Arch Imperial templates is set to this by default:

A. 1', 9'
B. 288'-0", 192'-0"
C. 144'-0", 96'-0"
D. None of the above

7. The default scale factor for symbols and dimensions of the AEC_Arch Imperial template is:

A. ¼" = 1'
B. ½" = 1'
C. 1" = 1'
D. 1/8" = 1'

8. Walls have all of the following properties EXCEPT:

A. Width
B. Height
C. Justification
D. Thickness

9. The command to Convert a Polyline to a Wall is:

A. WALLCONVERT
B. POLYLINECONVERT
C. CONVERT
D. WALLPROPS

10. Select the entity type that CAN NOT be converted to a wall:

A. LINE
B. POLYLINE
C. SPLINE
D. ARC

11. The command to insert a Door into a drawing is:

A. DOORADD
B. ADDDOOR
C. INSERT
D. DOORINSERT

12. The Style of a door is defined using this command:

A. DOORSTYLE
B. STYLEDOOR
C. DOORTYPE
D. STYLETYPE

13. A single hinged door will display at an angle of 90 degrees if the opening percent is set to:

A. 90
B. 50
C. 25
D. 0

14. The _____ of an Arch Window defines its radius.

A. RADIUS
B. RISE
C. HEIGHT
D. DIAMETER

15. Roofs are created with the command:

A. ADDROOF
B. ROOFADD
C. INSERT
D. INSERTROOF

16. The following entity type CAN NOT be used by the CONVERT TO ROOF command:

A. Roof slab
B. Line
C. Polyline
D. Walls

17. The command to insert stairs into a drawing is:

A. STAIRADD
B. ADDSTAIR
C. INSERT
D. INSERTSTAIRS

18. Straight stairs are created using this Shape setting:

A. Multi-Landing
B. Spiral
C. Straight
D. U-Shaped

19. The Shape options for stairs include all those listed below EXCEPT:

A. U-SHAPED
B. Multi-Landing
C. Floating
D. Straight

20. To create HOUSED stairs, you need to select a specific:

A. SHAPE
B. STYLE
C. PROPERTY
D. ENTITY

21. Setting the Drawing Scale establishes the _____ of dimensions.

A. units
B. text height
C. text style
D. placement

22. Mass Elements can be created using the _____ Command:

A. MassElementAdd
B. ElementAdd
C. MassAdd
D. None of the above

23. Mass Elements and Mass Groups are placed on this layer (using the AIA standard):

A. A-MASS
B. A-SOLIDS
C. A-ELEMENTS
D. A-MASS-ELEMENT

24. Which components make up the display system in Architectural Desktop?

A. Scale, objects, and linetype settings.
B. The viewports, layer settings, and scale.
C. The plotscale, objects, and linetype settings.
D. A viewport, objects, and a viewing direction for the objects.

25. What does the priority for components represent when creating or modifying wall styles?

A. Importance values for use in wall cleanup at intersections.
B. Relative structural strength values for the components of the wall style.
C. Priority controls the display order of the individual wall components for the style.
D. The assembly order of the individual wall components for construction purposes.

ANSWERS:

True-False
1) T; 2) T; 3) T; 4) T; 5) T; 6) T; 7) T; 8) T; 9) F; 10) T; 11) T; 12) F; 13) T; 14) T; 15) T; 16) T; 17) F; 18) F; 19) F; 20) T; 21) T; 22) T; 23) F; 24) T; 25) T;

Multiple Choice
1) C; 2) C; 3) D; 4) A; 5) B; 6) B; 7) D; 8) D; 9) A; 10) C; 11) A; 12) A; 13) B; 14) B; 15) B; 16) A; 17) A; 18) C; 19) C; 20) B; 21) B; 22) A; 23) A; 24) B; 25) A

Appendix A
Toolbars

	Create AEC Content	CreateContent
	Add MultiView Block	MvBlockAdd
	Modify MultiView Block	MvBlockModify
	MultiView Block Definitions	MvBlockDefine
	Add Mask Block	MaskAdd
	Modify Mask Block	MaskModify
	Mask Block Definitions	MaskDefine
	Attach Mask to Objects	MaskAttach
	Profile Definitions	ProfileDefine
	Insert Profile as Polyline	ProfileAsPolyline

Appendix A
Toolbars

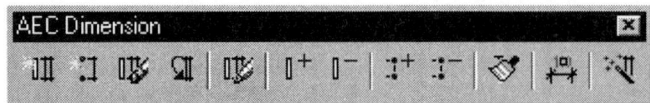

	Add AEC Dimension	DimAdd Brings up dialog box.
	Add Manual AEC Dimension	DimManAdd
	Modify AEC Dimension	DimModify
	Convert to AEC Dimension	DimConvert
	AEC Dimension Styles	DimStyle
	Attach Objects	DimAttach
	Detach Objects	DimDetach
	Add Dimension Points	DimPointsAdd
	Remove Dimension Points	DimPointsRemove
	Match AEC Dimension	DimMatch
	Activate Dim Text Grip Points	DimSetOverride
	AEC Dimension Wizard	DimWizard

Appendix A
Toolbars

	Add AEC Polygon	PolygonAdd
	Modify AEC Polygon	PolygonModify
	Convert to AEC Polygon	PolygonConvert
	AEC Polygon Styles	PolygonStyle
	Divide	PolygonOpDivide
	Join	PolygonOpJoin
	Subtract	PolygonOpSubtract
	Intersect	PolygonOpIntersect
	Trim	PolygonOpTrim
	Add Vertex	PolygonAddVertex
	Remove Vertex	PolygonRemoveVertex

	Drawing Setup	DwgSetup
	Set Current Display Configuration	DisplayManagerConfigsSelection
	Display Manager	DisplayManager
	Style Manager	StyleManager

Appendix A
Toolbars

	Curve Anchor	CurveAnchor Attach objects to the base curve of other objects, such as lines, arcs, circles, mass elements, polylines, roofs, or walls.
	Leader Anchor	LeaderAnchor Attach objects to nodes on layout tools with leaders.
	Node Anchor	NodeAnchor Attach objects to nodes on layout tools.
	Cell Anchor	CellAnchor Attach objects to the cell positions on 2D layout grids and 3D volume grids.
	Volume Anchor	VolumeAnchor Attach objects to volumes on 3D volume grids
	Release Anchor Objects	AnchorRelease Release objects from anchors to remove the anchoring relationship. Any anchored object can be released, even a railing that is anchored to a stair object or a window that is anchored to a wall.
	Position Anchored Objects	AnchorSetOffset You can change the position of an anchored object relative to the curve in the X, Y, and Z directions. You can also rotate and flip the object along its axes.

Appendix A
Toolbars

	Add Area Group	AreaGroupAdd 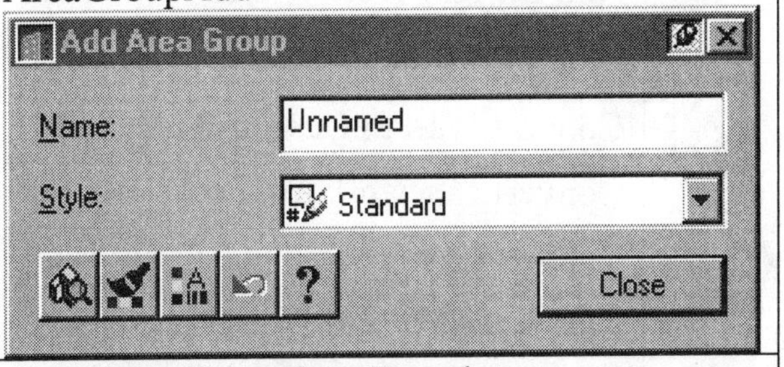
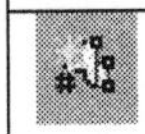	Create Area Groups from Template	AreaGroupCreatefromTemplate 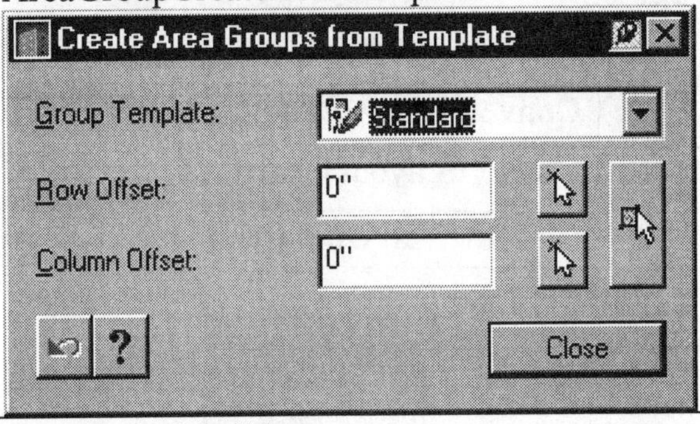
	Modify Area Group	AreaGroupModify
	Area Group Styles	AreaGroupStyle
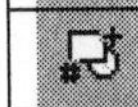	Attach	AreaGroupAttach
	Detach	AreaGroupDetach
	Create Polyline	AreaGroupCreatePline
	Area Groups Template	AreaGroupTemplate
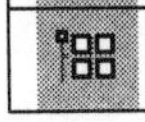	Area Group Layout	AreaGroupLayout

A-5

Appendix A
Toolbars

	Add Curtain Wall	CurtainWallAdd
	Modify Curtain Wall	CurtainWallModify
	Convert Wall to Curtain Wall	CurtainWallConvertWall
	Convert Linework to Curtain Wall	CurtainWallConvert
	Convert Layout Grid to Curtain Wall	CurtainWallConvertGrid
	Reference Curtain Wall	CurtainWallReference
	Curtain Wall Styles	CurtainWallStyle
	Add Curtain Wall Unit	CwUnitAdd
	Convert Linework to Curtain Wall Unit	CwUnitConvert
	Convert Layout Grid to Curtain Wall Unit	CwUnitConvertGrid
	Curtain Wall Unit Styles	CwUnitStyle

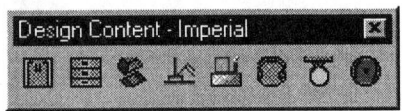

	Appliances	SetImpAppliances
	Casework	SetImpCasework
	Ceiling Fixtures	SetImpCeiling
	Electrical Fixtures	SetImpElectric
	Equipment	SetImpElectric
	Furniture	SetImpFurniture
	Plumbing Fixtures	SetImpPlumbing
	Site	SetImpSite

	Bathroom Fittings	SetMetBathroom
	Domestic Furniture	SetMetDomestic
	Electrical Services	SetMetElectric
	Kitchen Fittings	SetMetKitchen
	Office Furniture	SetMetOffice
	Piped and Ducting Services	SetMetPipeAndDuct
	Site	SetMetSite

![icon]	Break Marks	AnnoBreakMarkAdd	Bar, Bar (Filled), Cut Line (1), Cut Line (2), Cut Line (Curved), Pipe, Pipe (Filled)
![icon]	Detail Marks	AnnoDetailMarkAdd	Detail Boundary A, Detail Boundary B, Detail Boundary C, Detail Mark A1, Detail Mark A1T, Detail Mark A2, Detail Mark A2T

	Elevation Marks	AnnoElevationMarkAdd	Elevation Mark A1, Elevation Mark A2, Elevation Mark B1, Elevation Mark B2, Elevation Mark C1, Elevation Mark C2
	Leaders	AnnoLeaderAdd	Spline (Circle), Spline (Diamond), Spline (Hexagon), Spline (Square), Spline (Text), Straight (Circle), Straight (Diamond), Straight (Hexagon), Straight (Square), Straight (Text)

Appendix A
Toolbars

	Miscellaneous	AecDcSetImpMiscellaneous	Dimensions	
			Aligned	Angular
			Baseline	Continue
			Linear	Radius
			Fire Rating Lines	
			1 Hr	2 Hr
			2 Hr - Smoke	4 Hr
			Smoke	
			Match Lines	
			Match Line	Match Line (Swiss)

A-10

	Miscellaneous	AecDcSetImpMiscellaneous	North Arrows
			North Arrow A, North Arrow B, North Arrow C, North Arrow D, North Arrow E, North Arrow F, North Arrow G, North Arrow H, North Arrow I, North Arrow J, North Arrow K, North Arrow L, North Arrow M
	Revision Clouds	AnnoRevisionCloudAdd	Large Arcs, Large Arcs & Tag, Medium Arcs, Medium Arcs & Tag, Small Arcs, Small Arcs & Tag
	Section Marks	AnnoSectionMarkAdd	Section Mark A1, Section Mark A1T, Section Mark A2, Section Mark A2T

Appendix A
Toolbars

	Title Marks	AnnoTitleMarkAdd	Bar Scale (Inches)	Title Mark A1
			Title Mark A1 (Swiss)	

	Elevation Labels	AnnoElevationLabelAdd	2D Section	
			+10" Elevation Label (1)	+10" Elevation Label (2)
			+10" Elevation Label (3)	+10" Elevation Label (4)
			+10" Elevation Label (5)	+10" Elevation Label (6)
			+10" Elevation Label (7)	+10" Elevation Label (8)
			Model	
			3D Elevation Label (1)	3D Elevation Label (2)
			3D Elevation Label (3)	3D Elevation Label (4)
			3D Elevation Label (5)	3D Elevation Label (6)
			3D Elevation Label (7)	3D Elevation Label (8)

	Elevation Labels	AnnoElevationLabelAdd	Plan — Plan Elevation Label (1), Plan Elevation Label (2), Plan Elevation Label (3)
	Chases	AECCannomVblockInterferenceAdd	Chase (1), Chase (2), Chase (3), Chase (4), Chase (5), Chase (6)

	Add Door	DoorAdd
	Modify Door	DoorModify
	Door Styles	DoorStyle
	Add Window	WindowAdd
	Modify Window	WindowModify
	Window Styles	WindowStyle
	Add Opening	OpeningAdd
	Modify Opening	OpeningModify
	Add Window Assembly	WinAssemblyAdd
	Modify Window Assembly	WinAssemblyModify
	Convert Linework to Window Assembly	WinAssemblyConvert
	Convert Layout Grid to Window Assembly	WinAssemblyConvertGrid
	Window Assembly Styles	WinAssemblyStyle

Appendix A
Toolbars

	Add Elevation Line	BldgElevationLineAdd
	Elevation Line Properties...	BldgElevationLineProps
	Create Elevation	BldgElevationLineGenerate
	Update Elevation	BldgSectionUpdate
	Elevation Properties...	BldgSectionProps
	Elevation Styles...	2dSectionStyle
	Edit Linework	2dSectionResultEdit
	Merge Linework	2dSectionResultMerge

	Add Column Grid	ColumnGridAdd 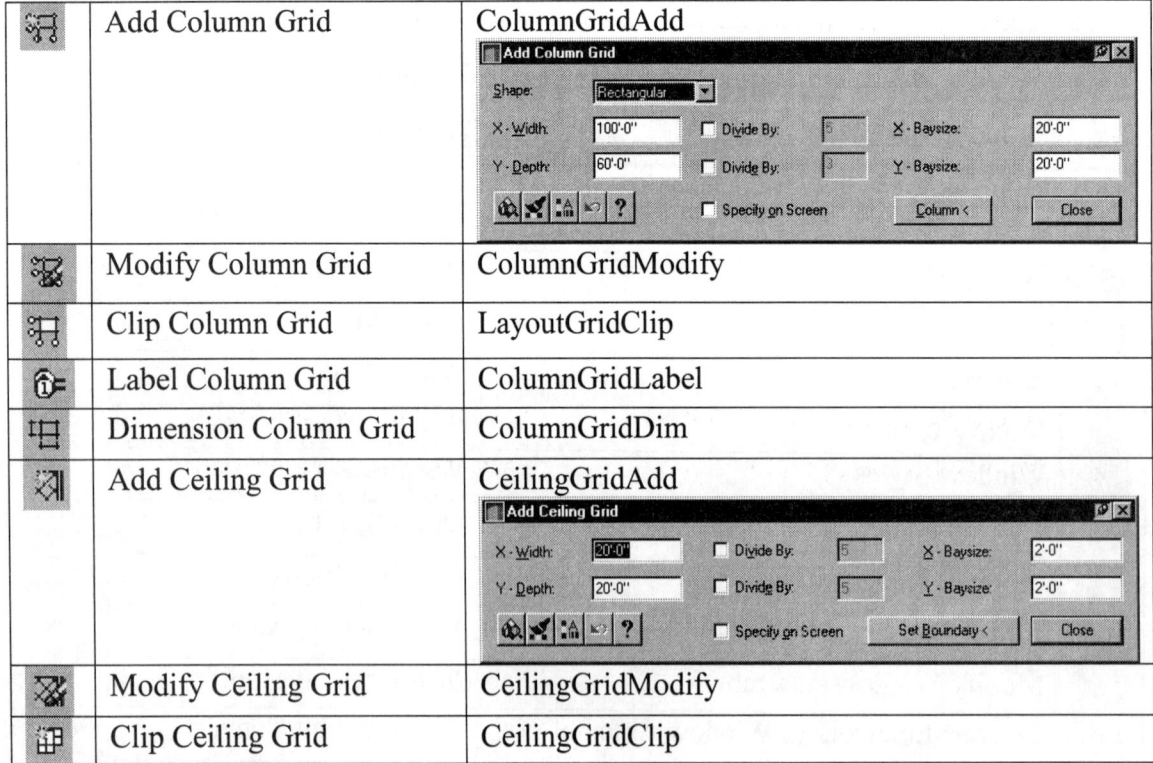
	Modify Column Grid	ColumnGridModify
	Clip Column Grid	LayoutGridClip
	Label Column Grid	ColumnGridLabel
	Dimension Column Grid	ColumnGridDim
	Add Ceiling Grid	CeilingGridAdd
	Modify Ceiling Grid	CeilingGridModify
	Clip Ceiling Grid	CeilingGridClip

A-14

Appendix A
Toolbars

	Layer Manager	LayerManager
	Select Layer Standard	DwgLayerSetup
	Layer Key Styles	LayerKeyStyle
	Layer Key Overrides	LayerKeyOverride
	Overrides ON/OFF	This is a TOGGLE
	Remap Object Layers	RemapLayers

	Add Layout Curve	LayoutCurveAdd
	Add Layout Grid (2D)	LayoutGridAdd
	Add Layout Volume (3D)	GridVolumeAdd
	Modify Layout Grid (2D)	LayoutGridModify
	Modify Layout Volume (3D)	GridVolumeModify
	Clip Layout Grid	LayoutGridClip

	Add Live Section Configuration	SectionConfigurationAdd
	Modify Live Section Configuration	SectionConfigurationModify

Appendix A
Toolbars

	Arch	arch
	Barrel Vault	barrel
	Box	box
	Cone	cone
	Cylinder	cylinder
	Dome	dome
	Gable	gable
	Pyramid	pyramid
	Sphere	sphere
	Iso. Triangle	isoc
	Rt. Triangle	right
	Add Mass Group	MassGroupAdd
	Modify Mass Element	MassElementModify

	Add Mass Group	MassGroupAdd
	Mass Group Properties	MassGroupProps
	Make Elements Additive	MassElementOpAdd
	Make Elements Subtractive	MassElementOpSubtract
	Make Element Intersection	MassElementOpIntersect
	Attach Elements	MassGroupAttach
	Detach Elements	MassGroupDetach

A-16

Appendix A
Toolbars

	Add Camera	CameraAdd
	Camera Properties	CameraProps
	Create Camera View	CameraView
	Adjust Camera View	CameraAdjust
	Create Video	CameraVideo

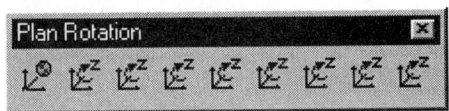

	Rotate to World UCS
	Rotate to 30d
	Rotate to 45d
	Rotate to 60d
	Rotate to 90d
	Rotate to 105d
	Rotate to 120d
	Rotate to 135d
	Rotate to 150d

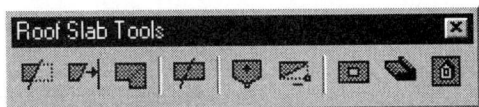

	Trim Roof Slab	RoofSlabTrim
	Extend Roof Slab	RoofSlabExtend
	Miter Roof Slab	RoofSlabMiter
	Cut Roof Slab	RoofSlabCut
	Add Roof Slab Vertex	RoofSlabAddVertex
	Remove Roof Slab Vertex	RoofSlabRemoveVertex
	Roof Slab Hole	RoofSlabHole
	Boolean Add/Subtract	RoofSlabBoolean
	Roof Dormer	RoofSlabDormer

Appendix A
Toolbars

	Add Roof	RoofAdd
	Modify Roof	RoofModify
	Convert to Roof	RoofConvert
	Edit Roof Edges/Faces	RoofEditEdges
	Add Roof Slab	RoofSlabAdd
	Modify Roof Slab	RoofSlabModify
	Convert to Roof Slabs	RoofSlabConvert
	Edit Roof Slab Edges	RoofSlabEdgeEdit
	Roof Slab Styles	RoofSlabStyle
	Roof Slab Edge Styles	RoofSlabEdgeStyle

	Door & Window Tags	SetImpDoorWindowTags
	Room & Finish Tags	SetImpRoomAndFinishTags
	Object Tags	SetImpObjectTags
	Wall Tags	SetImpWallTags
	Attach/Edit Schedule Data	PropertyDataEdit
	Renumber Data	PropertyRenumberData
	Add Schedule Table	TableAdd
	Update Schedule Table	TableUpdateNow
	Export Schedule Table	TableExport
	Schedule Table Styles	TableStyle

	Add Section Line	BldgSectionLineAdd
	Section Line Properties	BldgSectionLineProps
	Create Section	BldgSectionLineGenerate
	Update Section	BldgSectionUpdate
	Section Properties	BldgSectionProps
	Section Styles	2dSectionStyle
	Edit Linework	2dSectionResultEdit
	Merge Linework	2dSectionResultMerge

	Trim Slab	SlabTrim
	Extend Slab	SlabExtend
	Miter Slab	SlabMiter
	Cut Slab	SlabCut
	Add Slab Vertex	SlabAddVertex
	Remove Slab Vertex	SlabRemoveVertex
	Slab Hole	SlabHole
	Boolean Add/Subtract	SlabBoolean

	Add Slab	SlabAdd
	Modify Slab	SlabModify
	Convert to Slab	SlabConvert
	Edit Slab Edges	SlabEdgeEdit
	Slab Styles	SlabStyle
	Slab Edge Styles	SlabEdgeStyle

Appendix A
Toolbars

	Generate Slice	SliceCreate
	Set Slice Generation	SliceElevation
	Convert to Polyline	SliceToPline
	Attach Objects	SliceAttach
	Detach Objects	SliceDetach

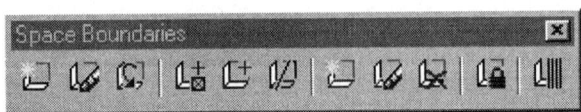

	Add Boundary	SpaceBoundaryAdd
	Modify Boundary	SpaceBoundaryModify
	Convert to Boundaries	SpaceBoundaryConvert
	Attach Spaces to Boundary	SpaceBoundaryMergeSpace
	Merge Boundaries	SpaceBoundaryMerge
	Split Boundary	SpaceBoundarySplit
	Add Boundary Edges	SpaceBoundaryAddEdges
	Edit Boundary Edges	SpaceBoundaryEdge
	Remove Boundary Edges	SpaceBoundaryRemoveEdges
	Anchor to Boundary	SpaceBoundaryAnchor
	General Walls	SpaceBoundaryGenerateWalls

Appendix A
Toolbars

	Add Spaces	SpaceAdd 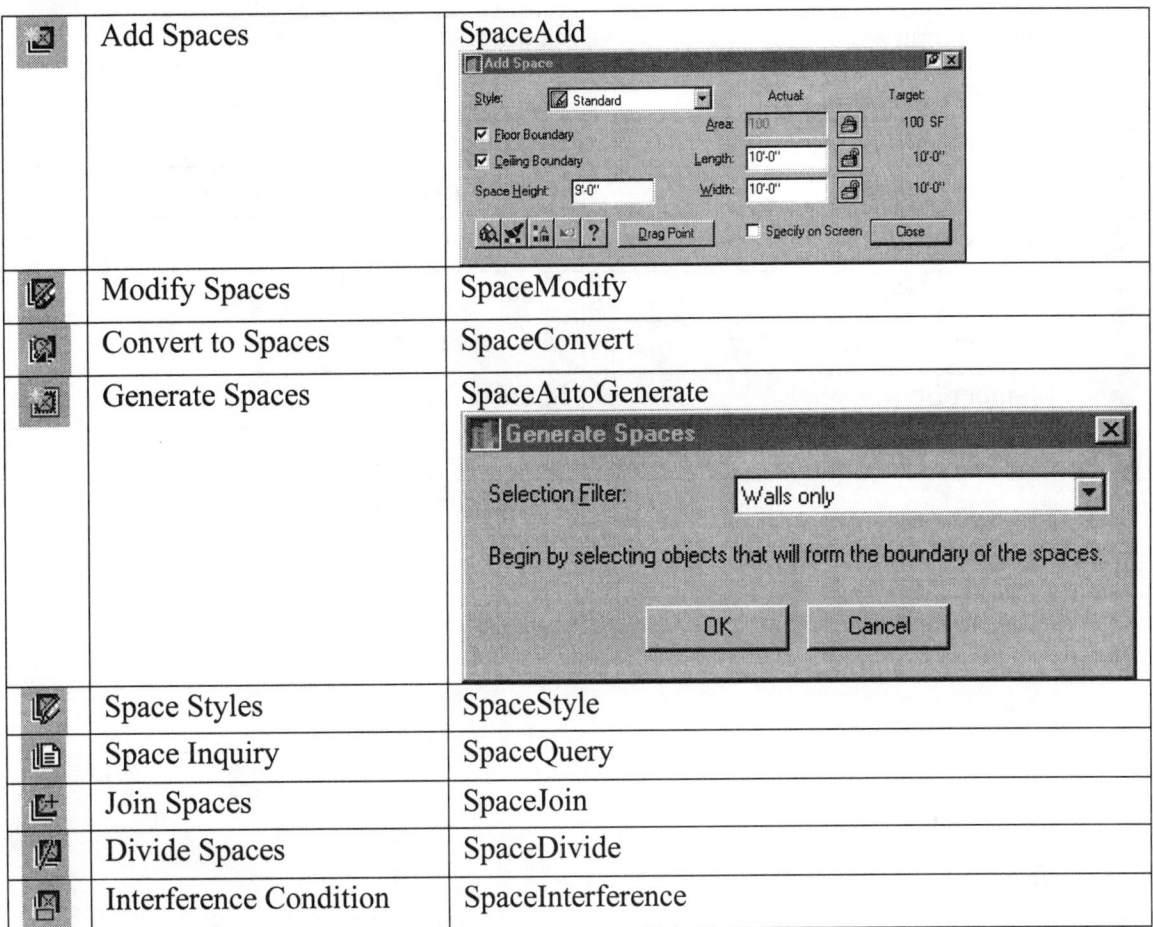
	Modify Spaces	SpaceModify
	Convert to Spaces	SpaceConvert
	Generate Spaces	SpaceAutoGenerate
	Space Styles	SpaceStyle
	Space Inquiry	SpaceQuery
	Join Spaces	SpaceJoin
	Divide Spaces	SpaceDivide
	Interference Condition	SpaceInterference

	Add Stair	StairAdd
	Modify Stair	StairModify
	Stair Styles	StairStyle
	Customize Edges	StairCustomizeEdge
	Add Railing	RailingAdd
	Modify Railing	RailingModify
	Convert to Railing	RailingConvert
	Railing Styles	RailingStyle
	Anchor to Stair	RailingAnchorToStair

Appendix A
Toolbars

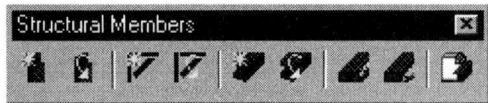

	Add Column	ColumnAdd
	Convert to Column	ColumnConvert
	Add Brace	BraceAdd
	Convert to Brace	BraceConvert
	Add Beam	BeamAdd
	Convert to Beam	BeamConvert
	Member Properties	MemberProps
	Member Styles	MemberStyle
	Structural Member Catalog	MemberCatalog

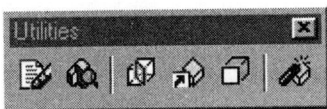

	Notes	Notes You can attach notes and reference documents to an object.
	Object Viewer Brings up the Object Viewer – similar to the Birds Eye View window found in AutoCAD.	ObjectViewer
	Quick Slice	QuickSlice If you have a 3D model of stairs in your current drawing, you can use a quick slice to create a polyline outline of the stair in a section view instead of using the section object. You can use the Quick Slice command to produce profile shapes, such as roof trusses, or to slice through mass groups.
	Reference AEC Objects	EntRef The objects that you can create reference objects from include all AEC objects and AutoCAD polylines. These reference objects are automatically updated when you change the original object. When you select an object to reference, you must specify a point on the original object as an insertion point for the marker of the reference object. A reference marker is displayed in the drawing to mark the insertion point.
	Hidden Line Projection	CreateHLR When you create hidden line projections, the 3D objects that you select are copied and collected into a unnamed (also called anonymous) 2D block. The unnamed block can be placed in your drawing parallel to the XY plane or in the current 3D view. You can edit or explode the inserted block.

Appendix A
Toolbars

	Explode AEC Objects You can convert or explode AEC objects from Architectural Desktop (ADT) into AutoCAD primitive entities such as lines, arcs, and 3D faces. This may be helpful when exchanging drawings with another person who wants to work on a copy of the drawing using only the primitive entities.	ObjExplode 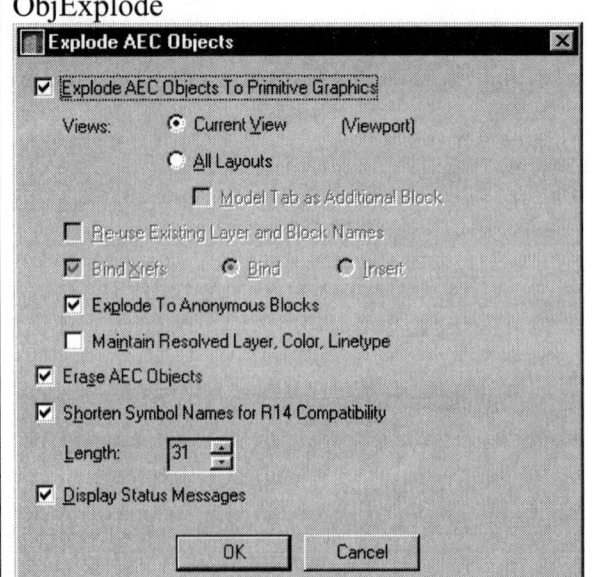

Appendix A
Toolbars

	Add Wall Modifier	WallModifierAdd
	Remove Wall Modifier	WallModifierRemove
	Convert Polyline to Wall Modifier	WallModifierConvert
	Wall Modifier Styles	WallModifierStyle
	Insert Modifier Style as Polyline	WallModifier
	Override Endcap Style	WallApplyEndCap
	Wall Endcap Styles	WallEndCapStyle
	Insert Endcap Style as Polyline	WallEndCap
	Merge Walls	WallMerge
	Override Cleanup Radius	WallApplyCleanupRadiusOverride
	Toggle Wall Graph Display	WallGraphDisplayToggle
	Roof Line	RoofLine
	Floor Line	FloorLine
	Interference Condition	WallInterference
	Sweep Profile	WallSweep
	Sweep Profile Miter Angles	WallSweepMiterAngles
	Body Modifier	WallBody
	Join Walls	WallJoin
	Reverse Wall Start/End	WallReverse
	Anchor to Wall	WallAnchor
	Dimension Walls	WallDim

Appendix A
Toolbars

	Add Wall	WallAdd 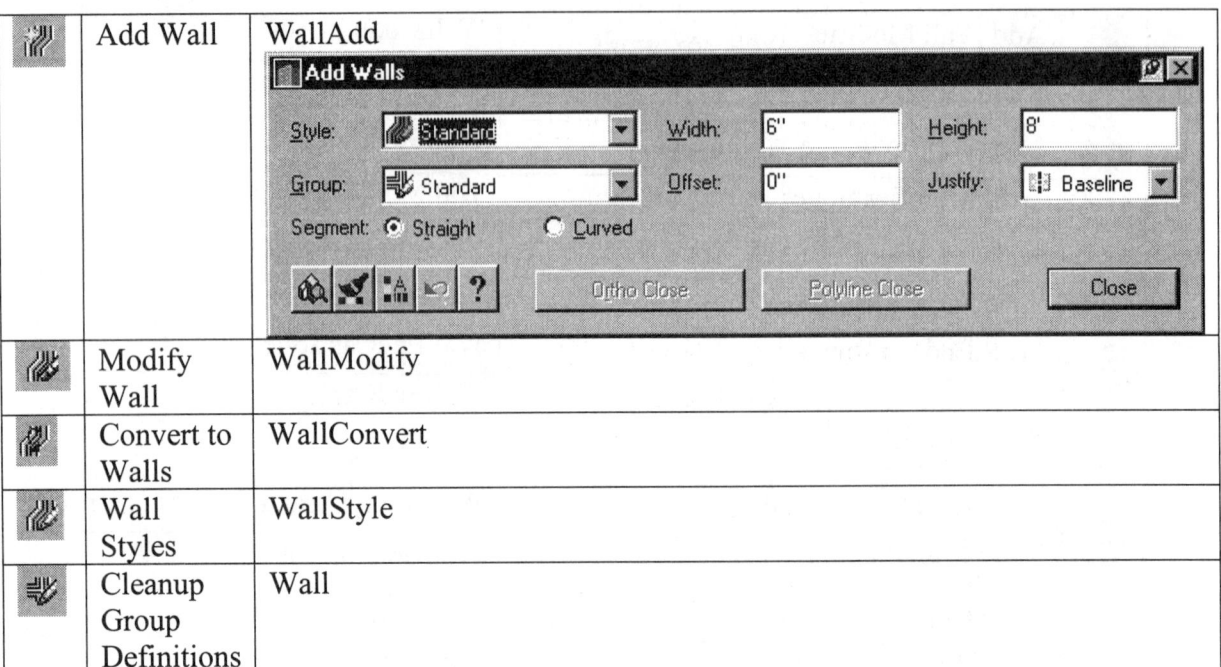
	Modify Wall	WallModify
	Convert to Walls	WallConvert
	Wall Styles	WallStyle
	Cleanup Group Definitions	Wall

Appendix B
Exercise Time Chart

	Time (mins)
Lesson 1	
Exercise 1: Creating a New Drawing	15
Exercise 2: Creating a New Geometric Profile	15
Exercise 3: Creating a New Wall Style	15
Exercise 4: Assigning Wall Properties	15
Exercise 5: Downloading Styles from Autodesk's pointA	15
Exercise 6: Creating a Fireplace/Chimney	30
Lesson 2	
Exercise 1: Creating Custom Line Types	15
Exercise 2: Creating a Custom Text Style	15
Exercise 3: Creating New Layers	15
Exercise 4: Creating a Site Plan	15
Exercise 5: Creating a Layer User Group	15
Lesson 3	
Exercise 1: Convert to Slab	15
Lesson 4	
Exercise 1: Layer Filters	15
Exercise 2: Snapshots	15
Exercise 3: Creating Exterior Walls	15
Exercise 4: Converting Polylines to Walls	15
Exercise 5: Wall Cleanup	15
Exercise 6: Adding Closet Doors	30
Exercise 7: Adding Interior Doors	30
Exercise 8: Add Opening	30
Exercise 9: Add Window Assemblies	30
Exercise 10: Adding a Fireplace	30
Lesson 5	
Exercise 1: Creating AEC Content	15
Exercise 2: Furnishing the Bedrooms	30
Exercise 3: Equipping the Bathrooms	30
Exercise 4: Furnishing the Common Areas	30
Exercise 5: Adding to the Service Areas	30
Exercise 6: Adding Decorator Touches	30
Lesson 6	
Exercise 1: Creating a Roof using Existing Walls	15
Lesson 7	
Exercise 1: Creating Member Styles	15
Exercise 2: Adding Members	30
Exercise 3: Add Floorboards	15
Exercise 4: Add Railing	15
Exercise 5: Add Stairs	15

Appendix B
Exercise Time Chart

	Time (mins)
Lesson 8	
Exercise 1: Creating Elevation Views	15
Exercise 2: Creating Section Views	15
Exercise 3: Creating 3D Section Views	15
Exercise 4: Modifying Section Views	15
Lesson 9	
Exercise 1: Dimensioning a Floor Plan	30
Exercise 2: Adding Wall Dimensions	30
Exercise 3: Add Drawing Scale	30
Lesson 10	
Exercise 1: Adding a Schedule Table	15
Exercise 2: Adding Window Tags	15
Exercise 3: Editing a Schedule	15
Exercise 4: Creating a New Table Style	15
Exercise 5: Exporting a Schedule	15
Lesson 11	
Exercise 1: Adding a Path	15
Exercise 2: Adding a Camera	15
Exercise 3: Creating a Video	15
TOTAL CLASS EXERCISE TIME:	**16 HOURS**

Index

A

Add AEC Dimension	9-8, 9-13
Add AEC section object	8-29
Add Door	4-22
Add Elevation	2-15
Add Elevation Line	2-20
Add Opening	4-24
Add Wall	1-45
Attributes	8-19, 9-26
AEC Content	1-18, 5-11
AEC Dimension	1-20
AEC DwgDefaults	1-11
AEC Editor	1-9
AEC Performance	1-12
AEC Stair Defaults	1-13
AEC Template	1-5
Annotation Plot Size	9-1
Array	7-14, 7-29
Attach Elements	1-60
Attribute, Edit	9-26

B

Beam, Add	7-9, 7-27
Brace, Add	7-18
Break Marks	9-2
Building Section Line	8-16

C

Camera	11-3
Camera Adjust	11-11
CameraProps	11-6
CameraVideo	11-13
Camera Viewa	11-6, 11-8
Chases	9-7
Column, Add	7-16
Convert to Walls	4-10
Create AEC Content	5-3
Custom Linetype	2-2
Custom Paper Sizes	8-9

D

Design Content	5-7
Desktop Display Manager	1-20, 9-12
Detail Marks	9-2
Dimensions	9-4, 9-9, 9-13
Display Locked	8-3
Display Props	8-18
Display Viewport Objects	8-27
Divide	2-14
Documentation Content	9-7
Documentation toolbar	2-20, 9-2, 9-17
Door Styles	1-25
Drawing Scale	9-1, 9-23, 9-25
Drawing Options	1-8
DWF eView	8-9

E

Edit Display Props	2-21, 8-18
Edit AEC Dimension Style	9-14
Edit Table Style	10-18
Elevation Labels	2-18, 2-22, 9-7
Elevation Label Modify	2-20
Elevation Line Properties	2-17
Elevation Marks	2-20, 9-3
Elevation Tools	2-15
Entity Display	2-20
Express Tools	2-2
Extrude	1-65

F

Filter Groups	4-1
Fire Rating Lines	9-4
Fireplace, Add	4-43
Flip Hinge	4-19
Foundations	3-1

G

Generate Member Style	7-4
Generate View	11-4

I

idrop	1-51
Isolate Group	7-8, 8-24

L

Layer, Current	1-70
Layer Key	5-5
Layer Key Defaults	1-73
Layer Key Overrides	1-69, 1-72
Layer Key Styles	1-68, 1-71, 7-26
Layer Manager	1-67, 2-7, 2-32, 4-47, 7-7, 7-24
Layer, New	2-6, 7-24, 8-26
Layer Standards	1-68, 1-71, 7-25, 8-25
Layer User Group	2-26, 4-47, 7-25
Layout Rename	8-13
Leaders	9-3
Linetypes	2-2

M

Mass Element	1-22, 1-31, 1-53
Mass Group	1-22, 1-60
Mass Model	1-22
Match Lines	9-4
Member Properties	7-20
Mirror	7-12
Miscellaneous Tools	2-24
Multi-View Block Properties	2-19, 8-19

N

New Filter Group	1-70, 4-2
New Layer	1-69, 2-6
New Layout	8-7
New User Group	1-70
North Arrows	2-24

O

Opening, Add	4-24
Opening, Modify	5-28
Options	1-8
Overrides	1-72

P

Page Setup	8-8
Paper Space	8-1
Path	11-1
Perspectives	11-3
Plot Device	8-9
Plotter Configuration	8-8
Point Style	2-14
Presentation Format	7-42
Profiles	1-31
Properties	1-36, 1-64, 11-2
Property Sets	10-26

Q

Quick Select	2-29, 7-23

R

Railing, Add	7-30, 7-43
Railing Modify	7-37
Railing Properties	7-36
Railing, Style	7-32
Remap Object Layers	1-72
Remove from Group	4-48

Reposition Along Wall	4-35
Reposition Within Wall	4-42
Revision Clouds	9-5
Roof, Add	6-1
Roof, Convert	6-1, 6-4
Roof Slabs	6-7
Roof Slab Properties	6-8

S

Scale	9-10
Schedules	10-1
Sections	8-5
Section, Create	8-16, 8-30
Section Marks	8-28, 9-5
Scale, Viewport	8-14
Schedule Tables	1-29, 10-12, 10-22
Section Line, Add	8-15
Site Plan	2-1, 2-12
Snapshots	1-69, 4-3
Space Styles	1-28
Stairs	7-38
Stairs, Add	7-41
Stair Styles	1-27
Start-Up Dialog	1-5
Structural Members	7-6
Structural Member Catalog	7-3, 7-4, 7-15
Style Manager	1-30, 10-2
Surveyor's Units	2-12

T

Table, Add	10-12
Table Export	10-29
Table Properties	10-24
Table Style	10-21
Text	2-24
Text Styles	2-5, 2-21, 2-26
Title Marks	9-6, 9-24
Trim Planes	7-21

U

UCS, Move	7-40
UCSICON	8-5
Units	2-12

V

Video	11-1, 11-13
Viewports	8-1, 8-24
Visibility	8-21

W

Walls, Add	4-7
Wall Cleanup	4-14
WallDim	9-13
Wall Styles	1-23, 1-38, 1-46
Wall Style Manager	1-24
Window, Add	4-34
Window Tags	10-14
Window Styles	1-26
Work-3D Tab	4-32

About the Author

Elise Moss has worked for the past twenty years as a mechanical designer in Silicon Valley, primarily creating sheet metal designs. She has written articles for Autodesk's Toplines magazine and AUGI's PaperSpace. She is President of Moss Designs, a Registered Autodesk Developer, creating custom applications and designs for corporate clients. She is also President of Silicon Valley AutoCAD Power Users, the largest AutoCAD user's group in the United States. She has taught CAD classes at DeAnza College, Silicon Valley College, and for Autodesk resellers. She holds a baccalaureate degree from San Jose State.

She is married with two sons. Her older son, Benjamin, is currently studying electrical engineering at UC Santa Cruz. Her younger son, Daniel, is studying architecture at West Valley College. Her husband, Ari, is a computer scientist.

Elise is a third generation engineer. Her father, Robert Moss, is a metallurgical engineer in the aerospace industry. Her grandfather, Solomon Kupperman, was a civil engineer for the City of Chicago.

She can be contacted via email at elise_moss@mossdesigns.com.

More information about the author and her work can be found on her website at www.mossdesigns.com.

Other books by Elise Moss

Autodesk Inventor R4 Fundamentals: Conquering the Rubicon
Autodesk Inventor R4 Intermediate: Mastering the Rubicon
AutoCAD 2000i Mechanical Drafting for Beginners

About the Co-Author

Lay Christopher (Chris) Fox works as an architectural drafter, rendering specialist and educator in western New York state, where he has lived for 22 years. He is president of ArchImage CAD Services, providing drafting, rendering and consulting services to homeowners, designers, architectural firms and institutions. He teaches classes in Autodesk products at the AutoCAD Training Center on the campus of Rochester Institute of Technology (RIT), and currently holds an adjunct professor position at RIT, teaching AutoCAD for the Civil Engineering Department.

A longtime officer of the Rochester Area AutoCAD Users Group, he has contributed articles to AUGI's PaperSpace newsletter. Professionally, his recent work has included 3D modeling for large-scale metal art sculpture and CAD Standards development for local colleges.

Chris holds a bachelor's degree in English Literature from Harvard. He has previously worked as a teacher, audio engineer, carpenter, building remodeler and construction company owner.

Chris is married and lives with two (sometimes three) dogs in the house he and his wife built overlooking the small town of Springwater. His Australian wife, Sally, works in theater, teaches acting at SUNY Buffalo and the University of Rochester, and flies downunder at least once a year for her sunshine fix.

More information about the co-author of this book can be found on his website at www.archimagecad.com. He can be contacted via email at lcfox@archimagecad.com.